BEING WEATHERWISE

INFORMATION ON GENERAL METEOROLOGY

PC DUTTA

notionpress.com

INDIA • SINGAPORE • MALAYSIA

ISBN 979-8-89233-448-8

CONTENTS

FOREWORD

Dr. Sanjay O'Neill Shaw

1. 'Being Weatherwise' stands as a beacon in the realm of meteorological literature, offering a comprehensive and accessible journey into the fascinating world of weather and climatology. It is a well written book by Sri P.C. Dutta which is a must for professionals, students and competitive examination aspirants. This book excels in presenting complex meteorological concepts with clarity and precision. The book's strength lies in its meticulous organization with chapters thoughtfully distributed to facilitate a seamless learning experience. It covers a wide spectrum of topics ranging from the basic principles of weather to the significance of weather elements. 'Being Weatherwise' leaves no stone unturned in its exploration of general meteorology and climatology. This inclusive approach makes it an invaluable resource for readers at different levels of expertise. One of the book's notable attributes is its clarity in explaining meteorological topics. The author has succeeded in breaking down intricate concepts into easily understandable portions, ensuring that even readers unfamiliar with the subject can grasp the information effortlessly.

2. A noteworthy aspect of 'Being Weatherwise' is its affordability. Despite its comprehensive coverage and high-quality content, the book remains budget-friendly. This makes it an attractive option for those seeking a reliable guide to meteorology. The book's affordability, combined with its educational value, positions it as an essential addition to any personal library.

3. In conclusion, 'Being Weatherwise' is a triumph in the genre of meteorological literature. Its well-organized structure, comprehensive coverage, clear language, and budget-friendly nature make it an indispensable companion for professionals, students and enthusiasts alike. If you're looking for a reliable guide that covers the elements of weather, the significance of weather elements, and general meteorology and climatology, 'Being Weatherwise' is an excellent choice that will enrich your understanding and serve as a valuable resource for years to come.

Dr. Sanjay O'Neill Shaw
Scientist-F
Regional Meteorological Centre, Guwahati
India Meteorological Department
Ministry of Earth Sciences
Govt. of India

MR MUKUL KUMAR SAIKIA, IAS

1. As the day breaks, it comes to our mind how the day would be on that particular day. Sometimes the day becomes very pleasant and exciting, the way we expected, and sometimes it turns out to be quite the opposite. Most of the time, this situation arises because of the weather conditions, which we fail to predict correctly. Further, there have been visible changes in the weather pattern across the world due to global warming which makes prediction of weather more challenging.

2. I am, therefore, very happy to know that Sri Paramesh Chandra Dutta, with his nearly four decades experience in Aviation Meteorology, has come forward to share his knowledge by writing a book named **'Being Weatherwise'** with an endeavor to help the readers gain a clear understanding of weather and climatology.

3. **'If you want to see the sunshine, you have to weather the storm'**... Yes, Sri Dutta is indeed very successful in his effort to help the readers understand the complex meteorological concepts effortlessly through his book 'Being Weatherwise,' and I believe the book will be handy for all sections of readers, including students, research scholars, and aspirants for various competitive examinations.

4. My best wishes to the writer for more such good work in the days to come.

Shri Mukul Kumar Saikia, IAS
Principal Secretary, Karbi Anglong Autonomous Council
Diphu, Karbi Anglong, Assam -782460
E-mail: ps-ka@nic.in / pskaac1952@gmail.com

MR DIPAK MALLA BUZAR BARUAH, SCIENTIST-E

1. "Being Weatherwise" stands out in the world of meteorological literature as a guiding light. Authored by Sri P.C. Dutta, this well-crafted book is a must-have for both professional students and those preparing for competitive examinations. The strength of this book lies in its ability to unravel intricate meteorological concepts with simplicity. The chapters are thoughtfully organized, ensuring a smooth learning journey from the basics of weather to the significance of weather elements.

2. Covering a broad spectrum of topics, "Being Weatherwise" leaves no stone unturned in exploring general meteorology and climatology. Its inclusive approach caters to readers at various expertise levels. The author excels in breaking down complex concepts into easily digestible portions, making it accessible even for those unfamiliar with the subject. Remarkably, this invaluable resource maintains its affordability while delivering high-quality content, making it an attractive choice for those seeking a dependable guide to meteorology.

3. In essence, "Being Weatherwise" triumphs in the realm of meteorological literature. It's well-organized structure, comprehensive coverage, clear language, and budget-friendly nature make it an essential companion for both professional students and enthusiasts. Graphical and pictorial representation are very nice and will help the students in understanding the concept. If you seek a reliable guide encompassing the elements of weather, the significance of weather elements, and general meteorology and climatology, "Being Weatherwise" is an excellent choice that promises to enrich your understanding for years to come.

Dipak Malla Buzar Baruah
Scientist-E & Director (IT)
NIC Assam
Govt of India

PREFACE

1. The most common and widely discussed subject in the world is 'Weather'. This is due to its pivotal role in the day-to-day life of individuals and its substantial impact on global business and commerce. The agricultural sector across the globe heavily depends on weather, regardless of how modern or developed a country is.

2. However, despite being one of the most common subjects of discussion, the general populace has meagre knowledge about weather, climate, and their associated features. People often talk about extreme weather conditions like freezing cold or scorching heat but lack the deeper understanding or curiosity to explore about the causes behind such weather conditions or how long these conditions might persist. This lack of interest might stem from the fact that an uncomfortable weather condition is often followed by a period of weather that either neutralizes the adverse condition or brings about a pleasant one.

3. It's not that people are entirely unaware of weather conditions. Many farmers around the world possess remarkable ability to sense weather changes in advance and act accordingly, whether it's clearing crops from fields or preparing them for the upcoming weather conditions. The cultivation of different crops in different seasons is based on the climatology of a specific area.

4. It would be beneficial if individuals possessed adequate knowledge about weather and climatic conditions, especially concerning the area where they reside or plan to visit. Understanding the factors responsible for conditions such as heatwaves, cold snaps, heavy rainfall, floods, etc., can be genuinely interesting and helpful. Having general knowledge about weather also aids in planning outdoor activities like long drives, outings, and tours. Lack of knowledge about weather has led to disasters in some tourist places, particularly in hill stations, due to occurrences like landslides and heavy snowfall. Furthermore, there has been a high incidence of accidents caused by lightning strikes on people who were outdoors during thunderstorms.

5. However, the subject of 'Weather' is intricate and multifaceted. Weather elements such as temperature, wind, atmospheric pressure, etc., are governed by fluid dynamics, geography, topography, astronomy, and numerous other factors. Despite being governed by scientific rules, the science of weather, Meteorology, is not at par with other scientific subjects. Weather elements are highly variable and interdependent.

6. Recognizing the importance of meteorological knowledge, various academic-level books cover the subject. Additionally, many competitive examinations include a considerable number of questions related to weather and climate.

7. Numerous publications and scientific literature prepared by meteorologists and other professionals are available worldwide in various languages, with many accessible in the public domain on the internet. However, these books or research papers are primarily intended for meteorological professionals. For a layman or a student, gathering information often requires searching through different publications. With this in mind, this book on meteorological information has been introduced, compiling essential information about weather, climate, seasonal patterns, etc. Most of the included information pertains to India and neighboring countries.

8. Furthermore, I want to reiterate that this book is a compilation of information referred from various literatures, books available in the public domain, web pages of different meteorological organizations, and other publications. None of the information presented in the book is a product of my own research. Therefore, this book may be regarded as a compilation work.

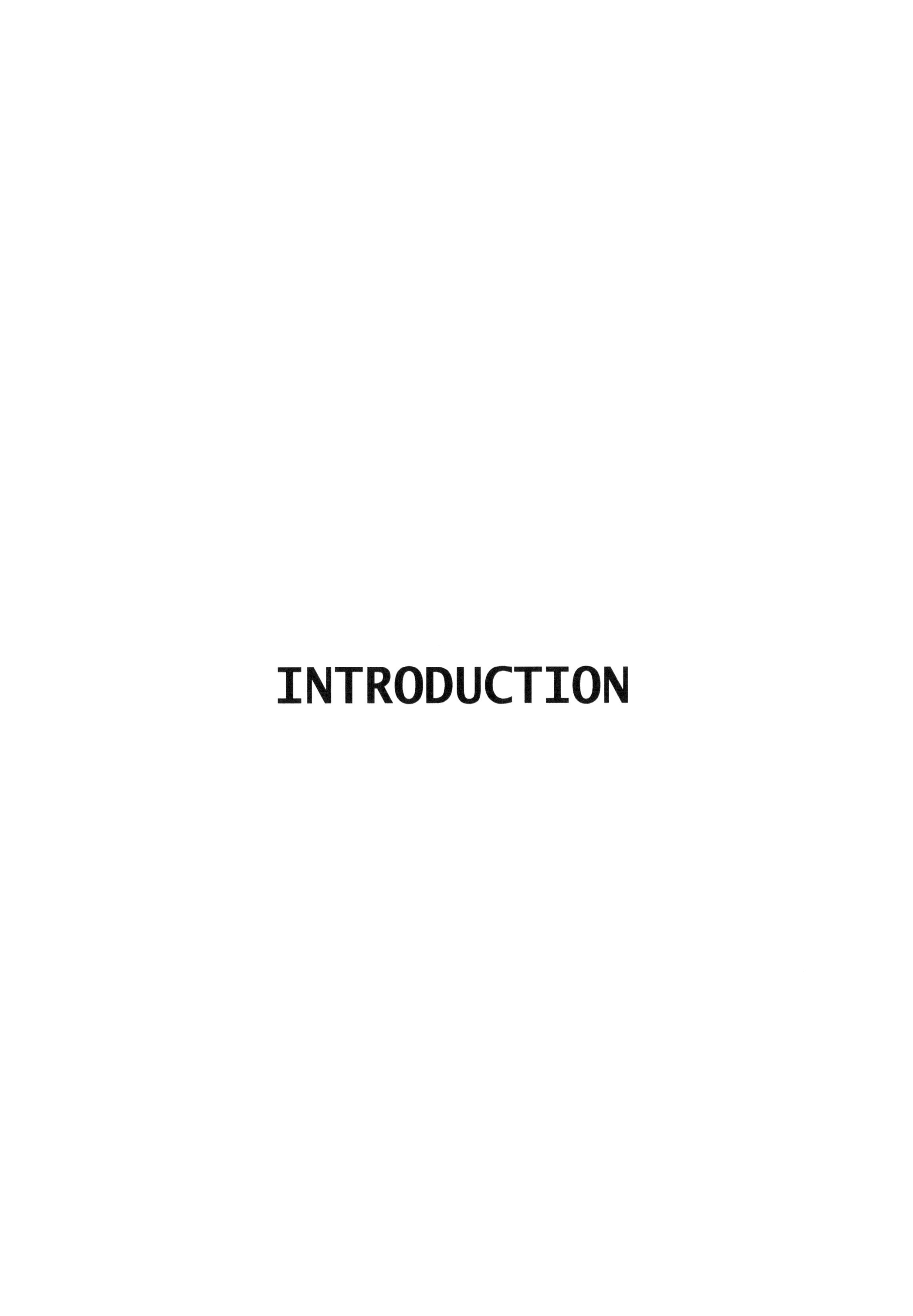

INTRODUCTION

CHAPTER **1**

ELEMENTS OF WEATHER

1. **Weather and Climate.** To delve into the topic of weather, it's crucial to first understand the distinctions between the terms 'weather' and 'climate', which are commonly interchanged despite their unique meanings. These terms, though they appear similar, are distinct and different from each other.

2. **Weather.** Weather describes short-term natural events such as fog, rain, snow, wind, thunderstorms, etc. over a specific place at a specific time. It can be defined as the 'state of the atmosphere at a specific time over a specific place.' The state of the atmosphere includes current weather, temperature, atmospheric pressure, wind, clouds, etc., which will be discussed elaborately. What is important to note for now is that weather pertains to a particular place at a particular time.

3. **Climate.** Climate is the long-term average of weather conditions or weather elements over a particular place, a region, a country, or a larger area of the globe. Statements like 'India experiences a monsoon type of climate,' 'Dry and hot climatic conditions prevail over northwest India during the pre-monsoon season,' and 'NE India is the wettest part of the country' are based on climatological interpretation. After conducting a study of the weather conditions over a given area during different parts of the year and calculating the averages of different weather elements for a long period, such conclusions have been reached. The India Meteorological Department considers data for at least 30 years for the evaluation of climatological summaries.

4. **Elements of Weather.** For a common person, weather means the currently prevailing weather conditions that are seen or felt, such as rain, a hot day, fog, snow, etc. However, in meteorology (Meteorology), 'weather' has a broader meaning. The terms mentioned above actually refer to the prevailing conditions that are observed or felt. There are other elements that are included in weather and are inseparable. The various elements observed by a meteorologist to prepare a weather report are discussed below.

5. **Surface Wind (direction and speed).** The wind direction and speed measured on surface of the earth including the sea surface. measuring instruments or sensors are installed 10 M above the ground and adequately away from any structures like buildings, walls, high rise topographical features, etc. so that the free flow of wind is not disturbed. Direction of wind is the direction from which the wind blows and not towards which direction it blows. Direction is measured in degree using 36 point of compass and speed is measured in Kt (Nautical mile per hour) or kmph.

6. **Visibility**. In Met, visibility means transparency of the atmosphere. In other words, it is the maximum distance at which an object can be seen as such along the horizon with naked eyes under proper illumination. Visibility is affected by the obscuring weather conditions like haze, dust haze, precipitation, fog, mist, smoke haze, etc. In Met, visibility of 5000 M or above is considered good visibility condition. Visibility is reported in Kilo-metre or metre.

7. **Current weather**. The prevailing weather phenomena or the weather events like fog, rain, snow, thunderstorm, mist, etc. This is the element which is felt the most and play vital role in agriculture, aviation, event management, etc. It also has great influence in our day-to-day life.

8. **Clouds**. This element consists of type, amount and base of clouds. Clouds form by condensation of water vapour. There are various types of clouds with distinct characteristics and associated with different atmospheric conditions. Clouds give very good indication of the prevailing meteorological situations over an area. Amount of cloud is reported in Octa, type of cloud is reported as per the approved cloud atlas and height is reported in Feet.

Amount of cloud *– The sky is divided in eight equal parts and one part is called Octa (1/8). A particular cloud (say low cloud) which are scattered in the sky are brought together (assumed). If it is estimated that the total amount of cloud is about one-fourth of the sky, then the amount will be 2 Octa (2/8). Observation is recorded in respect of all type clouds (low/med/ high) and separately for vertically growing clouds (Cb or TCu) if any. Following is the procedure: -*

Amount	*Reported as*
1 – 2 Octa	*FEW (Few)*
3 – 4 Octa	*SCT (Scattered)*
5 – 7 Oct	*BKN (Broken)*
8 Octa	*OVC (Overcast)*

9. **Dry Bulb Temperature (DB)**. It is the air temperature measured using a mercury thermometer or sensor. The measuring instrument is generally installed at a height of 1.25 M from the ground level in such a way that the thermometer (Or the sensor) receives free flow of air and ensured that direct sun rays do not fall upon it. Standard unit for reporting temperature is degree Celsius. The cage (shelter) which is used to house different types of thermometers in an observatory is called Stevenson Screen.

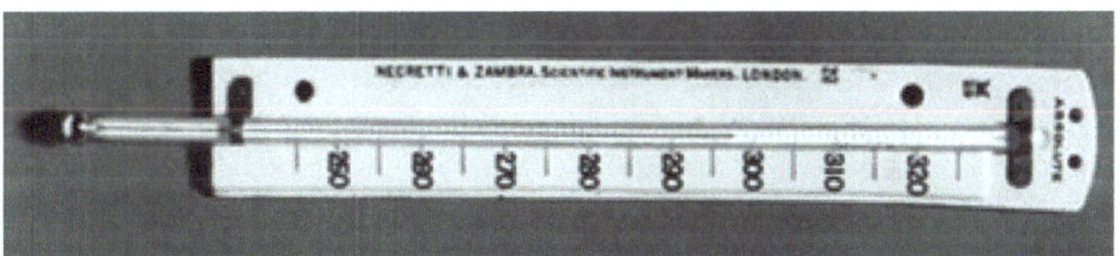

10. **Wet Bulb Temperature (WB)**. It is the temperature attained by air when water is allowed to evaporate freely. The bulb of a mercury-in thermometer is covered with a layer of muslin cloth. A wick from the muslin dips into a bottle of water and keeps the muslin constantly wet. Thus, free evaporation takes place from the thermometer bulb and it registers wet-bulb temperature. Difference of dry bulb and wet bulb temperatures $(T – Tw)$ is used to calculate Dew Point temperature, humidity and vapour pressure.

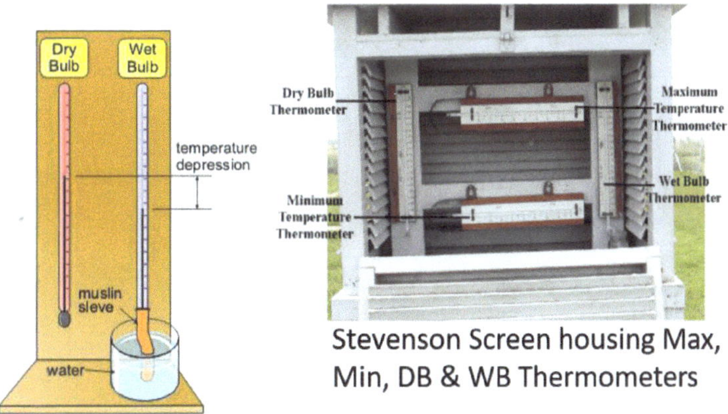

Stevenson Screen housing Max, Min, DB & WB Thermometers

11. **Dew Point Temperature (DP)** The Dew Point is the temperature to which air must be cooled for condensation (Cloud formation) to take place. In other words, this is the temperature at which air become saturated, that is, water in gaseous form (vapour) can be held by air at only below this temperature. We all know that temperature reduces with increase in height. If a parcel of air is lifted to a height where the parcel will attain temperature equal to the Dew Point temperature, then process of cloud formation will take place. Hence, Dew Point temperature is an important element not only to forecast clouds and associated weather, but also to understand the characteristics of the prevailing the air mass.

12. **Relative Humidity (RH).** Relative Humidity Is a ratio, expressed in percentage, of the amount of atmospheric moisture present relative to the amount that would be required to saturate it.

$$RH = \frac{\text{Actual amount of water vapour}}{\text{Amount required for saturation}} \times 100$$

13. **Atmospheric Pressure at the altitude of the place of observation (QFE).** It is the pressure exerted by the column of atmosphere at a point over the observatory. Pressure is measured using a barometer or a sensor. If the measuring tool is kept at a level higher or lower than the station level, necessary correction is required to be applied to the reading.

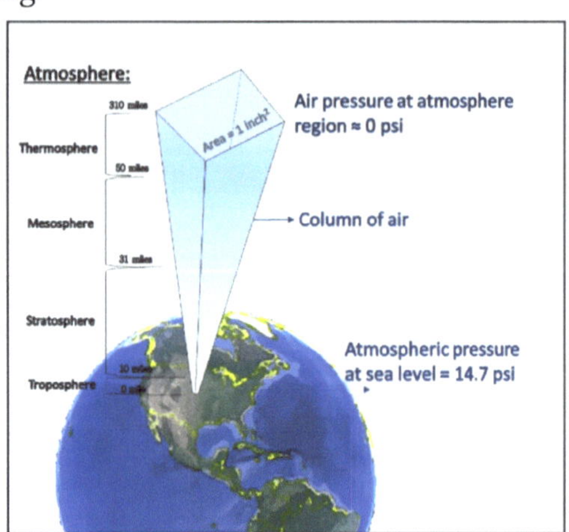

14. **Atmospheric Pressure reduced to Mean Sea Level (MSL) as per ISA (QNH).** QFE, as mentioned in above paragraph, actually represent the atmospheric pressure over a particular place. Hence, it cannot represent a wider region. It is because, the stations are located at different altitude (height above sea level) and the column of air above the stations are not equal and hence, their pressure readings will have difference. To counter this issue, there is a requirement to reduce all the station level to a common level, and most suitable for the purpose is the Mean Sea Level (MSL). Thus, all stations apply correction to reduce the pressure to MSL. While applying the correction it is assumed that column of air from the station level to the MSL is the air which holds the characteristics as per the

International Standard Atmosphere (ISA).

15. Consider a small locality with hilly topography. Also consider the locality has six Met observatories with different altitude (height above MSL). The QFE reading at each observatory will be different due to different altitude. Now, when all the observatories reduce their QFE values to the MSL, the readings become same if other situations are identical for all the places.

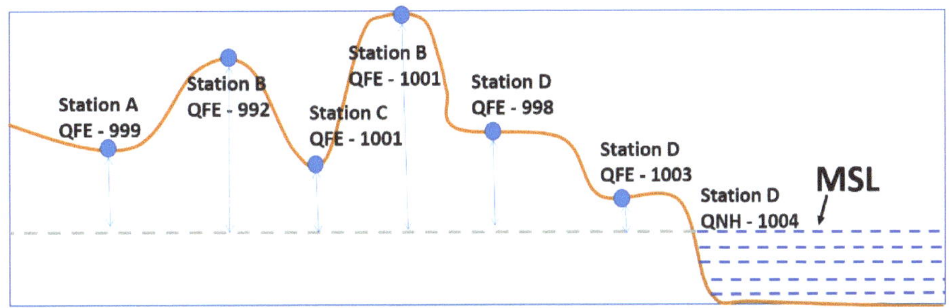

16. **Atmospheric Pressure reduced to MSL as per standard Met practices (QFF).** This form of atmospheric pressure is obtained for plotting of charts for the purpose of analysis by the meteorologist. The equal pressure values in the charts are joined to draw the isobars and find out location of pressure systems line low pressure, high pressure, depression, etc.

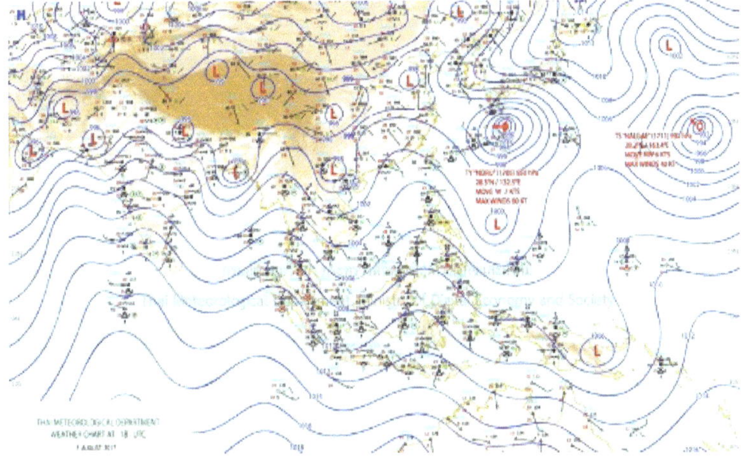

17. The other elements like maximum & minimum temperature, rainfall information, etc. also form part of a weather report.

CHAPTER 2

MET OBSERVATION

1. Observation or measuring/estimating the elements discussed in the previous chapters is carried out by a weatherman. The observations, however, are not taken as per the wish or need of the observer. Besides, observation taken in isolation will make little sense in coming to a conclusion on the pattern of weather situation prevailing over a region or over a larger domain. Hence, certain procedures are laid down for taking observation. The most important being the time of observation. It is important for the observers around the world to take observation at a particular time so that the actual atmospheric condition in respect of various elements can be picturized and the behavior of the pattern can be assessed.

2. Another important aspect is a robust communication system to operate globally so that the data can be exchanged smoothly among the various regions irrespective of political boundaries so as to reach the user ends, that is, each meteorological agency in the world for the purpose of timely analysis and issue of forecast for the respective areas of responsibility.

3. The global responsibility of exchange of data is carried out by WMO through World Weather Watch (WWW) and it functions through three main components which are linked to each other. These are Global Observing System (GOS), Global data Processing and Forecasting System (GDPFS) and Global Telecommunication System (GTS).

4. GTS has been designed by WMO Information System (WIS) to establish a common ICT infrastructure for all WMO programmes. WIS consists of three components, namely Global Information System Centres (GISCs), Data Collection or Production Centres (DCPCs) and National Centres (NCs).

5. GTS is the backbone system for the exchange of data and information. It is an integrated network of surface-based and satellite-based communication links interconnecting Met telecommunication Centres (MTCs) operated round the clock by the member countries. The data exchanged through GTS includes information on weather; water and climate analysis and forecast; tsunami, seismological data, tropical cyclones, flood, etc.

6. **Organization of GTS**. Following are the three levels of GTS: -

a) **Main Telecommunication Network (MTN)**. MTN is linked to the three Global Meteorological Centres located at Melbourne, Moscow and Washington and fifteen Regional Telecommunication Hubs (RTH), one of which is located at New Delhi.

b) **Regional Meteorological Telecommunication Network**. This network system interconnects all regions of WMO and Meteorological Telecommunication Centres (MTCs). There are six regions of WMO which are as follows: -

i.	Region-I	Africa
ii.	Region-II	Asia
iii.	Region-III	South America
iv.	Region-IV	North & Central America and the Caribbean
v.	Region-V	Southwest Pacific
vi.	Region-VI	Europe and Atlantic

c) **National Meteorological Telecommunication Network (NMTN)**. This network is responsible to NMCs for receiving and exchanging observational data of the respective country. NMTN of India is located at New Delhi.

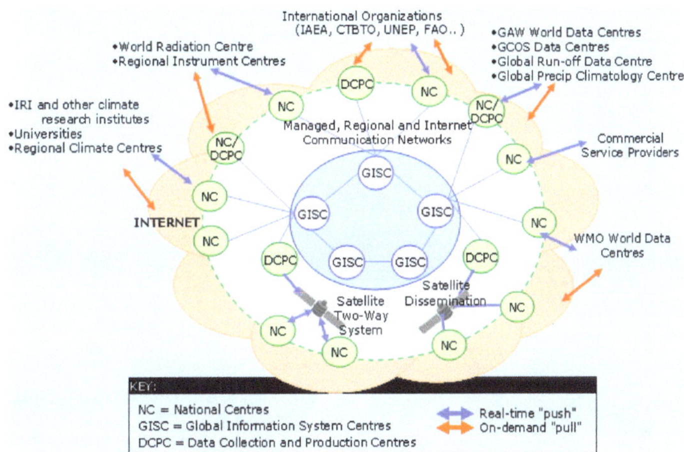

d) **India Meteorological Department (IMD)**.   The organization of IMD has been discussed separately at the end of this chapter.

7. As known by now, the world has a dense network of Met observatories. In India, almost all districts have at least one observatory. Besides, each airport mandatorily has an observatory. The Met observatories are categorized as Synoptic observatories, Aviation Met Observatories (Current weather observatories) and upper air observatories.

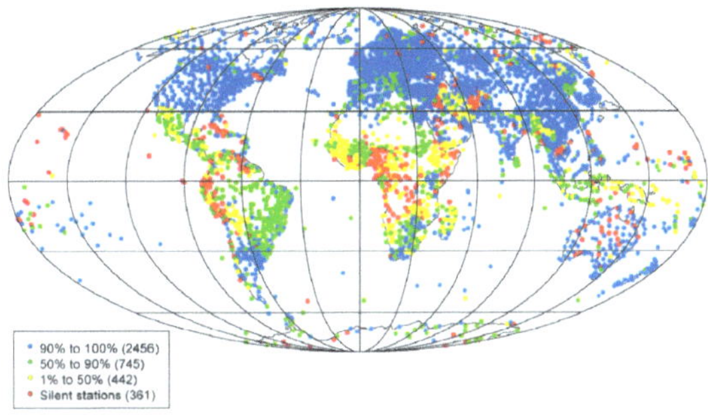

8. As per WMO guidelines, the synoptic observatories take observation every three hours beginning at 0530h (00 UTC), the current weather observatories take observation on hourly/half hourly basis and the upper air observatories measure upper winds at various levels two to four times a day.

9. Each observatory, after taking observations of the various weather elements at the specified time, prepares a report in a given format and pass it to the respective regional centres which, in turn, forward it to NMTN New Delhi for further global exchange.

10. To prepare a weather report all the elements listed above are required to be included. An example of a current weather observation report is given below: -

Consider that the following weather conditions are observed at Met observatory, Guwahati airfield: -

Station	:	Guwahati
Date and Time of Observation	:	25 May 22, 0700h IST (0130 UTC)
Surface Winds	:	Blowing from west (270^0), speed 05 Kt
Visibility	:	4000M
Present Weather	:	Rain
Clouds	:	2/8 Stratus at 1000'
		3/8 Stratocumulus at 3000'
		7/8 Altostratus at 9000'
		Total amount – 8/8
Surface Temperature	:	32.4^0 C
Dew Point Temperature	:	29.0^0 C
Relative Humidity (RH)	:	089 %
Atmospheric Pressure (QNH)	:	1009 hPa

11. After having made the above observation, there will be a requirement for communicating the above observation to the other users in a manner that the statement is brief but complete and is easy for the user to understand. Thus, the above report is framed in the prescribed coded format as follows: -

METAR 25/0130Z VEGT 270/05KT 4000M -RA FEW010 SCT030 BKN090 32/29 Q1009=

The above codded report denotes the elements as follows: -

METAR	Identification of Code	**25/0130Z**	Date and time of observation in UTC
VEGT	ICAO Location Indicator for Guwahati airfield	**270/05KT**	Wind direction 270^0 and Speed 5 Kt
4000M	Visibility in M	**-RA**	Prevailing weather is 'Light Rain
FEW010	1-2 Octa of low cloud (Stratus) at 1000 Ft height	**SCT030**	3-4 Octa of Low Cloud (Sc/Cu) at 3000 Ft
BKN090	5-7 Octa Medium Cloud (Ac/As) at 9000 Ft	**32/29**	DB Temperature is 32^0 C and DP Temperature is 29^0 C
Q1009	Atmospheric Pressure Reduced to Mean Sea Level is 1009 hPa	**=**	End of message indicator

12. It is worth mentioning that Meteorological observations and forecasts are most widely used by the personnel associated with the Aviation Industry including the Military Aviation. Most of the elements and weather features including the upper air pattern, information on monthly and seasonal climatological characteristics of an area, etc., are of vital importance to the Aviation industry.

13. The other type of surface observation is the Synoptic Observation and is taken on a three-hourly basis. The report of synoptic observation is prepared using a different code called SYNOP. These reports are also exchanged worldwide and are used to prepare Met Charts (Surface) for further analysis by the meteorologist. In surface charts, isobars are drawn to locate the pressure systems like low pressure, depression, high pressure, etc. Surface charts also give good indication of the characteristic and movement of prevailing air mass, movement and intensity of low-pressure systems, temperature & humidity profile, etc. On the basis of the analyzed charts forecasts and other reports are prepared by the meteorologists. A typical analyzed surface chart is shown in the previous chapter.

14. **Upper Air Observation**.   Weather phenomena does not occur only due to the surface conditions. Weather takes place in the troposphere and hence, all the layers of troposphere are required to be observed to understand the actual pattern. In fact, the upper air conditions often play more important role in causing weather. Hence, it is important to understand the situation of the entire tropospheric levels. Upper air observations are taken in respect of wind (direction & speed), temperature and humidity.

15. Upper air observation is taken using a balloon filled with hydrogen. Hydrogen being lighter than air makes the balloon ascend upward. The balloon is filled with a calculated amount of hydrogen so that it ascends at a known rate. When the rate of ascent is known, it is possible to calculate the vertical position of the balloon at a certain time after release of the balloon. The balloon is followed using a theodolite and readings in respect of azimuth and elevation angles are noted after some interval of time (generally one minute). These readings are used to calculate the wind direction and speed at a certain height. Observations are taken by upper air observatories at 00 UTC, 06 UTC, 12 UTC and 18 UTC.

16. In addition to the above observations, certain upper air observatories take observations using radio sounding equipment attached to a large balloon. This balloon can be followed to great height and the attached sounding equipment provides temperature and humidity information along with the wind direction and speed. Such observations are taken twice a day, at 00 UTC and 12 UTC.

17. Both types of upper air observations mentioned above are exchanged worldwide using different codes. The upper air observations are plotted in upper air charts for different levels. Stream lines are then drawn for different levels. The upper air charts are used to understand the movement of air mass, location and movement of upper air systems like cyclonic circulation, ant-circulation, jet stream, trough & ridge, fronts, etc.

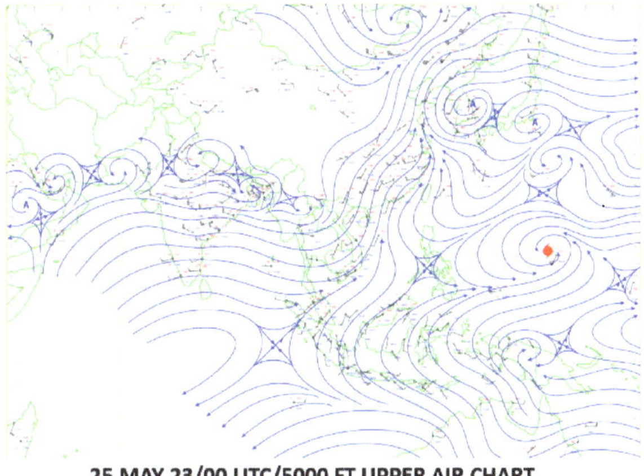

CHAPTER 3

ORGANIZATION STRUCTURE

INDIA METEOROLOGICAL DEPARTMENT (IMD)

1. The ministry of Earth Science (MoES), under the govt of India, is mandated to provide services for weather, climate, ocean and coastal state, hydrology, seismology and natural hazards; to explore and harness marine living and non-living resources in a sustainable manner for the country and to explore the three poles of the Earth (Arctic, Antarctica and Himalayas)

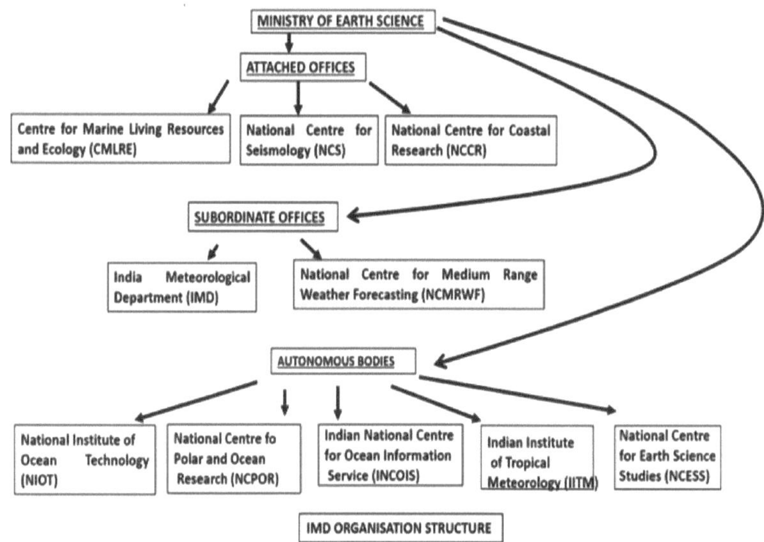

The Director General of Meteorology is the Head of the India Meteorological Department, with headquarters at New Delhi. There are 4 Additional Directors General at New Delhi and 1 at Pune. There are 20 Deputy Directors General of whom 10 are at New Delhi.

There are 6 Regional Meteorological Centres (RMC), each under a Deputy Director General.

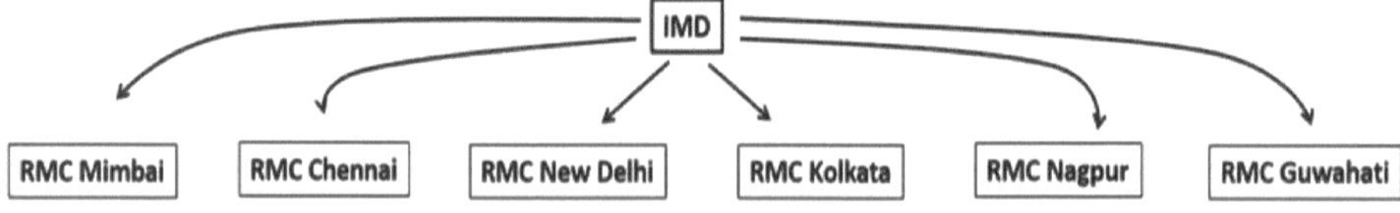

Under the administrative control of Deputy Director General, there are different types of operational units

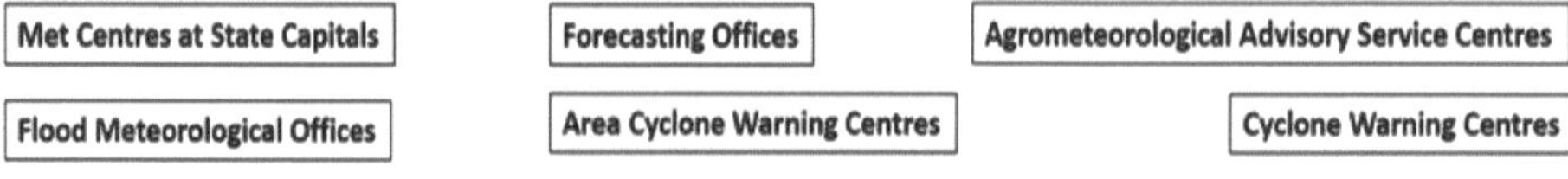

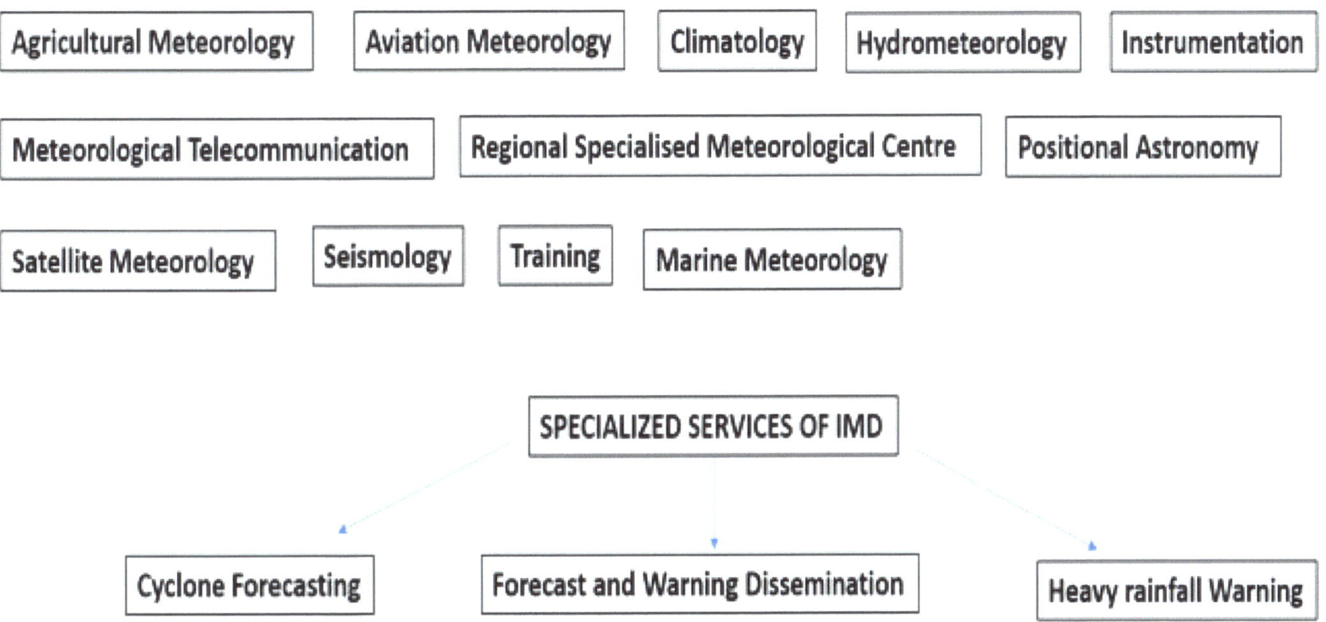

19. Brief description of the services rendered by a few disciplines of IMD are as follows:

    a)   **Hydrometeorological Services.** IMD renders assistance and advice on the meteorological aspects of hydrology, water management and multipurpose river valley projects management. These services are utilized by the Central Water Commission, Ministry of Agriculture, Ministry of Water Resources, Railways, Damodar Valley Corporation Flood Control Authorities and the State Governments. Hydromet Division of IMD caters the information on various rainfall products through its 'Customized Rainfall Information System (CRIS)', in form of reports and maps on the CRIS portal.

        i.   **Flood Meteorological Unit**. Flood Meteorological Offices (FMOs) have been set up by IMD at ten locations viz., Agra, Ahmedabad, Asansol, Bhubaneshwar, Guwahati, Hyderabad, Jalpaiguri, Lucknow, New Delhi and Patna. During the flood season, FMOs provide valuable meteorological support to the Central Water Commission for issuing flood warnings.

        ii.   **Rainfall Monitoring**. The real-time monitoring and statistical analysis of district wise daily rainfall is one of the important functions of the Hydrometeorological Division of IMD at New Delhi. Based on the real time daily rainfall data, weekly district wise, sub-division wise and state wise rainfall distribution summaries are prepared regularly by the Rainfall Monitoring Unit. Maps showing weekly and cumulative rainfall figures in 36 meteorological subdivisions of the country are prepared. This information is very important to many user agencies, particularly for agricultural planning.

    b)   **Meteorological Services for Agriculture in India**.   In order to provide direct services to the farming community of the country an exclusive Division of Agricultural Meteorology was set up in 1932 under the umbrella of India Meteorological Department (IMD) at Pune with the objective to minimize the impact of adverse weather on crops and to make use of favourable weather to boost agricultural production. The services of divisions are: -

i. Gramin Krishi Mausam Seva

ii. Dissemination of Agromet Advisories

iii. Feedback & Awareness of Agromet Service

iv. Training Programme to AMFUs

c) **Environmental monitoring and service.** This division of IMD conducts monitoring and research related to atmospheric constituents that are capable of forcing change in the climate of the Earth, and may cause depletion of the global ozone layer, and play key roles in air quality from local to global scales. EMRC also provides specific services to Ministry of Environment and Forest & Climate Change and other Government Agencies in the assessment of air pollution impacts. IMD contributes in the field of atmospheric environment to the World Meteorological Organization (WMO) Global Atmosphere Watch (GAW) programme.

i. **Ozone Monitoring Network.** National Ozone Centre of IMD is designated as secondary regional ozone centre for Regional Association II (Asia) of World Meteorological Organization. The centre maintains a network of ozone monitoring stations including Maitri and Bharati in Antarctica.

ii. **Precipitation and Particulate Matter Chemistry Monitoring.** IMD is monitoring Precipitation Chemistry through a network of eleven stations since 1970s. The precipitation chemistry network includes Allahabad, Jodhpur, Kodaikanal, Minicoy, Mohanbari (Dibrugarh), Nagpur, Port Blair, Pune, Srinagar, Visakhapatnam and Ranichauri. The rainwater samples collected from these stations are analyzed in Precipitation Chemistry Laboratory at IMD, Pune which is equipped with Ion-chromatograph, UV-VIS Spectrophotometer, Atomic Absorption Spectrophotometer, Semi-micro Balance, pH and Conductivity Meter. High Volume Samplers for collecting PM10, PM2.5 and Total Suspended Particulate Matter (TSP) have been installed at Delhi, Ranichauri, Pune and Varanasi. The filter papers are being analyzed for chemical characterization of aerosols.

iii. **Aerosol Monitoring Network.** Environment Monitoring and Research Center, India Meteorological Department has established Aerosol Monitoring Network by installing sky radiometer at twelve locations. The network is used to measure optical properties of aerosols such as Aerosol Optical Depth, Single Scattering Albedo, Size Distribution, Phase Function etc. The network stations are: New Delhi, Ranichauri, Varanasi, Nagpur, Pune, Port Blair, Visakhapatnam, Guwahati, Kolkata, Jodhpur, Rohtak, Thiruvananthapuram.

iv. **Black Carbon Monitoring Network.** Black Carbon Monitoring Network of 16 stations for measurement of Spectral Aerosol Absorption Coefficient, Equivalent Black Carbon Concentration and bio-mass burning component has been established during 2016. The names of the stations are New Delhi, Ranichauri, Varanasi, Nagpur, Pune, Port Blair, Visakhapatnam, Guwahati, Kolkatta, Jodhpur, Bhuj, Trivandrum, Ranchi, Amini, Chandigarh and Srinagar.

v. **Multi-wavelength Integrating Nephelometer Network.** IMD has established a network for measurement of aerosol scattering coefficient at twelve locations is under installation. The network stations are: New Delhi, Ranichauri, Varanasi, Nagpur, Pune, Port Blair, Visakhapatnam, Guwahati, Kolkata, Jodhpur, Bhuj, Thiruvananthapuram.

vi. **System for Air Quality Forecasting And Research (SAFAR).** The system for air quality forecasting and research (SAFAR) has been operationalized by IMD to monitor and forecast air quality in Delhi.

This is a joint project of IITM and IMD. The system is also operational at Pune, Mumbai and Ahmedabad. All major air pollutants (PM2.5, PM10, Ozone, CO, NOx (NO, NO2), SO2, BC, Methane (CH4), Non-methane hydrocarbons (NMHC), VOC's, Benzene, Mercury), solar radiation and meteorological parameters are measured at ten air quality station installed in each city. SAFAR provides location specific information on air quality in near real time and its forecast 1-3 days in advance

d)   **Meteorological Telecommunication Services in IMD.**   Information System and Services Division (ISSD) of IMD provides support functions needed for meteorological data and processed weather products to the users, both national and international, round-the-clock on near real-time basis and is known internationally as Regional Telecommunication Hub (RTH) under the aegis of WMO.

The Meteorological Telecommunication in IMD consists of an integrated network of point-to-point & point to multipoint (MPLS VPN) links and meteorological centers within the country and the world for receiving data and relaying it selectively. It is mainly organized on a two-level basis, namely: -

i.   The Meteorological Telecommunication Network (MTN) within the Global Telecommunication System (GTS) of World Weather Watch (WWW) program of World Meteorological Organization (WMO), and

ii.   The National Meteorological Telecommunication Network (NMTN).

In regard of Meteorological Telecommunication Networks on the GTS, New Delhi Telecommunication center is a designated Regional Telecommunication Hub (RTH) located on the Main Telecommunication Network (MTN). The MTN is the core network of GTS which connects three World Meteorological Centers (WMCs) and 14 other RTHs on the MTN. RTH New Delhi is also a National Meteorological Centre (NMC) for telecommunication purposes within the framework of GTS.

RTH New Delhi is directly connected with WMC Moscow, RTH Tokyo and RTH Cairo, RTH Beijing, RTH Toulouse, RTH Jeddah and WMC Melbourne located on the MTN; RTHs Bangkok and Tehran and NMCs Colombo, Dhaka, Karachi, Kathmandu, Male, Muscat and Yangon in the RMTNs.

e)   **SATMET Services.**   Satellite Meteorology Division started functioning in India Meteorological Department since early 70's. From 1972 to 1982, IMD used to receive the Satellite imageries of NOAA and NASA meteorological satellites through SDUC and images were printed on photographic paper for using in weather forecasting.

The Indian National Satellite (INSAT) programs a series of multipurpose geo-stationary satellites by ISRO to satisfy the telecommunications, broadcasting, meteorology, and search and rescue operations started in 1982. INSAT is one of the largest domestic communication satellite systems in Asia-Pacific region with nine operational communication satellites placed in Geo-stationary orbit. It is a joint venture of the Department of Space, Department of Telecommunications, India Meteorological Department, All India Radio and Doordar shan.

The first successful INSAT-1B satellite data receiving and processing system was established in 1983 in IMD as a full-flashed satellite Meteorological Division for providing satellite Metrological services to the nation. INSAT-1 series of satellites were Multipurpose satellites, having a meteorological payload, two channels Very High-Resolution Radiometer (VHRR) for imaging the Earth in Visible (0.55-0.75 μm) and Infra-Red (10.5-12.5 μm) channels having resolution of 2.75 X 2.75 kms in visible and 11 X 11 kms in IR channel. Since the launching of INSAT-2 series of satellites, the spatial resolution of VHRR enhanced to 2 X 2 kms in visible and 8 X 8 kms in IR channel. In INSAT-2E satellite VHRR payload also added Water Vapour (5.7-7.1μm) channel having resolution 8X8 kms along with new payload, charged couple device (CCD) of three channels

(Visible -0.62 to 0.69μm), Near Infra-Red 0.77 to 0.86μm) and Short Wave Infra-Red 1.55 to 1.77μm) bands, having a resolution of 1KM.

During both INSAT-1 and INSAT-2 series, the temporal resolution of scan acquisitions was 3 hourly. Since 1996, the satellite imageries were opened for public through IMD website. A dedicated meteorological satellite, Kalpana-1 was launched on 12th September, 2002 carrying VHRR payload and multipurpose INSAT-3A satellite was launched on 10th April, 2003 carrying VHRR and CCD payloads along with communication transponders. The temporal resolution of scan acquisitions by Kalpana-1 was 3 hourly with 3 triplets (00,06,12 UTC) during 2002 – 2005, hourly with 3 triplets (00,06,12 UTC) during 2005 – 2008 and Half Hourly during 2008-17. The temporal resolution of scan acquisitions by INSAT-3A, VHRR payload was 3 Hourly (2003 – 2008) and then Hourly (2008-16). The CCD payload used to take acquisitions six times during day (03,05,06,07,09 &11 UTC). Both INSAT-3A and Kalpana-1 had been decommissioned and discontinued since September 2016 and September 2017 respectively.

At present, there are two operational meteorological satellites; INSAT-3D & INSAT-3DR carrying 6 channel imager for imaging the earth in visible (0.55-0.75um),SWIR (1.55-1.70um) of resolution 1KmX1 Km, MIR (3.80-4.00um), TIR-1 (10.30-11.30um), TIR-2(11.50-12.50um) of resolution 4KmX4Km and WV (6.50-7.10um) of resolution 8KmX 8Km and 19 channel sounder consisting of 7 channels of LWIR (14.71-12.02um), 5 channels of MWIR (11.03-6.51um), 6 channels of SWIR (4.572-3.74um) and one channel of visible (0.695um) each of resolution 10X10 Km scan the atmosphere for derivation of profiles. INSAT-3D was launched on 26th July, 2013 and located at 82 degree east and INSAT-3DR was launched on 08th September 2016 and is located at 74 degree east. At Present 48 satellite passes are acquired daily from each INSAT-3D and INSAT-3DR IMAGER payload, are being used in stager mode so that after every fifteen minutes a new set of images/ product become available to the forecasters.

f) **Marine Weather Service**. IMD is the nodal agency to provide Marine Meteorological Services since 1966. Marine Meteorological Division, New Delhi and Climate Research Centre, Pune coordinate to provide the service. Through Marine Weather Service, IMD facilitates following:

   i. Round the clock watch the AoR (North of equator and west of 98.5⁰ E

   ii. Routine forecast for Indian Navy, shipping, fishermen, off-shore oil exploration, etc.

   iii. Carries out responsibilities of forecasting under Global Marine and Distress Safety (GMDSS) for Indian Ocean region north of the equator.

   iv. Forecast for tourism.

   v. Special warnings and advisories for severe weather conditions (cyclone) over north Indian Ocean.

   vi. Preparation of Marine Climatological Summary (MCS).

   vii. Support during Marine Pollution Emergencies.

g) **Cyclone Warning Services**. IMD is the nodal agency to provide cyclone warning services in the country. It has a three-tier network with cyclone warning divisions at new Delhi, three area cyclone warning centres at Mumbai, Chennai and Kolkata and four cyclone warning centres at Bhubaneswar, Visakhapatnam, Ahmedabad and Thiruvananthapuram. IMD also acts as Regional Specialized Meteorological Centre (RSMC) and Tropical Cyclone Advisory Centre, New Delhi.

Issue of warnings, advisories for various agencies, bulletins and press release on formation, intensification and movement of tropical cyclones, alerting civil agencies to take adequate precautionary measures, etc are the major functions of Cyclone Warning Services of IMD.

RSMC of IMD is also responsible to provide tropical cyclone and storm surge guidance to 13 member countries of WMO which are Bangladesh, India, Iran, Maldives, Myanmar, Oman, Pakistan, Quatar, Saudi Arabia, Sri Lanka, Thailand, UAE and Yemen.

h) **Heavy Rainfall Warning Services**. MD provide information to all concerned agencies on occurrence or expected occurrence of Heavy Rainfall over any part of the country. Following advisories/warnings/bulletins are issued in respect of heavy rainfall: -

   i.   Nowcast (lead time validity up to 03 hours) at station or district level.

   ii. Short to medium range forecast (lead time validity 01 to 05 days) at station, block, district, Meteorological sub-division level.

   iii. Extended range forecast (Lead time validity up to 04 weeks) at meteorological sub-division level during week 1 and week 2.

   The warnings/advisories/bulletins are disseminated to: -

   i.   National Disaster Management Authority

   ii. NDRF

   iii. State Disaster Management Authority

   iv. Other Central and State govt authorities

   v.   Print and Electronic media

   vi. Other users

j) **Heat and cold wave warning services**. IMD alerts all concerned agencies on occurrence or expected occurrences of heat wave and cold wave condition over any part of the country.

k) **Fog Information and Warning Services**. Fog is almost a regular phenomenon that engulf north Indian region during winter months and disrupt aviation, railway service and road transport service causing to inconvenience, economic loss and even loss of lives and property in accident. IMD has built up methodology to provide forecast and timely report of fog, that is, visibility. Warnings, advisories and reports are issued and disseminated timely to various agencies like disaster management authorities, airlines, airports, railway and highway authorities.

l) **Central Aviation Meteorological Division**. Aviation industry is perhaps the agency which needs the services and guidance of meteorological services at every stage of the flights. Air operation cannot be successful without adequate information like current weather of aerodrome of origin, destination and enroute weather conditions; information on upper air temperature, winds, clouds, turbulence, etc. IMD provides a crucial service to the national and international civil aviation sector in fulfilment of the requirements prescribed by the International Civil Aviation Organization (ICAO) and the Director General of Civil Aviation of India (DGCA).

   Central Aviation Meteorological Division (CAMD) at DGM, New Delhi is the nodal office for the aviation services in the country. It also maintains the liaison with ICAO, WMO, DGCA, AAI and Airlines on technical aspects of aviation. The installation and maintenance of Airport Meteorological Instruments are done by IMD (SI Division), Pune. The telecommunications requirements for aviation are managed by the IMD (Telecommunication Division) functioning at New Delhi and by the telecommunication unit of Airport Authority of India.

   The principal requirements in the aviation point of view are:

i.   Supply of current weather observations to all aeronautical users

ii.   Issue of forecast and warnings on meteorological hazards to aviation

iii.   Adherence to procedures and formats for dissemination of products to aviators

Aerodrome Meteorological Offices functioning at Mumbai, Kolkata, Delhi and Chennai airports also serve as Meteorological Watch Offices (MWOs) catering to flights in respective Flight Information Regions (FIR). These services are provided through 17 Aerodrome Meteorological Offices (AMO) and 72 Aeronautical Meteorological Stations (AMS) located at various national and international airports of the country.

m)   **Positional Astronomy Centre, Kolkata**. The Indian Astronomical Ephemeris is published annually by the Positional Astronomy Centre, Kolkata for providing astronomical data to observational astronomers and other users. The ephemeris includes:

i.   Sun/Moon rise/set timings

ii.   Phase of Moon

iii.   Eclipse events of the year

iv.   Astronomical phenomenon

v.   Sidereal longitude of Sun, Moon and planets

vi.   Mean places of stars

vii.   Night sky charts

The Rashtriya Panchang is being published by the Positional Astronomy Centre since 1879 Saka Era (1957-58 A.D.). This is being done with the objective of unifying the divergent practices of calendar systems existing in different parts of the country and also to promote panchang calculations on a modern scientific basis. The Rashtriya Panchang is being published in all the major Indian languages besides Hindi and English.

n)   **Data Supply**. On various occasions various agencies like educational/research institutes or other govt/private agencies need meteorological data for a certain period or a specific time. IMD enhanced the efficiency & transparency of data supply and has reduced the service delivery time by opening up an online portal for the purpose. Various users can log in and receive data as per the existing norms. This function is carried out by National Data Center, IMD, Pune.

SIGNIFICANCE OF WEATHER ELEMENTS

CHAPTER 4

SURFACE WINDS

1. The horizontal motion of air is called wind. Winds at the surface of the earth plays an important role in the process of weather formation. Besides, it also has remarkable impact on human comfort and discomfort. We always welcome a pleasing cool breeze and on the other hand, the hot and dry loo makes our life highly uncomfortable.

2. Every place experiences a pattern in the wind flow which generally changes from season to season. For example, the Indo Gangetic plain experiences dry and hot westerly wind flow during the pre-monsoon (April to June) season while during the monsoon season it is swayed by the cool and moist easterly to south-easterly monsoon current. Knowledge of such wind pattern is of vital importance in building large structures like large factories, runways, bridges, etc. Pattern of wind flow is also given adequate weightage in constructing our own houses. When we say pattern of any element, it's the long time average of that element that prevails over a place during a particular period of the year. In other words, climatological data pertaining to the element.

3. In meteorology the direction from which surface wind flows is considered as the direction of wind. It is reported in 36 point of compass in three digit. Example 360^0, 050^0, 270^0, etc. Speed is generally reported in Knot (Kt) or Km per hour (Kmph).

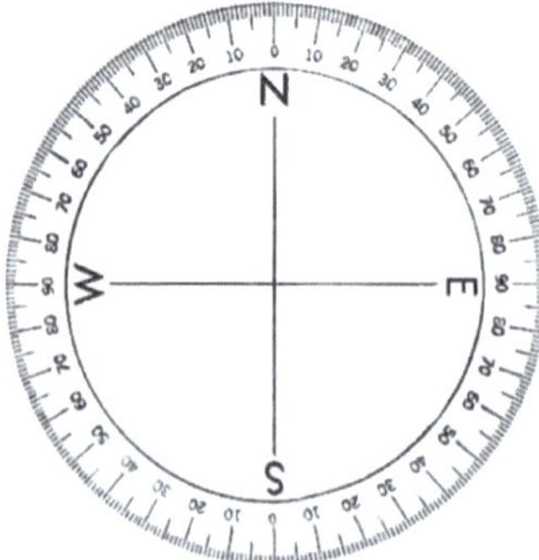

4. Direction and speed of wind are governed by the theory of fluid dynamics. We will keep it simple to understand and will not discuss the scientific theory in depth. Air being a fluid, moves from the area of higher atmospheric pressure towards an area of lower pressure.

5. Before discussing further on the subject, it would be necessary to discuss about the global wind pattern. Many of you must aware of the pressure belts on the earth. There are distinct high and low pressure zones on the earth. Winds generally flow from the high pressure belts to the low pressure belts. It is called the global circulation. There are some seasonal changes in the position of these belts due to position of the Earth in respect to the Sun during various parts of the year.

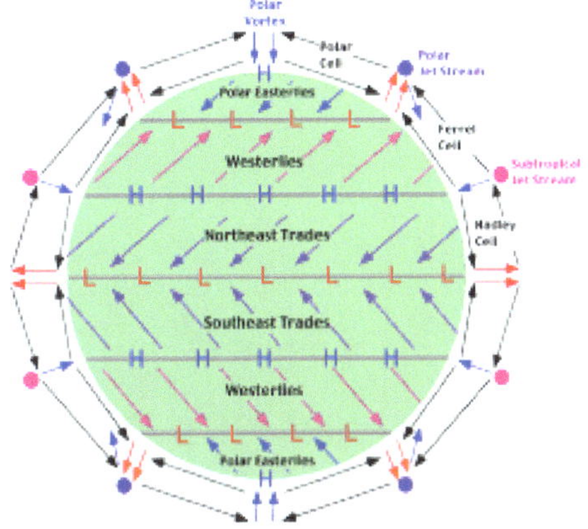

6. In both the hemisphere of the earth following pressure zones prevail:-

    Polar High (b) Sub-Polar Low (c) Sub-tropical High (d) Equatorial Low

7. It is seen in the figure that the winds flow from the Highs to the Lows. But what is important to note here that the winds are not flowing in a perpendicular direction as it should have. Rather, the movement of air is in a slant direction. This is because, the winds tending to flow perpendicularly from the Highs to the Lows are deflected to the right and left in the southern and northern hemisphere respectively by a force called the Coriolis force which is produced due to the rotation of the Earth.

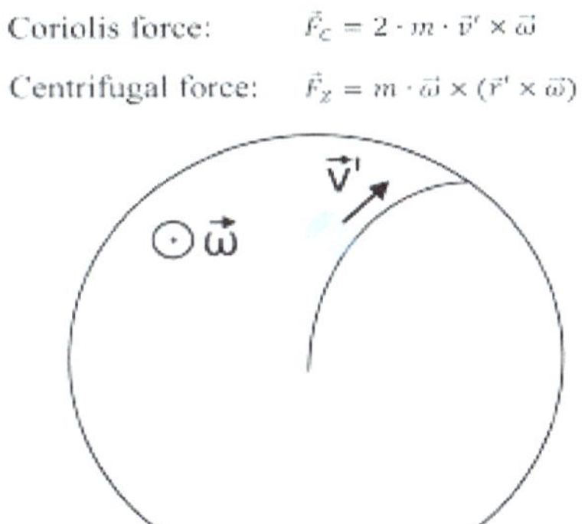

Coriolis force: $\vec{F_C} = 2 \cdot m \cdot \vec{v}' \times \vec{\omega}$

Centrifugal force: $\vec{F_Z} = m \cdot \vec{\omega} \times (\vec{r}' \times \vec{\omega})$

8. In other words, the deflection is anti-clockwise in the southern hemisphere and clockwise in the northern hemisphere. The theory pertaining to the said pattern of wind flow is called **Buys Ballot Law** which states that *in the northern hemisphere, if you stand with your back to the wind, your left hand will indicate the area of low pressure.*

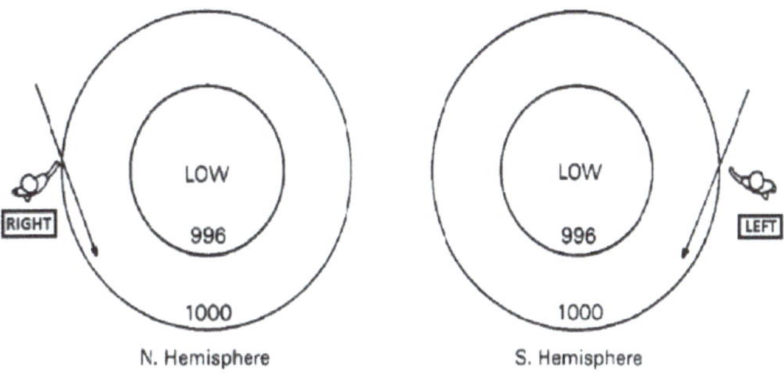

9. The pattern of wind flow can be better understood in the following figure which is a typical Isobaric (Isobar – line joining places of equal pressure) chart of India and neighborhood during southwest monsoon season.

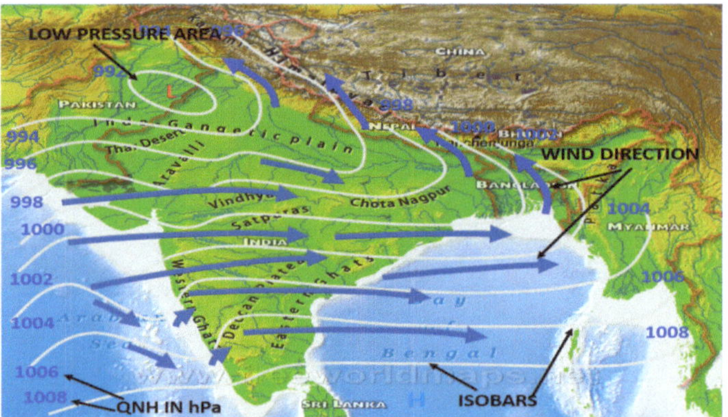

10. In the above Isobaric map, the white curved lines are the Isobars and the blue arrows indicate the direction of wind at the surface. It is to be noted that the direction of the winds are more or less parallel to the Isobars. Thus, during the monsoon the places over the Gangetic plains and many parts of NW India experiences easterly (Purvi) and the most of the other regions experience westerlies (Paschimi).

11. During Winter the pressure pattern is the reverse of the monsoon season with the High-Pressure area prevailing over the higher latitude (North) and Low-pressure area being located to the south (Indian Ocean). Hence, a reversal in the wind direction also takes place over the sub-continent during winter season.

12. Though the atmospheric pressure distribution is the major factor to determine the direction and speed of the winds, there are some other local factors which also influence the winds. These factors arise due to the topography of a locality. Hence, such winds generated by the local topographies are called the Local Winds. Such local winds will prevail when the effects of topographical features over-power the force generated due to pressure gradient.

Local Winds.

13. **Land Breeze and Sea Breeze**. These terms might be familiar to the most people. The winds that blow from land to sea are called the Land Breeze and that blow from the sea to the land are called the Sea Breeze. Land and sea breeze are the characteristics of the coastal areas and occurs on a daily basis, being more prominent in the summer months.

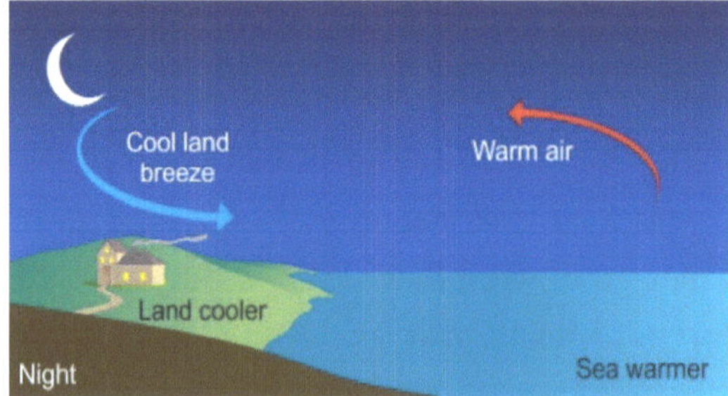

14. During night the rate of long wave radiation is higher over land than over sea. It means that land mass gets cooled down faster than the sea surface. And thus, the air adjacent to the land surface is colder than that over the sea. Cold air is denser and heavier than the warm air. Hence, the heavier air mass will flow towards the warmer airmass, that is, towards the sea. Such winds that flow from the land to the sea during the night time are called the land breeze. The air over the sea being warmer and lighter, rises up when replaced by the land air. Aloft, the air overlying the land descends to replace the air that has moved towards the sea thereby making a horizontal and vertical circulation.

15. During day the reverse of the land breeze process takes place. Due to incoming short-wave solar radiation the land gets heated up faster than the sea surface and the sea surface remains comparatively cooler than land. Hence, winds from the sea flow towards the land and tend to defuse the effect of solar heat over the land. For the same reason the coastal areas mostly do not experience heat wave condition.

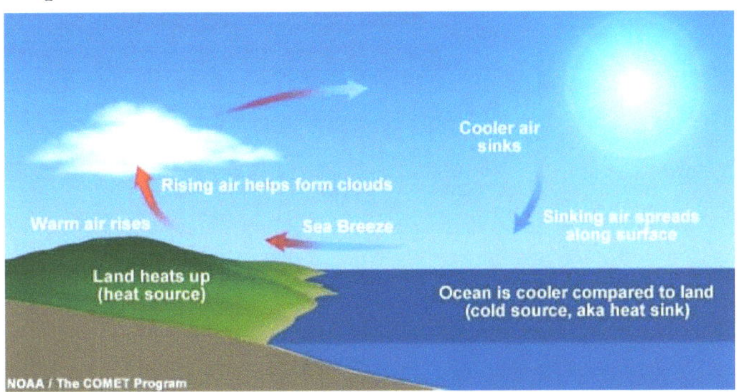

16. In general, land breeze is weaker than the sea breeze and affects only a few Kilo Metre from the coast. Sea breeze is much stronger and can penetrate into the land as deep as 50 to 80 Km. When the Isobaric pattern or/and local topography favours, the speed becomes stronger and can penetrate even deeper. The sea breeze that originates on a daily basis during the Pre-monsoon season (March to May) over north Konkan region (north Maharashtra coast) affects up to Pune which is above 100 Km (aerial distance) from the coast. During the monsoon season the isobars are compact over Saurashtra and Kutch region. During this season very strong winds often of the gale force (mean speed > 34 Kt) prevail over that region.

17. **Katabatic and Anabatic Winds**. These winds affect the areas adjacent to the high hills. During night, the slopes of the mountains cool down more rapidly than the surrounding air due to radiation. With the slopes, the air very close to it also cools down more rapidly and hence, become denser and heavier than the air away from the slope. This results in movement of cool air down the mountain slope. Such winds are called the Katabatic Winds.

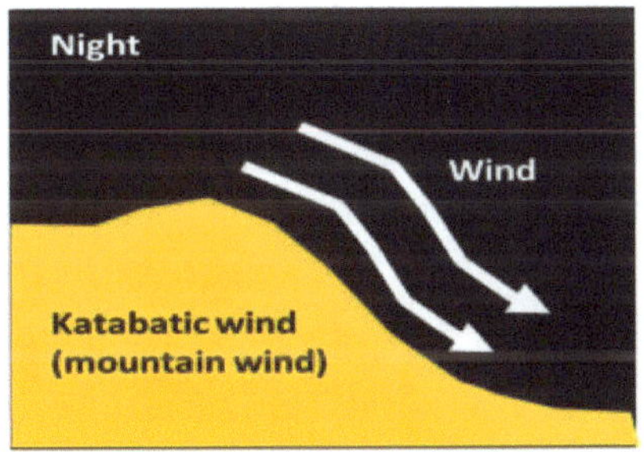

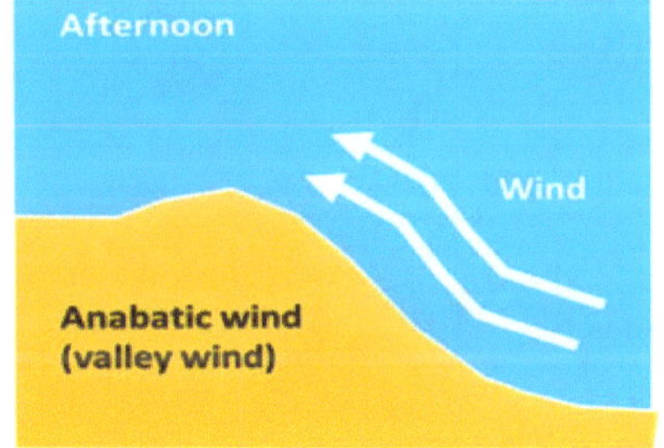

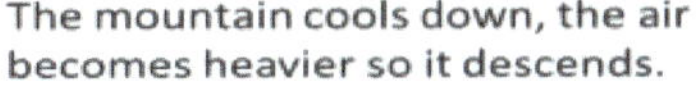

18. During day time, the hill slopes get heated up more rapidly due to the incoming solar radiation. So, the air adjacent to the slope becomes warmer than the surrounding and rises up and makes the surrounding air to move upwards along the slope. These upsloping winds are called the Anabatic Winds.

19. Katabatic winds start affecting from mid night to morning hours and the Anabatic winds are the afternoon/evening phenomena. Both types of winds are more prominent when the sky is clear so that there is adequate sunshine during day to heat up the air near the hill slope and during night there is sufficient radiation from these slopes. These winds, however, are not as stronger as the land/sea breeze and can hardly be experienced in prevalence of steep pressure gradient in the locality. However, these winds play important role in causing thunderstorm activities over the foothill areas. This is one of the factors for making the Brahmaputra valley one of the most thunder prone areas in the world.

20. Local winds, which are known by different names in different part of the world, plays important role in the lives the people. They also hold a major space in the literature and has inseparable role to play in art and culture of the locality.

21. **Estimating Wind Speed**. It would certainly be interesting if one can estimate the wind speed without the use of any instrument. While going for a walk or while working outdoor, prevailing wind speed matters and influences the output. For instance, prevalence of gentle breeze while walking would be refreshing. On the other hand, when strong breeze prevails, we may have to suspend walking or it would be difficult for them to manage who are working outdoor or playing in the ground.

22. Sir Francis Beaufort, who was a hydrographer and an officer in Royal Navy, devised a scale in 1805 to estimate wind speed. The scale was later named after him as the **Beaufort Scale**. The scale is as shown below: -

Description	Speed (Knots)	Mean Speed (Knots)	Beaufort Force	Terms used for description	Wind effect on land
Calm	<1	0	0	Calm	Calm, Smoke rises vertically
Light air	1—3	2	1	Light	Wind motion visible in smoke
Light breeze	4—6	5	2		Wind felt in exposed skin, leaves rustles
Gentle breeze	7-10	9	3		Leaves and small twigs in constant motion
Moderate breeze	11—16	13	4	Moderate Fresh	Dust and loose papers are raised. Small branches begin to move.
Fresh breeze	17—21	19	5		Small trees begin to sway
Strong breeze	22—27	24	6	Strong Gale	Large branches are in motion. Whistling is heard in overhead wires. Use of umbrella becomes difficult.
Moderate gale	28—33	30	7		Whole tree in motion. Difficulty Experienced walking in the wind.
Fresh gale	34—40	37	8		Twigs and small branches break from trees. Cars veers on road.

Strong gale	41—47	44	9	Severe Gale	Large branches break from trees. Light structural damage.
Whole gale	48—55	52	10		Trees broken and uprooted. Considerable structural damage.
Storm	56—63	60	11		Widespread damage to structures and vegetation.
Hurricane	64—71	68	12		Considerable and widespread damage to structures and vegetation

Sir Francis Beaufort

23. Knowledge of the force and description given in the table above is very important on the present day also. It has helped an observer preparing a Met report from a remote location for the purpose like operation of a helicopter for any operational reason or rescue operations. It is also helpful to the common men and those work outdoor including the fishermen out in the sea.

CHAPTER 5

VISIBILITY

1. The transparency of the atmosphere is called visibility. In other words, it is the maximum distance at which an object located at surface can be seen with naked eye and can be recognized as such. In meteorology horizontal visibility is only considered as visibility. Slant visibility will always be better than the horizontal visibility (if not seen through the clouds). Visibility is reported in metre or kilo-metre.

2. Poor visibility does not contribute in making a weather condition but it is a product of certain weather conditions like fog/mist, haze, dust haze, dust storm, dust whirl, strong surface winds, smoke haze, volcanic ash, precipitation, etc.

3. Poor visibility condition affects our lives and the economy is even nailed. Persistent and widespread poor visibility condition, besides restricting the vehicular movements, also leads to cancellation/diversion of flights and trains. Visibility plays a very important role in aviation. It is very crucial during landing or take off of an aircraft. The pilots are rated in respect of their capability to make a landing or take-off in poor visibility condition. For example, if the visibility, say less than 2000 M, then all pilots are not permitted to execute a landing or take off. Also, in certain runways, where instrumental landing system (ILS) is not available, are not clear for undertaking air operations in certain range of poor visibility condition. In other words, constant monitoring of visibility and subsequently reporting the correct visibility by a Met observer plays a vital role in air operations and aerospace safety.

4. Different weather conditions affect visibility in different range. Same are tabulated below:-

Elements	Visibility Range	Remarks
Haze	5000M to 2000M	
Smoke Haze	5000M or less	
Dust Haze	5000M or less	
Mist	5000M to 1000M	
Fog	Less than 1000M	
Precipitation (Rain, Drizzle, Snow, Sleet, etc.)	More than 5000 M to Nil	Occasionally fog/mist in combination with precipitation adds to further decrease in visibility.
Shower	2000 M to Nil	
Dust storm/Sand storm	Less than 1000M to Nil	
Dust raising wind	Less than 5000M	Dust and loose particles raised by strong surface winds with mean speed 20 Kt or more.

Poor visibility condition in dust storm and rain respectively

5. **Haze**. Suspension of minute solid particles which reduces transparency of atmosphere. Generally, haze reduces visibility to 3 to 5 Km but may reduce visibility further down. The minute solid particles may be dust, salt, aerosols, or photochemical smog that are so small (with diameters of about 0.1 micron (0.00001 cm) that they cannot be felt or seen individually with the naked eye, but the in aggregate reduces horizontal visibility. Haze is reported only when the prevailing relative humidity is less than 75%.

6. **Smoke Haze**. In smoke haze transparency of the atmosphere is affected by suspension of very small smoke particles. Visibility may reduce to as less as a few hundred metre. In India smoke haze commonly occurs over north India during post monsoon and winter season due to industrial smoke, and burning of agricultural wastes. Smoke haze, in combination with mist is known as smog. Many flights get cancelled from/to the cities like Delhi due to smoke haze/smog during winter season. Besides affecting the flying activities, smoke suspended in atmosphere has also been a significant health hazard of the urban population. Smoke haze gives a light bluish or grayish look to the surroundings.

7. **Dust haze**. Dust haze is reported when visibility reduces to 5 Km or less due to suspension of minute dust particles and prevailing relative humidity is less than 75%. Raising of loose dust to the atmosphere is caused due to occurrence of dust storm (Andhi) or dust raising winds at surface. It generally occurs over dry areas during pre-monsoon season (India) and commonly experienced in northwest and north India. However, it is observed in other parts of the country when dust raised to great height by dust storm over northwest India and the same get migrated with the strong zonal westerlies at mid tropospheric levels. There are instances when dust raised over west UP or Rajasthan by dust storm are observed as dust haze after a few days over Brahmaputra valley. Occasionally, dust raised due to strong surface winds over south Pakistan region are carried to as far as Gujarat, Maharashtra and Telangana by lower tropospheric north westerlies. Visibility may drop to 5 Km to a few hundred metre in dust haze and aviation may be seriously affected. What is more concerning with dust haze in aviation is that, the poor visibility condition prevails for a long period leading to suspension of landing/take off of aircrafts and cancellation of flights. Improvements take place when most of the dust particles settles down or the surface wind direction changes significantly to carry away the dust particles or after a spell of rain. The surroundings look brownish in dust haze.

8. **Mist**. Mist occurs due to condensation of water vapours into minute water droplets in the air column adjacent to the surface. Such condensation takes place when relative humidity is 75% or more and prevailing temperature is sufficiently cool for the condensation to take place. Visibility may drop to 1 Km in mist and often flights have to be cancelled due to poor visibility condition in mist.

9. **Fog**. When mist is thick, that is, condensation takes place at a greater rate, we call it fog. Visibility in fog is less than 1000M and relative humidity is close to 100%. Essentially, fog is a type of cloud resting on the ground. The mechanism of occurrence of fog is the same as that of mist. However, the temperature needs to drop by a greater scale to saturate the atmosphere with the available moisture so that condensation takes place rapidly and vigorously for formation of fog.

10. Fog has been a major aviation hazard in India. Persistent fog often leads to cancellation flights. In addition, it also leads to cancellation of trains or delay in the schedules. In India it is a common phenomenon during the winter season. Some of the airfields in north India come to a standstill for hours during the winter season. Widespread fog also affects transportation and vehicular movement affecting day-to-day life.

11. Great efforts have been made across the world carrying out research on dispersal of fog to minimize its hazardous effect on aviation. But little has been achieved so far. However, appreciable results have been attained from the present-day satellite technology and through Numerical Weather Prediction (NWP) system in respect of location, characteristics, movements and dissipation of fog.

12. Fog is categorized into two types. Radiation fog and advection fog. The most common type in India is the radiation fog which is generally experienced over the land areas in the winter season mainly over north, central and eastern parts. It occurs due to radiational cooling of the earth surface during night.

13. **Radiation Fog**. Over plains of north India, the earth surface cools rapidly during night. Once the surface gets cooled, the air layer close to the surface also cools. When this air layer attains dew point temperature, the moisture available in the air starts condensing and ultimately causes formation of fog.

14. Radiational cooling is an every-day's phenomenon over north India during the winter season. However, fog does not occur on a daily basis in this season. This is because, some other favourable conditions need to prevail for formation of fog. These factors are briefly discussed below:-

   a)   The surface needs to be cooled sufficiently so that the air layer close to the surface attains dew point temperature and condensation of the water vapours can take place. For adequate cooling, the outgoing long wave radiation should be uninterrupted. This will be possible only when the sky remains clear during night. If the sky is covered with clouds, outgoing radiation is arrested to a significant extent. Hence, clear sky condition is an essential factor.

   b)   If the wind at surface is calm or very light, then cooling by conduction will extent only to a few centimetre above the ground. In such case the condensed moisture in the very thin layer of air will settle down on surface as dews. If light winds (say about 03 to 05 Kt) prevail over the surface, then the cooling is transmitted to some height above the ground and the dew point is maintained in all levels within this layer and formation of fog takes place. On the other hand, if the winds are stronger, cooling is spread by vigorous turbulence to a much deeper layer.

   c)   The other important factor is prevalence of high relative humidity. When relative humidity is high, dew point temperature can be attained through lesser cooling. In other words, formation of fog can take place in relatively higher temperature.

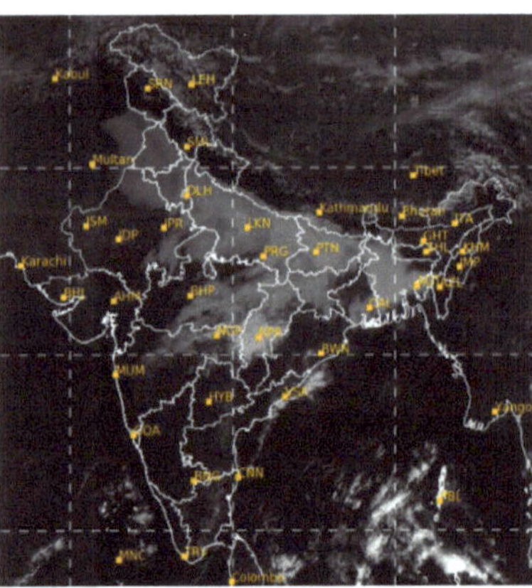

Fog can be detected by Satellite
This is an INSAT imagery (Visible band) at 0945h on 04 Jan 23 Showing widespread fog extending from NW India to Bangladesh and over central part of the country.

15. **Influence of Local Topography**.   Radiation fog is sometime seen as a widespread phenomenon covering a large area like in north and northwest India during the winter season. However, in of of the cases it is a localized phenomenon and are largely influenced by the local topography. Some of the topographical effects are briefly discussed below:-

   a)   **Wet or water-logged ground**. Radiation is less in such areas. However, availability of abundance of moisture plays the major role and becomes the dominating factor.

   b)   **Coastal areas**. In coastal areas adequate moisture is available. Besides, there are abundant hygroscopic nuclei available from the sea spray around which condensation takes place. Advection fog from the sea surface may also get transported to the coast. In India however, fog over the coastal regions is not common due to prevalence of comparatively higher temperature due to geographical location.

c) **Valleys**. In the valleys, small or large in area, are assisted by katabatic winds for formation of fog during late night or early morning hours. In presence of a river flowing in the valley enhances the occurrence of fog. Occurrence of fog in the Brahmaputra valley is an ideal example. Sometime the river itself is free of fog as radiation from the water surface is comparatively less than the either sides.

16. **Advection Fog**. It is caused by the movement of warm and moist air to a cold land surface. The availability of moisture assists fog formation when the warm moist air comes in contact with the cold surface. Advection fog moving in from the sea also is a contributory factor in coastal belts. In India, this type of fog is reported by the coastal stations of Saurashtra and Kutchch region during winter season.

17. **Steaming Fog and Frontal Fog**. These are other two types of fog. Steaming fog is formed by the evaporation of water from a warm sea surface into cold overlying air. Frontal fog is caused by precipitation or very low cloud near a frontal surface. This is also referred to as upslope fog.

18. **Vertical Extent of Fog**. Fog generally does not extend beyond 1000M. Once fog gets lifted from the ground, it is seen as low clouds (stratus). Basically, mechanism of formation of stratus cloud is similar to that of fog. On penetration of sunlight to the surface, fog gets lifted due heating of the land. Once lifted up, fog (low clouds) dissipates rapidly or are blown away by lower level winds.

CHAPTER 6

PRECIPITATION AND CLOUDS

1. **Precipitation**. Fall of liquid water drops or snow/sleet/hail on the surface of the earth is called precipitation. Thus, rain, drizzle, shower, hail, snow, sleet are all forms of precipitation. The form of precipitation received on the ground depends on the source cloud and upper air temperature profile of the area.

2. **Mechanism of Precipitation**. Initial condition for precipitation is formation of cloud. In other words, condensation of the water vapours is the basic requirement. Process of condensation above the ground is similar to that of fog or mist near the ground. Condensation in upper air causes formation of clouds. The condensed water droplets are so small that they remain suspended in air countering the gravitational force with the help of prevailing vertical current of air. However, in favourable conditions, when condensation around the nuclei continues to such an extent that the water droplets grow larger and heavier for the vertical current to hold it in air. In such condition the droplets (water or snow) fall on the ground as precipitation.

Condensation Nuclei. Condensation nuclei are **fine** *particles (particles of smoke, dust, etc.), suspended in the atmosphere. When the water vapours come into contact with these nuclei, vapour molecules condense and accumulate. When the weight of the nucleus and its condensate becomes too heavy, it becomes visible and falls in the form of precipitation.*

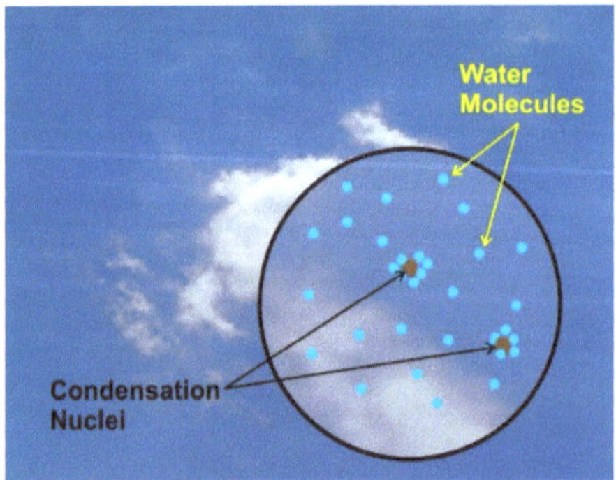

3. **Types of Precipitation.** Based on the nature of the falling liquid or solid precipitation, it is broadly classified into the following: -

a)   Drizzle

b)   Rain

c)   Shower (Rain or Snow)

d)   Freezing precipitation

e)   Snow

f)   Hail

Characteristics or types of precipitation broadly depends on one or more of the following: -

a)   Prevailing synoptic situation (Surface pressure or upper air pattern)

b)   Atmospheric temperature profile & atmospheric Instability

c)   Topography

d)   Geographical Location

e)   Source Cloud

4.   In fact, precipitation is preceded by formation of suitable (rain bearing) clouds; and formation of the suitable clouds depends on convergence (accumulation) of air having adequate moisture over a place at surface or at any level in upper air; favourable condition for lifting of the accumulated air (Triggering mechanism) to a level where the air mass attains saturation; and coldness (On attaining dew point temperature) to form clouds. Subsequently, when the water droplets in the clouds grows larger, they cannot counter the gravitational force and fall on the ground as precipitation (rain or snow, sleet or hail).

| Rainfall | Shower of Rain | Snowfall | Hail |

5.   **Clouds**. Precipitation of any form are closely associated with the nature of the source cloud. There cannot be precipitation without clouds. Hence, knowledge about different types of clouds and their characteristics is important.

6.   **WMO Definition of Cloud**. WMO defines cloud as follows: -

"A cloud is a hydrometeor consisting of minute particles of liquid water or ice, or of both, suspended in the atmosphere and usually not touching the ground. It may also include larger particles of liquid water or ice, as well as non-aqueous liquid or solid particles such as those present in fumes, smoke or dust".

7.   **Types of Clouds**. Broadly, clouds are categorized as Low, Medium, High and Convective Clods. The types of clouds that fall under these categories are as follows:-

a)   Low Clouds.

    i.   Stratus (St)

    ii.   Stratocumulus (Sc)

b)   Medium Clouds.

    i.   Altocumulus (Ac)

    ii.   Altostratus (As)

    iii.   Nimbostratus (Ns)

c)   High Clouds.

    i.   Cirrus (Ci)

    ii.   Cirrostratus (Cs)

    iii.   Cirrocumulus (Cc)

d)   Convective Clouds.

    i.   Cumulus (Cu)

    ii.   Towering Cumulus (Tc)

    iii.   Cumulonimbus (Cb)

8. Clouds are generally encountered over a range of altitudes varying from sea level to the tropopause (top of the troposphere). The troposphere can be vertically divided into three levels, high, middle and low. In the same order, the clouds are categorized as High, Medium and Low clouds based on the level of their formation. That is, each level is defined by the range of heights at which clouds of certain genera occur most frequently. The levels overlap and their limits vary with latitude.

Category	Types	Height over Polar region	Height over Temperate region	Height over Tropical region
Low Cloud	Stratus Stratocumulus	From the Earth's surface to 2 km (0 − 6 500ft)	From the Earth's surface to 2 km (0 − 6 500ft)	From the Earth's surface to 2 km (0 − 6 500ft)
Medium Cloud	Altocumulus Altostratus Nimbostratus	2 − 4km (6 500 − 13 000 ft)	2 − 7 km (6 500 − 23 000 ft)	2 − 8 km (6 500 − 25 000 ft)
High Cloud	Cirrus Cirrocumulus Cirrostratus	3 − 8 km (10 000 − 25 000 ft)	5 − 13 km (16 500 − 45 000 ft)	6 −18 km (20 000 − 60 000 ft)
Convective Cloud	Cumulus Towering Cumulus Cumulonimbus	From the Earth's surface to 2 km (0 − 6 500ft)	From the Earth's surface to 2 km (0 − 6 500ft)	From the Earth's surface to 2 km (0 − 6 500ft)

9. **Stratus Clouds**. Generally grey cloud layer with a fairly uniform base, which may give drizzle, snow or snow grains. When the Sun is visible through the cloud, its outline is clearly discernible. Sometimes Stratus appears in the form of ragged patches. Fog can also be described as stratus cloud resting on ground.

10. Stratus clouds sometimes may cause light drizzle. However, such drizzle may temporarily reduce visibility to a hazardous edge. This cloud, generally moves faster with the low-level winds. Hence, the weather phenomena associated with it are short lived. The sun rays are often blocked for a long duration by persistent stratus clouds.

11. **Stratocumulus Cloud**. Grey or whitish, or both grey and whitish, patch, sheet or layer of cloud that almost always has dark parts, composed of tessellations, rounded masses, rolls, etc., which may or may not be merged. Generally speaking, this cloud looks like the cotton patches floating in air.

12. Mechanism for formation of stratocumulus cloud is simple. When an air parcel near to the surface of the earth gets heated up, becomes lighter and gets lifted up with the available moisture in it. As the air parcel rises up, temperature decreases due to the environmental temperature (temperature in the surrounding air). Due to fall in temperature, the air parcel attains dew point temperature, becomes saturated and the water vapours get condensed thereby forming clouds. If the air is stable above the level of condensation, i.e. the vertical motion of air or the vertical current gets arrested, the cloud will not grow vertically and continue to float in the air.

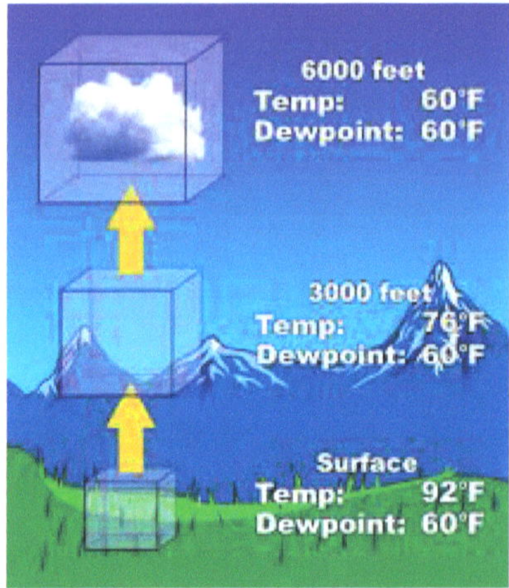

13. Stratocumulus clouds generally does not cause precipitation. However, during rainy season, say during monsoon season in Asian region or during winter season in association with extra tropical disturbance over higher latitude regions, this cloud may cause light rain or snow.

14. **Altocumulus Cloud**. White or grey, or both white and grey, patch, sheet or layer of cloud, generally with shading, composed of laminae, rounded masses, rolls, etc., which are sometimes partly fibrous or diffuse and which may or may not be merged.

 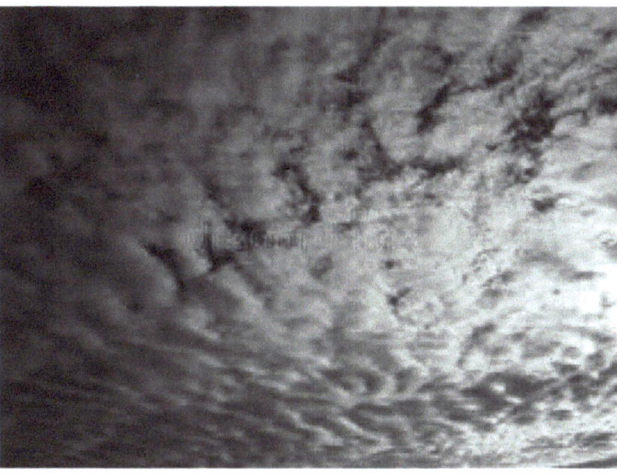

15. This cloud generally does not cause any weather. Ac cloud is mainly associated with anticyclonic maritime out flow, feeble middle level disturbance, or may be remnant of any significant weather system that moved away. During active rainfall condition, it may cause intermittent light rain but often not reducing the horizontal visibility at surface to less than about 4000M. However, Ac cloud is capable of causing significant rain/snow over the area of high altitude like Ladakh region in India. In prevalence of high temperature at surface or at lower tropospheric levels, most of the raindrops originating from this cloud may evaporate before they reach the surface. During the winter season, when there is low temperature profile in the troposphere, this cloud is generally composed of ice crystals and causes ice accretion on the aircraft frame. The modern aircrafts are well equipped to counter ice accretion. However, the aircrews are supposed to have information about the area and the flight level where there is possibility of ice accretion to take adequate de-icing measures. When possible, such level of flights is avoided.

16. Sun rays can easily penetrate through this cloud. But when the cloud is thick and compact, sun light can be seen partially and occasionally a corona, an optical phenomenon may form in the sky.

17. Corona is an optical phenomenon, a colorful ring formed due to reflection and internal refraction of sun rays passing through the water droplets of the clouds. One or more sequences (seldom more than three) of small-diameter coloured rings can be seen around the Sun or Moon. In each sequence, the inside ring is violet or blue, the outside ring is red and other colours such as green and yellow may occur in between. Some images of corona formed by medium clouds are shown below: -

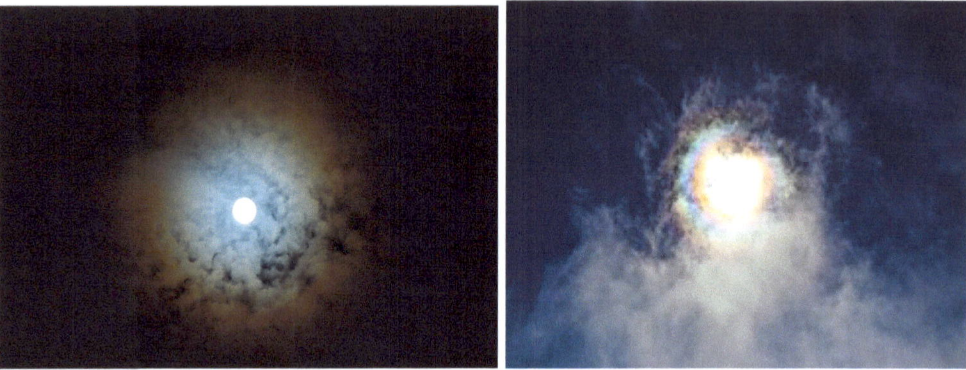

18. **Altostratus Cloud**. Greyish or bluish cloud sheet or layer of striated, fibrous or uniform appearance, totally or partly covering the sky, and having parts thin enough to reveal the Sun at least vaguely.

19. Altostratus is a middle-altitude cloud genus made up of water droplets, ice crystals, or a mixture of the two. Altostratus clouds are formed when large masses of warm and moist air rise, causing water vapor to condense. Altostratus clouds are usually grey or blueish featureless sheets, although some variants have wavy or banded bases. The sun can be seen through thinner altostratus clouds, but thicker layers can be quite opaque.

20. Altostratus clouds are associated with significant weather systems like active rainy condition, tropical & extra-tropical systems and other important localized convergence systems. This cloud is associated with abundant moisture content at mid-tropospheric level. Being associated with high moisture content, this cloud generally causes intermittent to continuous light to moderate rain or snow. Generally, it is not associated with shower or heavy rain. In India, this cloud is prominently observed during active monsoon conditions giving endless light to moderate continuous rain throughout the day and sometimes for days together.

21. While flying through this cloud, poor in-flight visibility and ice accretion are experienced. When the freezing level (0^0 C isotherm) is low, Altostratus cloud may also contain ice crystals and can produce some optical phenomena like coronas.

22. **Nimbostratus Cloud**. It is a multi-level, amorphous, nearly uniform and often dark grey cloud that usually produces continuous rain, snow or sleet but no lightning or thunder. Although it is usually a low-based cloud, it actually forms most commonly in the middle level of the troposphere and then spreads vertically into the low and high levels. Nimbostratus usually produces precipitation over a wide area.

23. Nimbostratus clouds form through the deepening and thickening of an altostratus cloud, often along warm or occluded fronts. These clouds extend through the lower and mid-layers of the troposphere bringing rain to the surface below. These mid-level clouds are often accompanied by continuous moderate rain or snow and appear to cover most of the sky. Nimbostratus will often bring precipitation which may last for several hours until the associated front passes over.

24. **Cirrus Cloud**. Detached clouds in the form of white, delicate filaments or white or mostly white patches or narrow bands. These clouds have a fibrous (hair-like) appearance, or a silky sheen, or both. Cirrus clouds are composed almost exclusively of ice crystals. These crystals are generally very small, which, together with their sparseness, accounts for their transparency. Dense Cirrus patches or Cirrus in tufts may contain ice crystals large enough to acquire an appreciable terminal velocity. Trails of considerable vertical extent fall from the base of Cirrus.

25. Thicker Cirrus clouds may cause halo phenomenon. However, the circular haloes mostly do not show a complete ring. Halo is an atmospheric optical phenomenon that results when the Sun or Moon shines through thin clouds. It is mainly due to reflection and internal refraction of light when it passes through the ice crystals of the clouds. The most common is the '22-degree halo'. They bear this name because the radius of the circle around the sun or moon is approximately 22 degrees. The white rays are split in such a manner that the colour of the innermost ring is red and the outer one is violet.

26. **Cirrocumulus (Cc)**. Thin, white patch, sheet or layer of cloud without shading, composed of very small elements in the form of grains, ripples, etc., merged or separate, and more or less regularly arranged; most of the elements have an apparent width of less than one degree.

27. Cirrocumulus clouds appears as the altocumulus clouds at greater height. However, Cirrocumulus can be distinguished from altocumulus in several ways, although both the clouds can occasionally occur together with no clear demarcation between them. Cirrocumulus generally occur at higher altitudes than altocumulus, thus the "cloudlets" appear smaller, as they are more distant from observation at ground level. They are also colder. Cirrocumulus clouds never cast self-shadows and are translucent to a certain degree.

28. Unlike other high-altitude tropospheric clouds like cirrus and cirrostratus, cirrocumulus includes a small amount of liquid water droplets, although these are in a supercooled state. Ice crystals are the predominant component, and typically, the ice crystals cause the supercooled water drops in the cloud to rapidly freeze, transforming the cirrocumulus into cirrostratus cirrocumulus clouds can be seen in the sky for a short duration. This cloud is not associated with any weather at surface.

29. **Cirrostratus Clouds (Cs)**.   Transparent, whitish cloud veil of fibrous (hair-like) or smooth appearance, totally or partly covering the sky, often producing halo phenomena.

30. Cirrostratus is a cloud of moderate vertical extent and is composed of a relatively sparse number of small ice crystals. This cloud is partially transparent and the disc (outline) of the Sun is visible visible through the cloud, at least when

the Sun is not too close to the horizon. Halo phenomena are often observed in thin Cirrostratus; sometimes the veil of Cirrostratus is so thin that a halo provides the only indication of its presence.

31. Cirrostratus clouds are basically sheet type of clouds and are generally not associated with any precipitation. However, in very high-altitude region like Ladakh or Siachen in western Himalayan region, this cloud sometimes causes light snowfall.

CHAPTER 7

CONVECTIVE CLOUDS

1. Convective clouds are the vertically developing clouds which, when grows to certain height causes lightning and thunder accompanied by precipitation which is sometime of appreciable intensity. The initial mechanism for formation of this cloud is same as that of the Stratocumulus cloud. That is, there is a requirement of lifting a parcel of air from the surface to a height where the parcel attains dew point temperature resulting in condensation. Now, from this level, if the instability or the vertical current continues, the cloud continues to grow higher. Initially, it becomes cumulus (Cu) cloud, on further growth it becomes Towering Cumulus (TCu) and finally, it forms Cumulonimbus (Cb) cloud.

2. Whether or not the atmosphere has stability depends partially on the moisture content. In a very dry troposphere, temperature decreases with height by less than $9.8\ ^0C$ per kilometer ascent indicates stability, while greater changes indicate instability. This lapse rate is known as the dry adiabatic lapse rate. In a completely moist troposphere, temperature decreases with height by less than 6^0C per kilometer ascent indicates **Stability**, while greater changes indicate **Instability**. In the range between 6^0C and 9.8^0C temperature decrease per kilometer ascent, the term **Conditionally Unstable** is used.

3. To put it in a simpler way,

   a) Atmosphere is said to be stable when vertical rise of an air parcel in the prevailing condition cannot take place or a great amount of energy would be required to raise the air parcel.

   b) The atmosphere is unstable when the prevailing conditions are sufficient to raise an parcel of air.

   c) There may be a situation when the prevailing condition at surface is stable but instability prevails at a slightly higher level. In this condition if through some mechanism the air parcel can be lifted to that height from where instability occurs, then the same parcel of air will attain instability. This condition is called Conditional Instability.

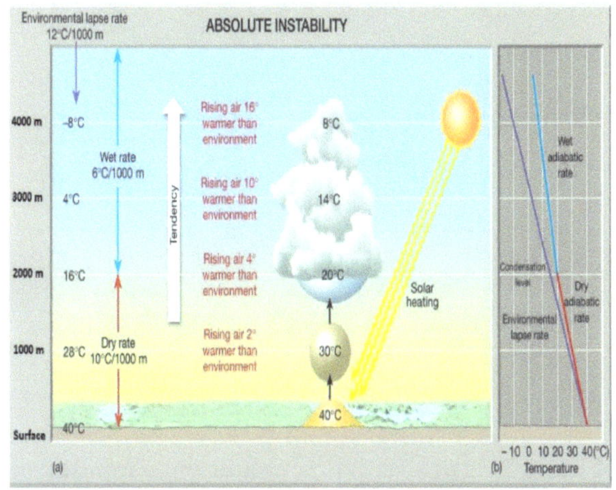

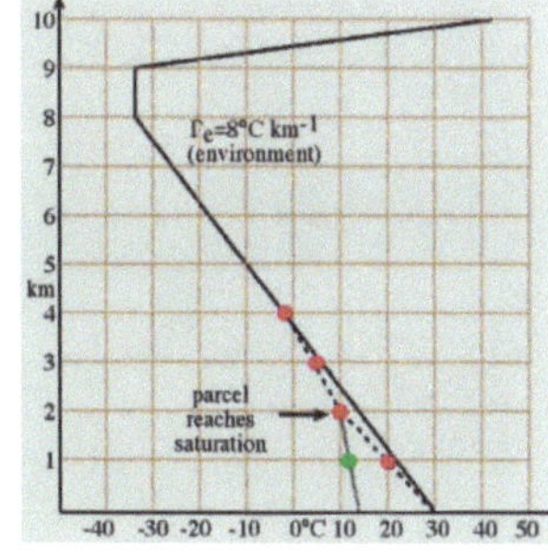

Conditional Instability - 4km

- The parcel continues to rise moist adiabatically (6°C/km)
  - Notice that now parcel temperature = Te
- What happens if the parcel is pushed upward just a little???

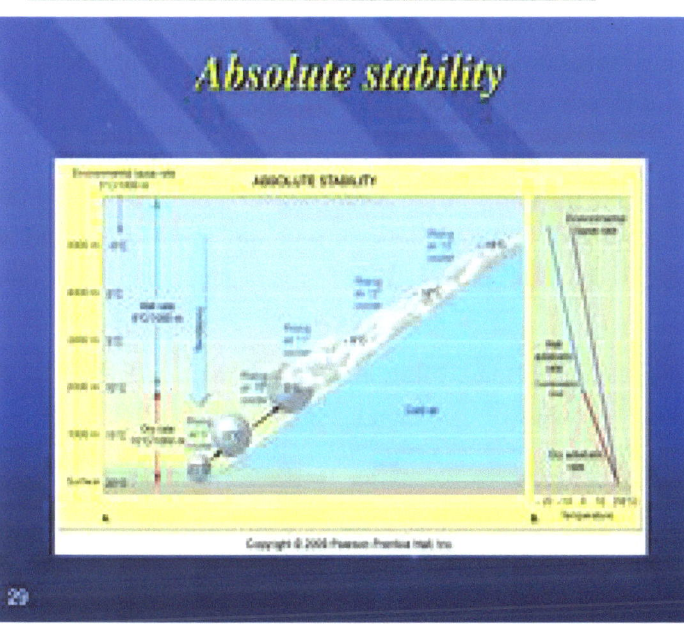

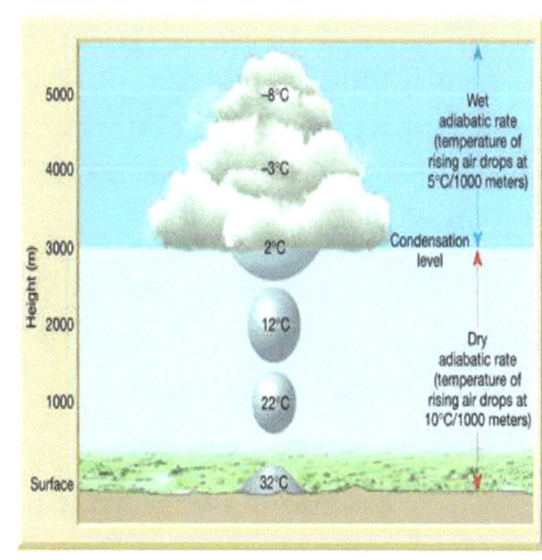

4. **Cumulus Clouds**. Detached clouds, generally dense and with sharp outlines, developing vertically in the form of rising mounds, domes or towers, of which the bulging upper part often resembles a cauliflower. The sunlit parts of these clouds are mostly brilliant white; their base is relatively dark and nearly horizontal.

5. Cumulus clouds form when there is instability but in lesser degree. The cloud after forming at the Lowest Condensation Level (LCL) grows little upward giving a look of a cauliflower. Further, the growth of the cloud is arrested due to prevalence of stability aloft.

6. Cumulus is composed mainly of water droplets. When of greater vertical extent, Cumulus may release precipitation in the form of showers of rain, snow or snow pellets. During active rainfall days, like in monsoon season over India, the smaller cumulus clouds are also capable of giving moderate shower. Such precipitations are short lived and passes by quickly with the prevailing winds. However, Cu clouds are often termed as the Fair-Weather Cumulus as they seldom cause weather outside the wet period.

7. Ice crystals may form in those parts of a Cumulus where the temperature is below 0 °C. The ice crystals grow at the expense of evaporating supercooled water droplets, transforming the cloud into Cumulonimbus. In cold weather, when the temperature in the entire cloud is well below 0 °C, this process leads to the degeneration of the cloud into diffuse trails of snow.

8. **Towering Cumulus (TCu)**. A vertically developed cumulus cloud, often a precursor to cumulonimbus. When a cumulus cloud is much taller, it is called a Towering Cumulus (TCU) cloud. The taller these clouds are, the more likely they are to produce showers.

9. Basically, TCu is a cloud which is larger and taller than the Cu clouds but smaller than the Cumulonimbus clouds. As the TCu clouds often grows further to become Cb cloud, the aircrews are cautioned about location of this cloud. This cloud is capable of producing lightning, turbulence and moderate to heavy shower. However, the individual TCu clouds can be avoided by the flyers by flying above the top of the cloud or little away from it.

10. **Cumulonimbus Clouds (Cb)**. Heavy and dense cloud, with a considerable vertical extent, in the form of a mountain or huge towers. At least part of its upper portion is usually smooth, or fibrous or striated, and nearly always flattened; this part often spreads out in the shape of an anvil or vast plume. Under the base of this cloud, which is often very

dark, there are frequently low, ragged clouds, either merged with it or not, and precipitation sometimes in the form of virga.

11. Virga is a mass of streaks of rain appearing to hang under a cloud and evaporating before reaching the ground.

12. Cumulonimbus, when overhead or close to the observatory, covers a large expanse of the sky and it can easily be confused with Nimbostratus, especially when identification is based solely on the appearance of the undersurface. The character of the precipitation may help to distinguish Cumulonimbus from Nimbostratus. If the precipitation is of the showery type, or if it is accompanied by lightning, thunder or hail, the cloud is Cumulonimbus.

13. Cumulonimbus is composed of water droplets and, especially in its upper portion, ice crystals. It also contains large raindrops and, often, snowflakes, snow pellets or hailstones. The water droplets and raindrops may be substantially supercooled.

14. The conditions under which Cumulonimbus clouds occur are similar to those that are favourable for the development of Cumulus clouds. However, in case of formation of Cb clouds, the vertical instability should persist to a greater height. The transformation of Cumulus into Cumulonimbus is due to the formation of ice particles in its upper part. The presence of ice particles is evident when some or all of the upper part loses the sharpness of its outlines or acquires a fibrous or striated texture.

15. Cumulonimbus may be described as a "cloud factory"; it can produce thick patches or sheets of cirrus or cirrostratus, altocumulus, altostratus or stratocumulus by the spreading out of its upper portions and by the dissipation of the lower parts. The spreading of the highest part usually leads to the formation of an anvil; if the wind increases strongly with altitude, the cloud top spreads only downwind, assuming the shape of a half anvil or, in some cases, a vast plume. Some images of Cumulonimbus clouds with anvil or plume top art shown below:-

16. A cumulonimbus anvil which is also known as an anvil cloud, is a cumulonimbus cloud which has reached the level of stratospheric stability and has formed the characteristic flat, anvil-top shape.

17. There are three distinct stages of a Cb cloud namely, formation stage, matured or developed stage and dissipating stage. In the formation stage, when the cloud continues to grow, it sucks air from the surrounding giving rise to updraft. The Cb cloud is a full grown cloud in its matured stage. In this stage the cloud has both up and down drafts. This is the stage when it generates lightning and thunder can be heard. In dissipating stage the cloud pours precipitation and mainly has downdraft. The downdrafts in this stage at times, generate very strong surface winds.

18. **Characteristics of Cumulonimbus**. Typically, Cumulonimbus clouds are accompanied by smaller cumulus clouds. The cumulonimbus base can be several kilometre across or as small as a few tens of metres. It occupies lower to upper tropospheric levels. It generally forms at altitudes ranging from 200 M to 4,000 M based on geographical location and prevailing seasonal characteristics of the environment. Peaks typically reach 12,000 M with extreme cases reaching 20,000 M or higher.

19. Cumulonimbus clouds with a well-developed top (anvil dome) are distinguished by a flat, anvil-like top caused by wind shear or inversion at the equilibrium level near the tropopause. The shelf of the anvil may precede the vertical component of the main cloud for many kilometres and be accompanied by lightning.

20. Rising air parcels occasionally exceed the equilibrium level (due to momentum) and form an overshooting top, culminating at the maximum parcel level. This largest of all clouds, when vertically developed, usually extends through all three cloud regions.

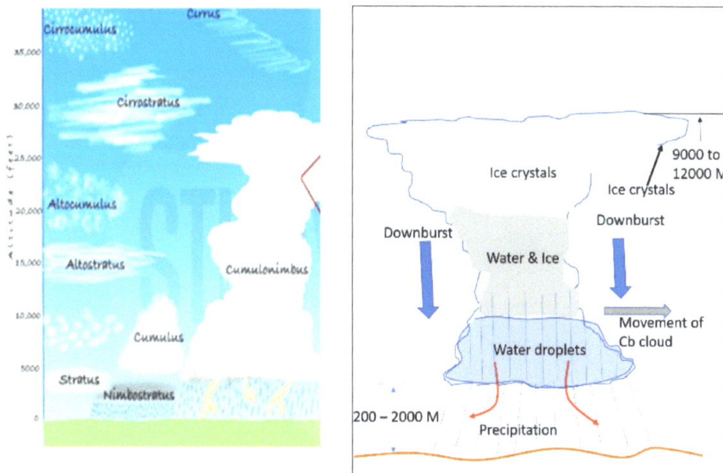

21. Cb clouds are associated with all types of significant weather systems that form due to instability like Tropical Cyclones, Western Disturbances, Fronts, etc. In such disturbances, generally air masses of different and contrasting characteristics converge along a boundary (sometimes referred as Fronts) or around a point. However, there are other disturbances where different air masses with not much difference in characteristics or disturbances in a single air mass also give rise to significant instability causing formation of Cb clouds and often of intense nature. The same are discussed in the subsequent paragraphs.

22. **Disturbance in Tropical Region**. The air masses which meet along a front over middle or higher latitudes, generally have significant difference in characteristics, that is, they have large variation in temperature and humidity which, ultimately give rise to weather over a wide area and last for a longer duration. However, the same is not the case in the tropical region. Here, the airmasses which near a common boundary, have little difference in temperature and humidity when they meet. Hence, the common boundaries in tropical region are not termed fronts but are called Tropical Discontinuity.

23. **Tropical Discontinuity**. The two types of tropical discontinuities are illustrated below; -

(a)   **Parallel Discontinuity**.

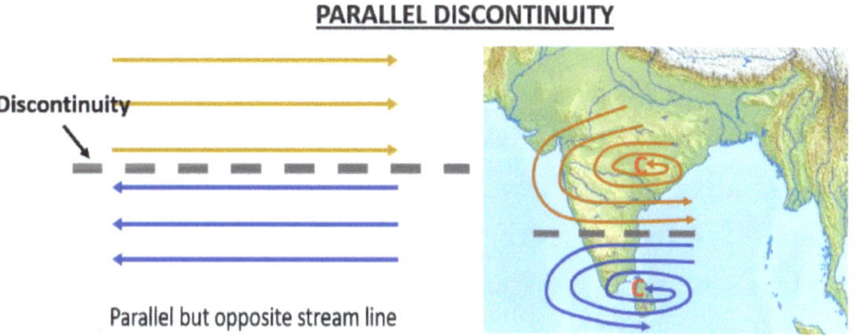

24. In above picture a case of parallel discontinuity is shown in which parallel stream line from opposite direction meet along a border line. Such discontinuity doesn't result in significant instability and therefore, convective build up is not associated with it. Hence, this type of discontinuity is also called Stable Discontinuity.

25. However, in case there is marked difference in the wind speed in either side of the discontinuity, the line of discontinuity will have the characteristics of shear line resulting cloud formation with some convective build up.

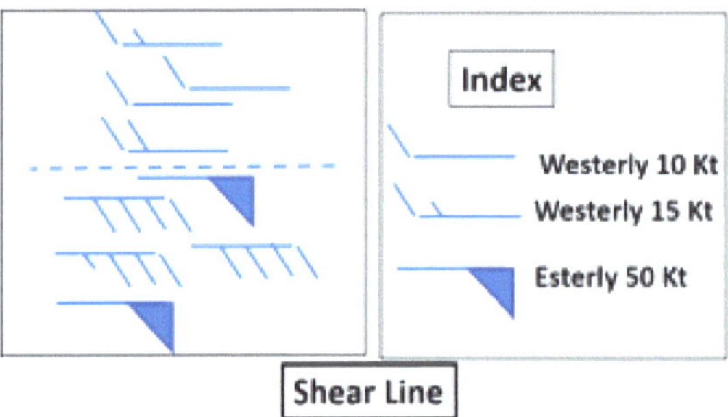

26. **Active or Unstable Discontinuity**.

   a)   **Confluence**. This occurs when two air masses converge on a line of discontinuity making an angle. In such situation these will be greater cloud formation making the sky condition from cloudy to overcast with convective buildup along a narrow belt covering either side of the discontinuity zone. Beyond the narrow zone (about 10 Km) clouding decreases gradually but persists beyond 60-70 Km. Confluence is common feature in Indian region in all the seasons.

   b)   **Discontinuity Between Winds of Opposite Direction**. This is the other type of active discontinuity and causes maximum instability. It is associated with large scale and intense convection. It is very common in the Indian Peninsular region during pre-monsoon season and is commonly known as peninsular discontinuity where air mass from the Bay of Bengal and the Arabian Sea meet. Occasionally this discontinuity may extend up to as north as Chhattisgarh or south/east Madhya Pradesh. The discontinuity is seen on a day to basis during the pre-monsoon season with varied location and orientation. Most of the thunderstorm and rainfall activity over the peninsular region are the result of this type of discontinuity.

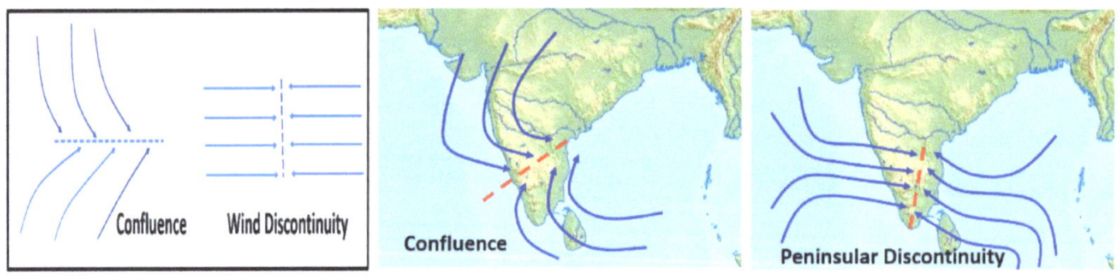

27. **Inter Tropical Convergence Zone (ITCZ)**. It is a zone where trade winds from both the hemisphere, that is north-easterlies from northern hemisphere and south-easterlies from the southern hemisphere meet along a narrow zone and results in instability along a large belt along the (solar) equator which shifts northward during summer and back to south during winter. ITCZ influences the weather worldwide and is associated with widespread clouding including Cb clouds and intense weather.

28. **Disturbance in a single airmass**. As mentioned earlier, the disturbances in the tropical regions discussed above involve different air masses. However, there are other phenomena which can cause significant instability and subsequent formation of Cb clouds. Some of such phenomena are insolation, uneven heating of surface, convergence of air due to surge, upper air cyclonic circulation, confluence, wind discontinuity, orographic lifting, etc. Instability due to insolation (local heating) and uneven heating of surface have been brought out earlier. The other mechanisms are briefly discussed below: -

29. **Surge or Speed Convergence**. This is a common type of disturbance during monsoon season which is mostly noticed over the Arabian Sea. It is also commonly known as the monsoon surge. When wind with comparatively higher speed meet an area of lesser wind speed, it gives rise to accumulation of air along the border of the two wind speed zones and results in lifting of the accumulated air. Under such situation normal monsoon rain is accompanied by moderate to heavy showers associated with convective phenomena.

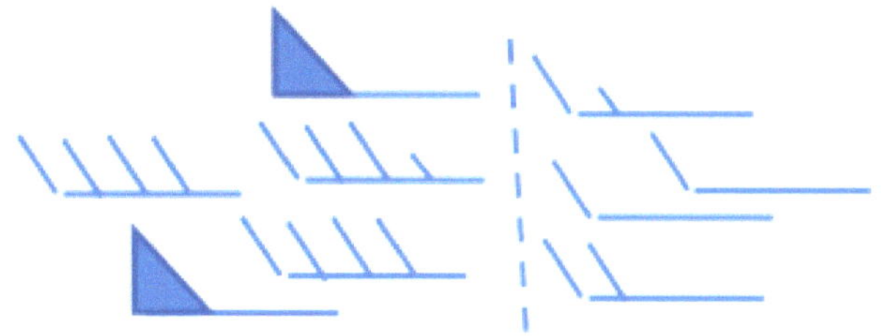

30. **Cyclonic Circulation**.   Due to varied regions the winds may undergo disturbances and form a wave. Under favourable conditions the waves deepen and winds follow a circular path around a centre. This type of features are called cyclonic circulations (Cycir) in which winds circulate in an anti-clock wise direction in the northern hemisphere and in clockwise direction in the southern hemisphere. Circulation around a centre results in accumulation of air and subsequent rise causing instability and formation of clouds including convective clouds. Under further favourable conditions such Cycir originating over a sea surface may intensify into a tropical storm. The Cycir along with the associated cloudings and weather move with the prevailing wind field.

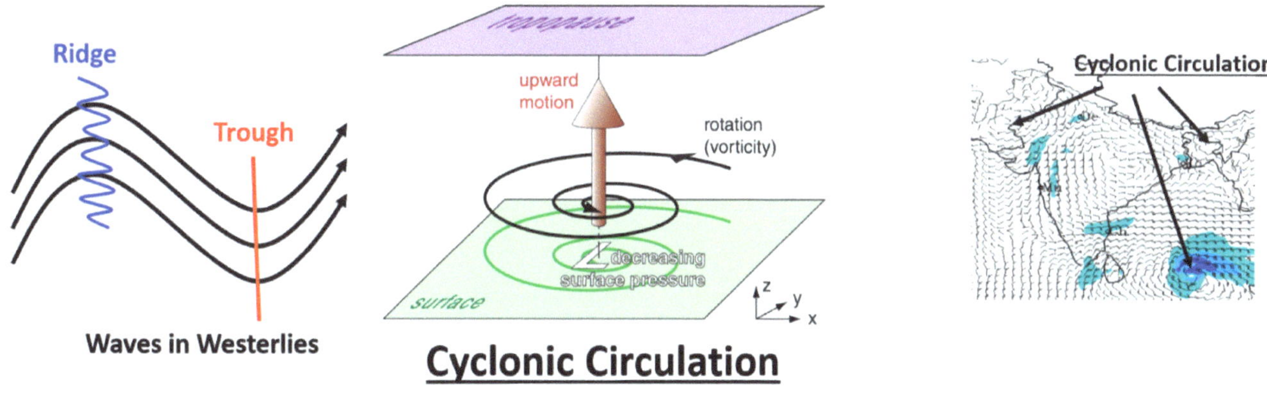

Cyclonic Circulation

31. The above disturbances may also develop involving a single air mass. Even then, convective build up and increase in clouding can take place as these features ultimately lead to convergence and trigger instability.

32. **Orographic lifting**. The winds, on meeting a hilly terrain ascend along the slopes of the hills. As they rise up, temperature decreases and saturation attained, which results in condensation and further vertical development of clouds. This phenomenon of lifting of air by the hill slopes is called orographic lifting. It is more effective during the summer months when difference in temperature is large between that on surface and aloft. However, adequate moisture is required along the vertical column for strong convective cloud to form.

33. Effect of orographic lifting is also prominent when afternoon sea breeze encounters hills during summer. Hence, pre-monsoon thunderstorms are common on the lee-ward side of the western Ghats (Mainly over Kerala, Coastal Karnataka and south Konkan even in absence of other instability phenomena).

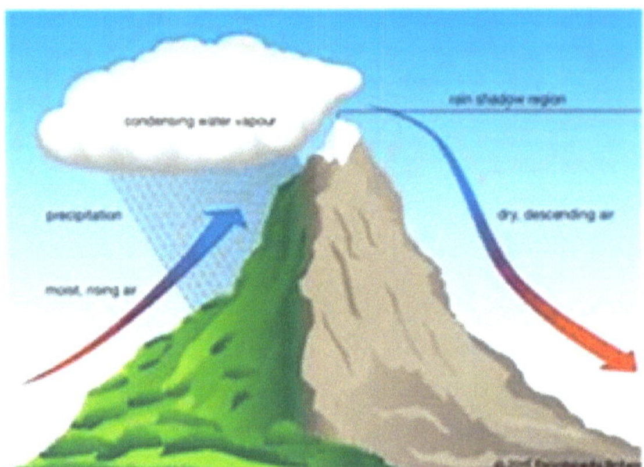

34. **Hazards associated with Cb Clouds**. Cb clouds are associated with intense and hazardous weather phenomena like heavy shower, cloud burst, strong winds, hail storm, thunderstorm & lightening, dust storm, tornado, etc. Most of losses & damages caused by weather are associated with Cb clouds. In aviation, it is regarded as the most dangerous cloud and is avoided by aircrews as far as possible.

Following are the aviation weather hazards associated with Cb clouds: -

a)   Strong Surface Winds generated by down draught of Cb clouds make it difficult for the pilot to control the aircraft while taking off or landing.

b)   Heavy shower from a Cb cloud reduces surface visibility drastically making it hazardous for making a landing or take off.

c)   Hail stones that fall from a Cb cloud may damage the frame of an aircraft and make the runway surface slippery making it nearly impossible for landing or taking off.

d)   While flying through the cloud an aircraft may encounter the following: -

   i.   Heavy turbulence caused by up and down drafts in the cloud. At times this may cause difficulty in controlling the flying machine. Besides, it may also make it uncomfortable for the passengers, sometime even leading to injuries.

   ii.   The middle and upper portions of the Cb clouds contain ice crystals and hail stones which may cause icing on the frame and hail storm may also damage the frame.

   iii.   There may be Clear Air Turbulence outside the Cb cloud.

   iv.   Wind shear (vast difference in the speed and direction along horizon and vertical.

   v.   Microburst caused by the down draught at surface or lower tropospheric levels is highly hazardous particularly while attempting to land and while or after takeoff. Because of their rapid onset and rapid changes in wind and aerodynamic conditions over short distances, microbursts are the most frequently implicated in crashes.

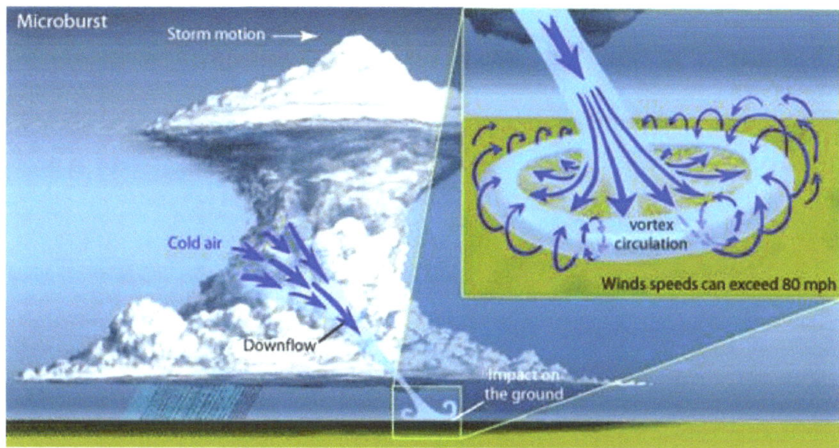

(A microburst is a localized column of sinking air (downdraught) within a thunderstorm and usually less than or equal to 2.5 miles in diameter. Most downbursts have visible precipitation shafts, but dry microbursts are generally invisible to the naked eye)

e) **Lightning**. The modern aircrafts are protected from lightning strike. However, sudden flash of extremely bright light may cause temporary blindness to the aircrews at night.

35. Formation, development, movement and expected intensity of Cb clouds are constantly monitored by the Met service agencies of aviation industry. Advisories and warnings are issued for safe air operations with the help of the modern infrastructures like weather radar, satellite imageries, lightening detection systems, available surface and upper air data, and above all, thorough professionalism and experience of the weathermen.

36. **Identifying Different Clouds in Satellite Imageries**. The images of various clouds shown above are actually as seen by the photographer or the observer. Thus, such observations can represent a small area which is like a dot on a map. Putting all the observations for a given time across a greater geographical area, say a country, will give a better picture of the prevailing clouds. It has been made possible through satellite observations. Cloud images received from satellite have been a vital tool in modern method of observation, analysis and forecasting.

37. Satellites use different method to capture cloud images. Most widely used for identifying clouds are as follows: -

a) **Visible Spectrum**. Use albedo (reflectivity) of the cloud to capture image. Low clouds (including fog) and convective clouds possess maximum reflectivity and are seen the brightest in a visible image. Medium clouds having lesser reflectivity are seen greyish. High clouds are practically not seen in visible image.

b) **Infrared (IR) Spectrum**. IR spectrum takes into consideration the temperature of the cloud tops. Lower the temperature, brighter will be the image of the clouds. Temperature in high clouds and top of the convective clouds (being at higher tropospheric levels), is very low. Hence, these clouds are seen bright in an IR image. Medium clouds are seen greyish and low clouds are not seen. However, during winter season, when the temperature of the low clouds are also very low during night and early morning, are seen in greyish colour.

c) **Water Vapour (WV) Spectrum**. This spectrum senses the middle level moisture content in clouds. The medium clouds (As, Ns) and convective clouds contain large amount of moisture (water). More moisture, brighter is the image. Cb clouds, again, are seen the brightest as it contains more water droplets than As and Ns clouds. High and low clouds are generally not seen in WV images.

38. Three images of the three spectrums of the same time are shown below. Effort has been made to show how different clouds are identified in the different images.

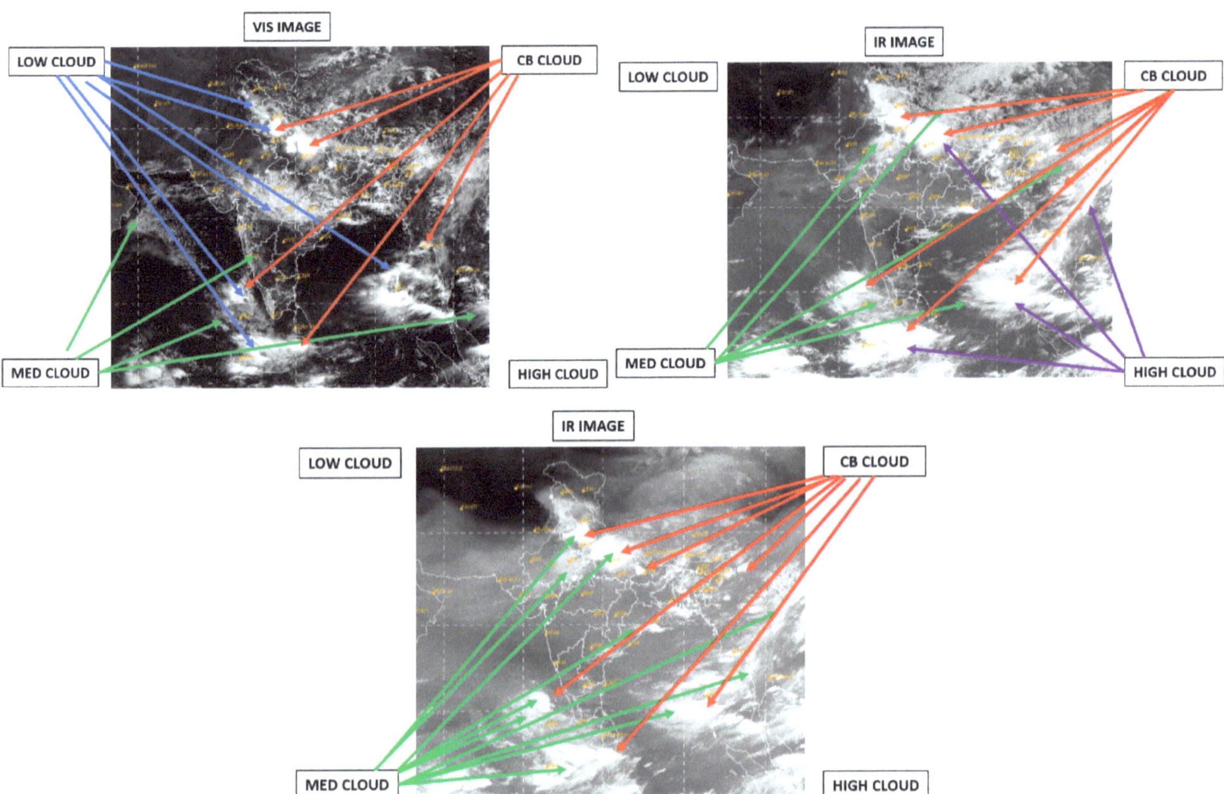
VIS IMAGE
LOW CLOUD
CB CLOUD
MED CLOUD
IR IMAGE
LOW CLOUD
CB CLOUD
MED CLOUD
HIGH CLOUD
IR IMAGE
LOW CLOUD
CB CLOUD
MED CLOUD
HIGH CLOUD

CHAPTER **8**

TEMPERATURE

1. Temperature is the measure of hotness or coldness expressed in terms of any of several scales, including Fahrenheit and Celsius. Temperature plays a vital role in maintaining the global circulation of air and in the process of making weather phenomena. Flow of air (wind) and development of weather involves transfer of heat both horizontally and vertically. Hence, it is important to understand about transfer of heat in the atmosphere. It is known to us that heat transfers from one body to the other and it takes place in different ways.

2. Transfer of heat takes place from one body to another due to differences in temperature between the bodies. With the increase in a body's temperature, molecules or atoms' vibrations increase. These vibrations are then transferred from one part of the body to another and from one body to another through a medium. When heat transfers from one body, the molecular movement or vibration in that body decreases gradually and the molecules tend to re-arrange to the original position. On further transfer or extract of heat from the body the molecular motion comes to a complete rest and there will be no further transfer of heat. The temperature at which this stage is attained is same for all matter. This temperature is called absolute zero and it is the lowest temperature which can be attained. The scale in which this temperature is set as zero is called the absolute scale. In Celsius scale this temperature is -273^0 C.

3. **Method of Heat Transfer.** If a hot body is brought in conducting contact with a cold body, the temperature of the hot body falls and that of the cold body rises, and it is said that a quantity of heat has passed from the hot body to the cold body.

   There are three modes of heat transfer, which can be described as follows: -

   a) **Conduction.** In this process transfer of heat takes place by conduction in solids or fluids at rest. It is the process of transmission of energy from one particle of the medium to another with the particles being in direct contact with each other. Conduction occurs more readily in solids and liquids, where the particles are closer together than in gases, where particles are further apart.

   b) **Convection.** Transfer of heat by convection in liquids or gases in a state of motion, combining conduction with fluid flow is called convection. In this process hot fluid from one part are bodily transferred to a colder part of the fluid. This is an important process of heat transfer within the atmosphere.

   c) **Radiation.** When transfer of heat takes place with no material carrier, it is called radiation. Unlike conduction and convection, heat transfer by thermal radiation does not necessarily need a material medium for the energy transfer. Each body, whose temperature is more than absolute zero, emits energy in form of electromagnetic waves which travel through space at the same speed as radio waves. Amount of energy radiated by a body depends upon its temperature. Hotter objects emit radiations of shorter wavelengths and cooler objects emit radiations of longer wavelength. The radiant energy is absorbed by a body, amount of absorption depending on nature of substance. Both short-wave radiation (solar) and long-wave radiation (terrestrial) exist in atmosphere. Atmosphere absorbs mainly long-wave radiations.

d) **Advection**.   It is a process in which hot air molecules move in a horizontal direction. In other words, transfer of heat takes place through horizontal wind motion or bodily movement of air masses.

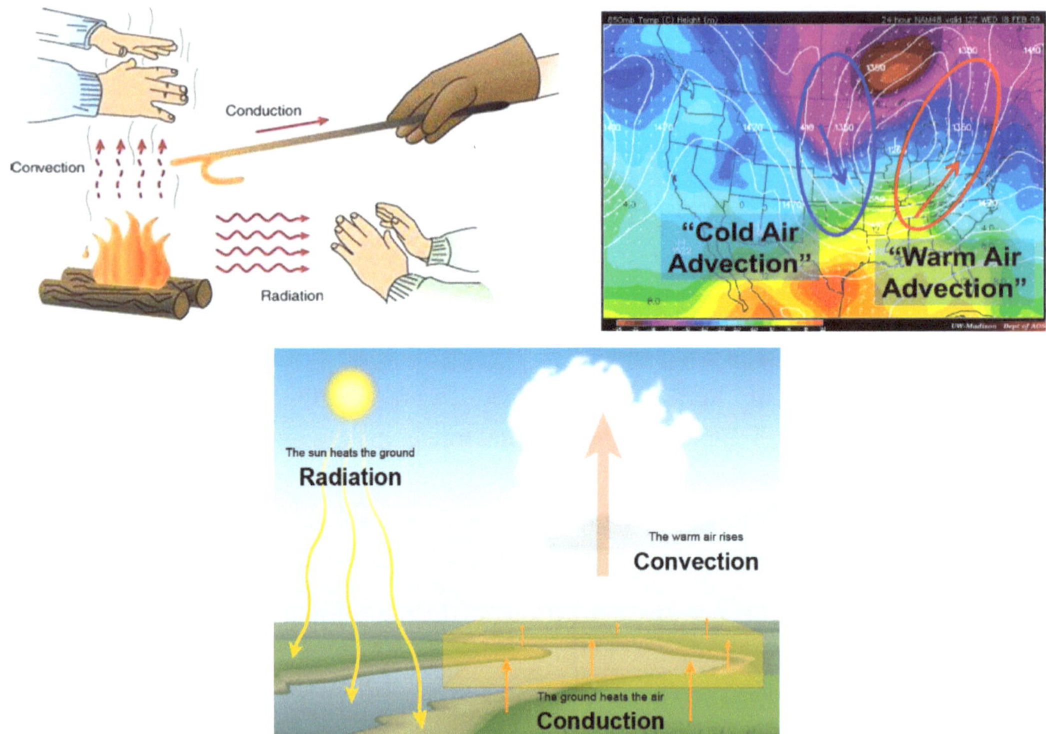

4. **Source of Heat in the Atmosphere**. The inner core of the Earth is extremely hot and is in a molten state. However, this heat cannot get transferred to the surface of the earth because the solid crust does not permit it. Hence, the Sun is the ultimate source of heat for the surface of the earth and the atmosphere aloft. The incoming solar radiation which reaches the surface of the Earth is called insolation. The Earth's surface is solid and absorbs the incoming solar radiation.

5. The Sun's temperature is estimated to be about 6000°C at its surface. At this temperature it emits an enormous amount of energy in the form of radiation. The radiation is mainly in the form of visible light, although a part of it is in the form of invisible energy in the ultraviolet and infra-red regions. The Sun is able to keep up this constant supply of energy, because within its interior, energy is being continuously produced by a process similar in principle to the thermonuclear fusion as in the hydrogen bomb.

6. When the sun's energy passes through the atmosphere several things happen to it. Around one fourth of this energy is directly reflected back by clouds and the ground. Around 8 percent is scattered by minute atmospheric particles and returned to space as diffuse radiation. Some 20 percent reaches the earth's surface as diffuse radiation after being scattered. Approximately 27 percent reaches the earth's surface as direct radiation and 19 percent is absorbed by the ozone layer and by water vapour in the clouds of the atmosphere.

7. The energy emitted by the sun which reaches the surface of the earth is called Insolation. The sun, having a mass of intensely hot gases, emits radiant energy in the form of waves, which consists of very short wave-length. The earth receives only about one two-thousand-millionth of the total insolation poured out by the sun, but this is vital to it. The amount received at the outer limit of the atmosphere is called Solar Constant. Thus, Solar Constant is the rate per unit area at which solar radiation is received at the outer limit of the atmosphere.

8. On an average, 47 percent of the solar energy arriving at the outer limits of the atmosphere eventually reaches the surface, and 19 percent is retained in the atmosphere. This 19 percent of direct solar radiation that is retained by the atmosphere is locked up in the clouds and the ozone layer and is thus not available to heat the troposphere. The

warmth of the atmosphere is due to the 47 percent of incoming solar energy reaching the earth's surface (that is, both land and bodies of water) and in the transfer of heat energy from the earth back to the atmosphere through such physical processes such as Long-Wave Radiation, Conduction and Convection. Some related phenomena such as advection and Latent Heat of Condensation also contribute to the warmth of the atmosphere.

9. Radiation is the process by which most energy is transferred through space from the sun to the earth. Radiation is given off by all bodies including earth and human being. The hotter is the body, shorter are the waves. We can simply say that the radiation from Sun comes to earth in the form of short waves and earth being cooler body, gives off energy in the form of long-wave. These are then radiated back to the atmosphere. This Long-Wave Radiation from the earth's surfaces heats the lower layers of the atmosphere. It is evident that the atmosphere is primarily heated from below by long wave radiation from the heated Earth surface.

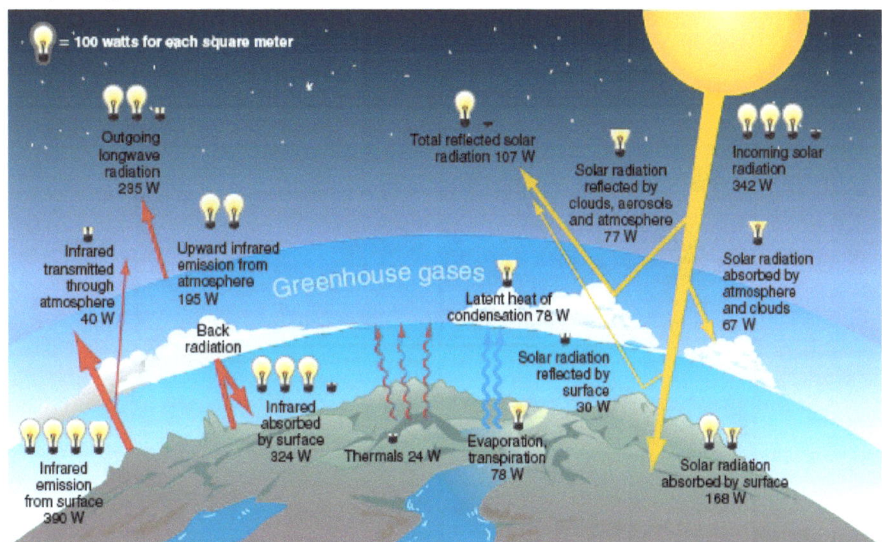

10. **Surface Temperature**. As understood from the above, the main source of temperature (energy) for the surface of the earth and the atmosphere (troposphere) are the incoming short-wave insolation from the Sun and the outgoing long wave radiation from the Earth's surface respectively. Amount of heating at the surface depends on the part of the day, that is, angle of the Sun. Heating is maximum when the sun is over head as the amount of heating received on unit area is the highest when the rays are incident vertically and through the shortest path.

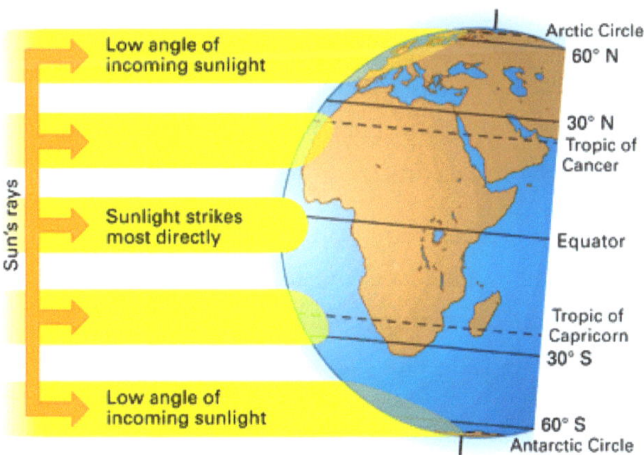

11. Temperature over the surface of the earth and the air layer very close to it under goes diurnal and seasonal changes. It is mainly because of the incident angle of the Sun rays and the position of the Earth in respect of the Sun. The geographical location and altitude of a place also obviously matter.

12. **What Causes the season**. Earth's tilted axis causes the seasons. Throughout the year, different parts of Earth receive the Sun's most direct rays. So, when the North Pole tilts toward the Sun, it is summer in the Northern Hemisphere. And when the South Pole tilts toward the Sun, it is winter in the Northern Hemisphere.

13. **The Earth's Tilt**. Many people believe that Earth is closer to the Sun in the summer and that is why it is hotter. And, likewise, they think Earth is farthest from the Sun in the winter. Although, this idea makes sense, it is somewhat incorrect. It is true that Earth's orbit is not a perfect circle. It is a bit lop-sided. During one part of the year, Earth is closer to the Sun than at other times. However, in the Northern Hemisphere, we are having winter when Earth is closest to the Sun and summer when it is farthest away. Compared with how far away the Sun is, this change in Earth's distance throughout the year does not make much difference to our weather.

14. There is a different reason for different season on the Earth. Earth's axis is an imaginary pole going right through the centre of the Earth from 'top' to 'bottom'. Earth spins around this pole, making one complete turn each day. That is why we have day and night. Earth has seasons because its axis doesn't stand up straight.

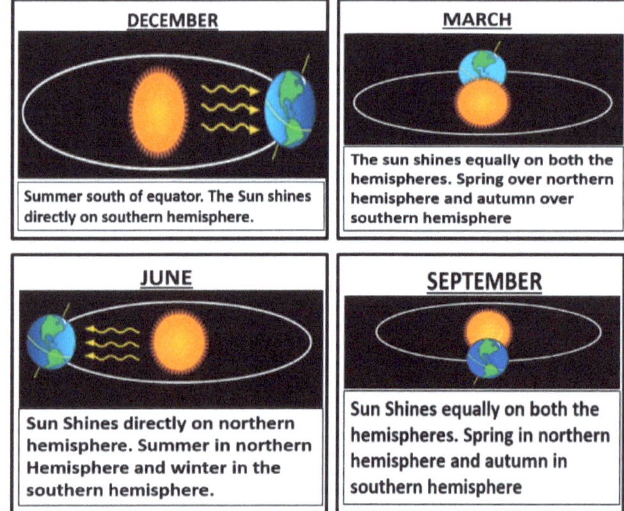

15. **Daily Change in Temperature**. Temperature at the surface depends on the nature of the surface. The rise in temperature by absorption of heat is inversely proportional to the specific heat of the substance. Water has the highest specific heat and hence, it experiences relatively smaller temperature changes. However, the earth's surface being solid have smaller specific heat and hence, temperature changes are greater on land. During day, ground temperatures may be much higher than the air temperature, sometimes by as much as 10°C. Maximum temperature is reached about two hours after midday. At night the ground cools because the earth emits longwave radiation. At the time of the minimum temperature, the ground is colder than the air close to it sometimes by about 5°C when the sky is clear and radiation effect is at its maximum. Minimum temperature is reached near about sunrise time.

16. Diurnal temperature variations are the greatest very near the Earth's surface. High desert regions typically have the greatest diurnal-temperature variations, while low-lying humid areas typically have the least. The Tibetan plateau presents one of the largest differences in daily temperature on the planet. During clear sky condition both incoming solar insolation during day and outgoing radiation during night are the maximum. Hence, the diurnal variation is also the greatest.

17. Air temperature also depends largely on the prevailing winds. During winter season if the prevailing winds are from land (Continental air), then the air temperature will be low and air originating from the sea will be warm. During summer, however, the situation would be the opposite. Winds blowing from the heated up land will be much warmer than the winds blowing from the sea.

Wind Patterns at lower levels

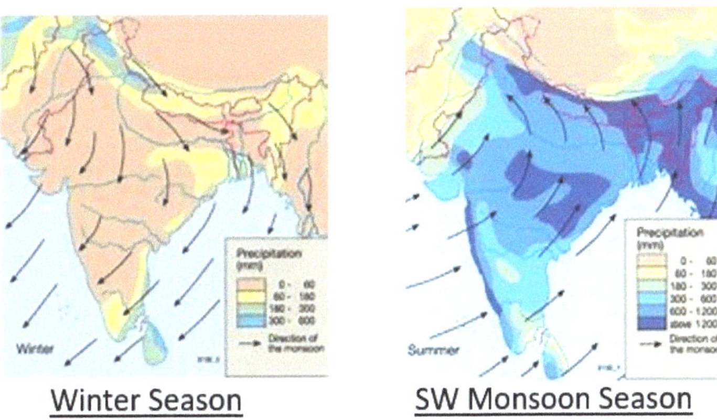

<u>Winter Season</u>　　　　<u>SW Monsoon Season</u>

18. In brief it can be stated that air temperature is affected by the following factors: –

    a)   The duration or length of time the sun shines (length of the day)

    b)   The angle or direction of the sun's coming (Angle of incident of Sun rays)

    c)   Sky condition (Cloudy or clear sky)

    d)   Altitude (Height of a place above sea level)

    e)   Latitude differences in an area (Geographical location)

    f)   Movement of ocean currents and wind (Warm or cold ocean current; dry continental or moist maritime wind)

    g)   Geographical conditions of a region (Topography)

19. **Upper Air Temperature**. Air at lower troposphere being very close to the surface undergoes both diurnal and seasonal variation. The middle level winds remain practically unaffected by diurnal variation. However, significant seasonal variation is seen in middle levels. The freezing level over north India comes down below 10000 Ft during winter whereas during summer months it remains above 15000 Ft. At higher tropospheric levels air temperature mostly remains stable varying only due to the prevailing wind pattern. The average temperature of atmosphere is illustrated in the following figure.

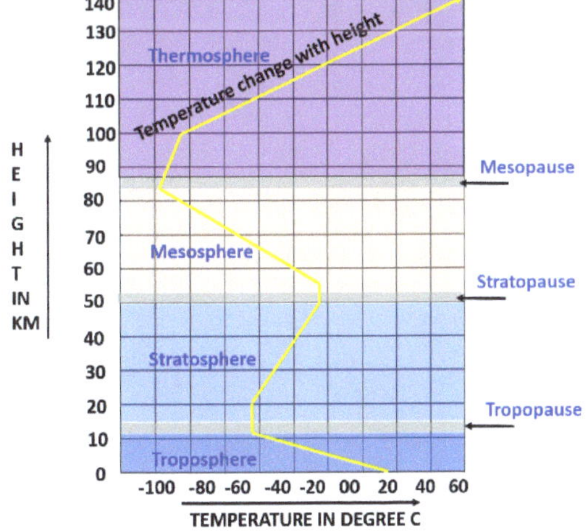

20. **Influence of Temperature in Weather**. Temperature is probably the most important factor influencing the weather. The most basic thing to understand is that warm air wants to rise while cold air wants to sink, because warm air is light and cold air is dense and heavy. As the sun heats the earth's surface it causes water on its surface to evaporate.

The evaporated water is absorbed by warm air and rises creating clouds. The greater the evaporation, the more the clouds created. At the same time, if conditions on the earth's surface are warm or hot, there is a rise in the ambient humidity. When you have high humidity and a front moving across it, the motion causes the formation of rain, thunderstorms, snow, etc.

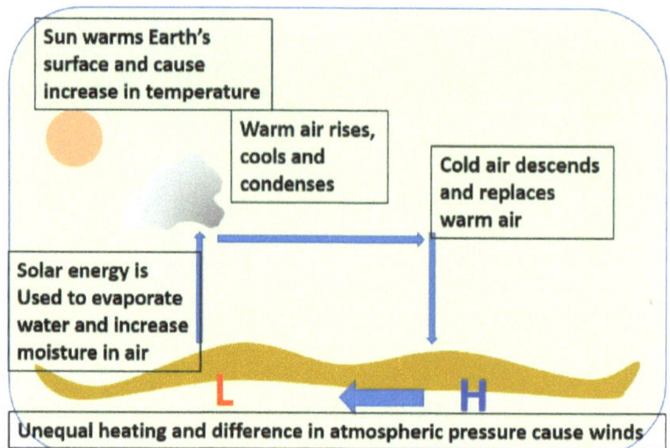

21. Air temperature also plays vital role on a day to day basis. Seasonal changes in surface pressure/wind and upper wind patterns are also broadly the result of change in temperature. Besides, it also plays important role in formation of significant weather systems like tropical cyclones, extra tropical systems, tornado, etc.

22. The spectacular examples of influence of temperature on weather systems across the globe are El Nino and La Nina effect on Southwest Monsoon. Occurrence of just 0.9^0 F above normal sea surface temperature at Peru coast (South America) during the Christmas period weakens the SW monsoon in that year over India and neighborhood. This situation is called El Nino. La Nina situation is the opposite to El Nino. In La Nina situation, the sea surface temperature over Peru coast is below normal during the Christmas period. This situation generally give rise to normal or above normal rainfall over Indian sub-continent during monsoon season in the same year.

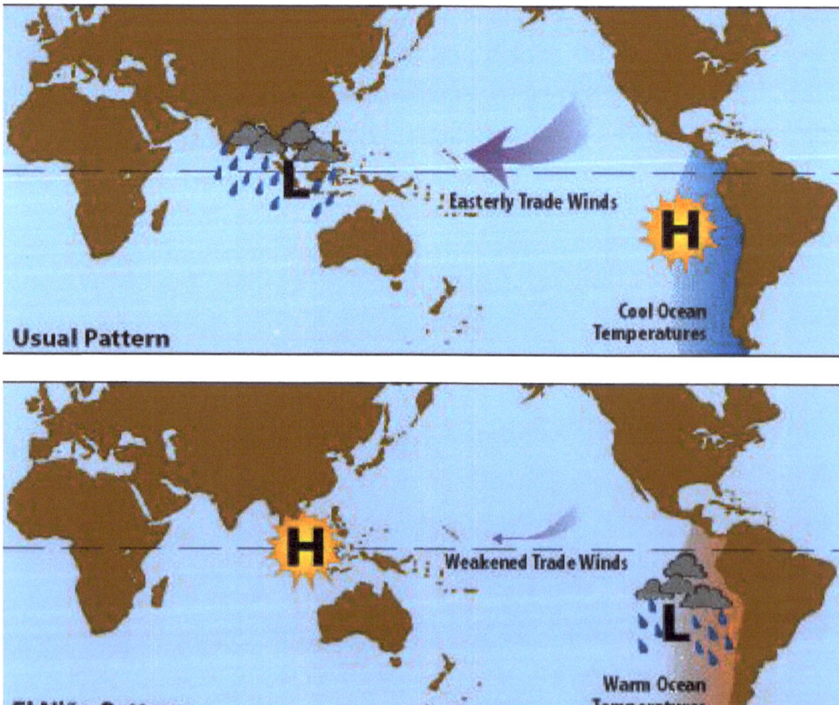

23. **Effect of Temperature in Aviation**. Apart from its importance in generation of weather phenomena, temperature can greatly affect the performance of an aircraft during takeoff, landing, and in-flight. Not only can the aircraft's engine performance be affected, but the aerodynamics involved with flying can vary as well. The performance of

an aircraft (both piston and jet type) is affected by the density of air, which in turn is inversely proportional to the temperature at constant pressure. Higher temperature implies lower density and so, has an adverse effect on engine performance. This effect is usually greatest during takeoff, but it should also be considered at other stages of flight, especially for jet aircrafts.

24. Higher the temperature, lower is the air density. Lower the density, lower is the engine performance. Lower the performance longer is the required runway length available for takeoff and lower is the load carrying capability of aircraft.

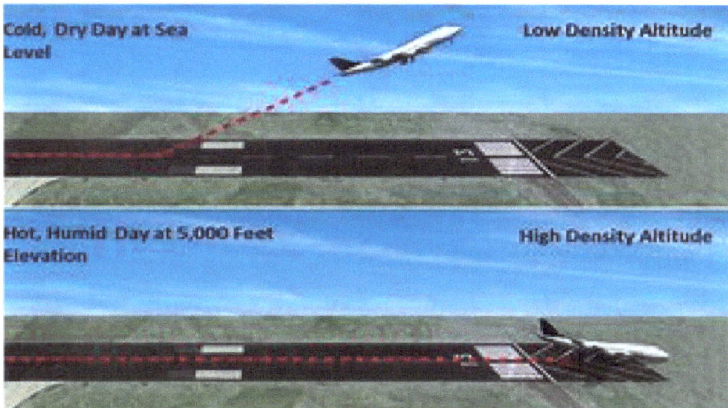

CHAPTER 9

ATMOSPHERIC PRESSURE

1. Although it is not visible, air pressure affects the weather pattern to a great extent. Rising air creates low pressure while sinking air creates high pressure. With high pressure, sinking air suppresses weather development. **High air pressure produces clear sky, dry and stable weather and low pressure makes air to rise and cause** weather. Besides, it is atmospheric pressure which controls the movement of air (Wind) which is yet another important factor for good or bad weather. Pressure distribution at surface or in upper air controls both wind direction and speed. As mentioned earlier, winds tend to flow in a direction parallel to the isobars keeping a lower pressure area to the left. All significant weather systems like cyclones, western disturbances, activity of the rainy season, all are associated with the distribution or pattern of atmospheric pressure.

2. Atmospheric pressure, also known as barometric pressure, is the pressure within the atmosphere of Earth. The standard atmosphere (symbol: atm) is a unit of pressure defined as 1013.25 hPa, which is equivalent to 760 mm Hg, 29.9212 inches Hg, or 14.696 psi. The atm unit is roughly equivalent to the mean sea level atmospheric pressure on Earth, that is, the Earth's atmospheric pressure at sea level is approximately 1 atm.

3. In most circumstances, atmospheric pressure is closely approximated by the hydrostatic pressure caused by the weight of air above the measurement point. As elevation increases, there is less overlying atmospheric mass, so atmospheric pressure decreases with increasing elevation. Because the atmosphere is thin relative to the Earth›s radius, especially the dense atmospheric layer at low altitudes, the Earth's gravitational acceleration as a function of altitude can be approximated as constant and contributes little to this fall-off. Pressure measures force per unit area, with SI units of pascals (1 pascal = 1 newton per square metre, 1 N/m2). On average, a column of air with a cross-sectional area of 1 square centimetre (cm^2), measured from the mean sea level to the top of Earth's atmosphere, has a mass of about 1.03 kilogram and exerts a force or weight of about 10.1 newtons, resulting in a pressure of 10.1 N/cm^2 or 101 kN/m2 (101 kilopascals, kPa). A column of air with a cross-sectional area of 1 in 2 would have a weight of about 14.7 lbf, resulting in a pressure of 14.7 lbf/in^2.

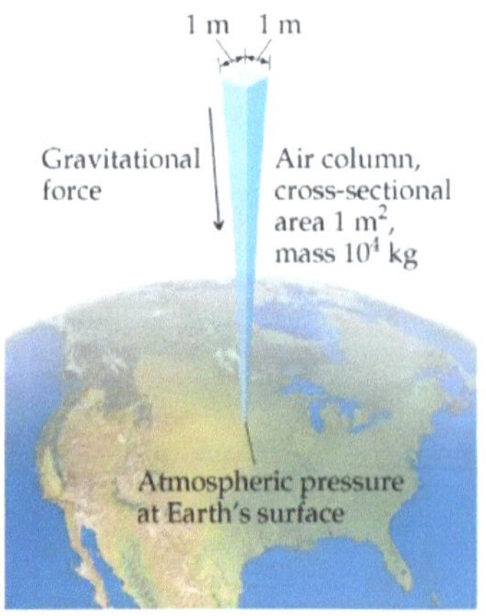

4. **Static and Dynamic Pressure**. Dynamic and static pressure are used in fluid dynamics to understand the pressure that is exerted on an object. These terms are used to define the state of a closed system of an incompressible, constant-density fluid. It is quite a simple concept. Static and dynamic pressure are often used to determine the pressure in a closed system. So, assuming that we are talking about a fluid (Water) in a pipe. This piece of pipe is placed horizontally and is filled with liquid.

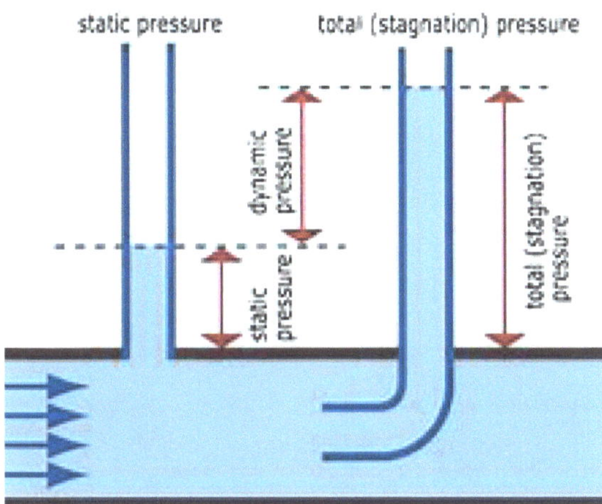

5. The liquid inside this pipe is stationary and not moving, but that doesn't mean that there is no pressure on it. The air provides pressure pushing it downwards, this is known as the static pressure when the liquid is at a state of rest. Now, imagine there is rushing of water in the pipe, the pressure exerted by the water pushing out the liquid, this is known as dynamic pressure.

6. This is commonly the only difference. The pressure exerted when there is no movement is static pressure, while pressure because of movement is dynamic pressure.

7. In respect of atmosphere, the pressure of the atmosphere is the force exerted on a surface of unit area by the air molecules impinging on it. When air is at rest, the motion of the molecules is entirely at random and the pressure exerted is uniform in all directions. This is known as static pressure, or simply as barometric pressure. If the air is in motion, an additional pressure is exerted on a surface of small area opposed to the direction of flow. This is known as dynamic pressure or wind pressure. This type of pressure has important applications in the design of pitot tubes of air speed indicators. However, in meteorology we are concerned mainly with static pressure. When the word pressure is used without qualification, it refers to static or barometric pressure.

8. **Global Pressure Pattern**. Due to prevailing temperature and position of the globe in respect of the Sun, there is a pattern of atmospheric pressure on the earth known as pressure belts. Pressure Belts of Earth refer to the regions that are dominated either by high-pressure cells or low-pressure cells. The high and low-pressure belts are arranged alternatively on the Earth's surface. The pressure belts are formed on the earth's surface because Warm air is light, and the air at the Equator rises, creating low pressure and the rising airmass descents over the high pressure belts. There are four pressure belts on the Earth.

   a) Tropical Low-Pressure Belts. This low-pressure belt interfaces from zero to 5° North and South of the Equator.

   b) Subtropical High-pressure Belt. The subtropical high-pressure belt is a series of atmospheric pressure zones that form a band located around 30 degrees north and south of the equator.

   c) Sub-Polar Low-Pressure Belts around 60 degrees north and south.

   d) Polar High-Pressure Belts north and south of 60 degrees north and south respectively.

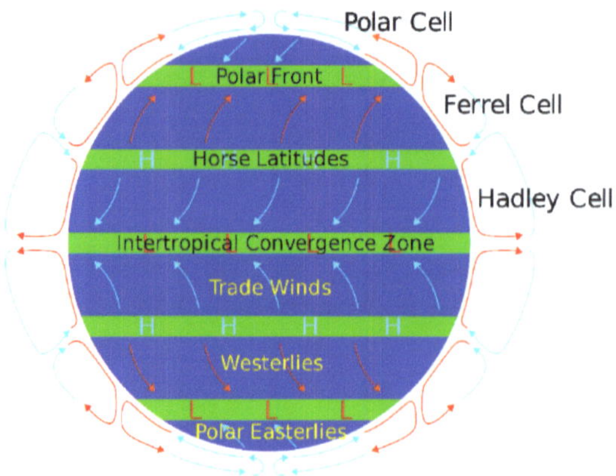

9. The normal wind pattern over the globe is in accordance to the pressure pattern. That is, wind flows from a high pressure area to a low pressure. However, the winds are reflected due to the Coriolis force generated due to rotation of the earth.

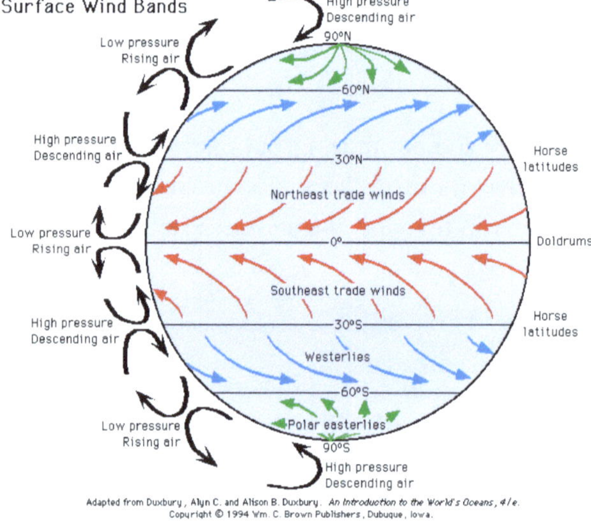

10. Due to difference in distribution of temperature due to change in the position of the Earth in respect of the Sun, there is seasonal/monthly variation in the pressure pattern. Apart from these, the pattern also undergoes temporary regional changes during formation and movement of significant weather systems like tropical cyclones.

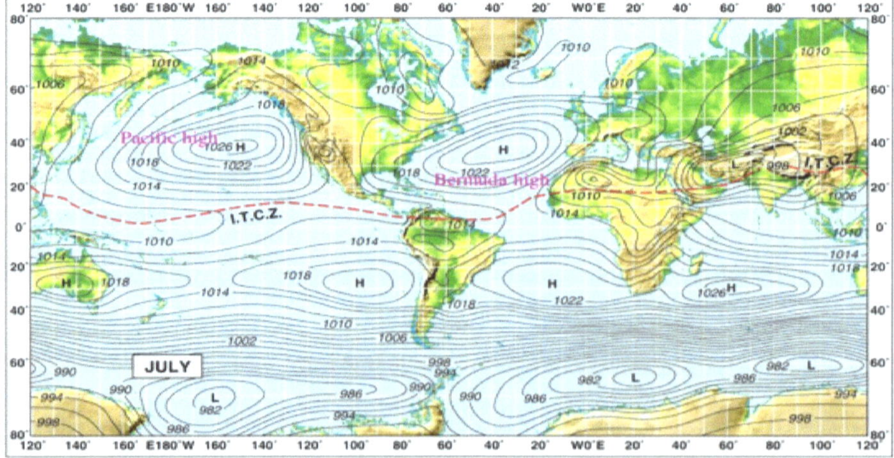

Analyzed Surface Charts Showing Pressure Pattern

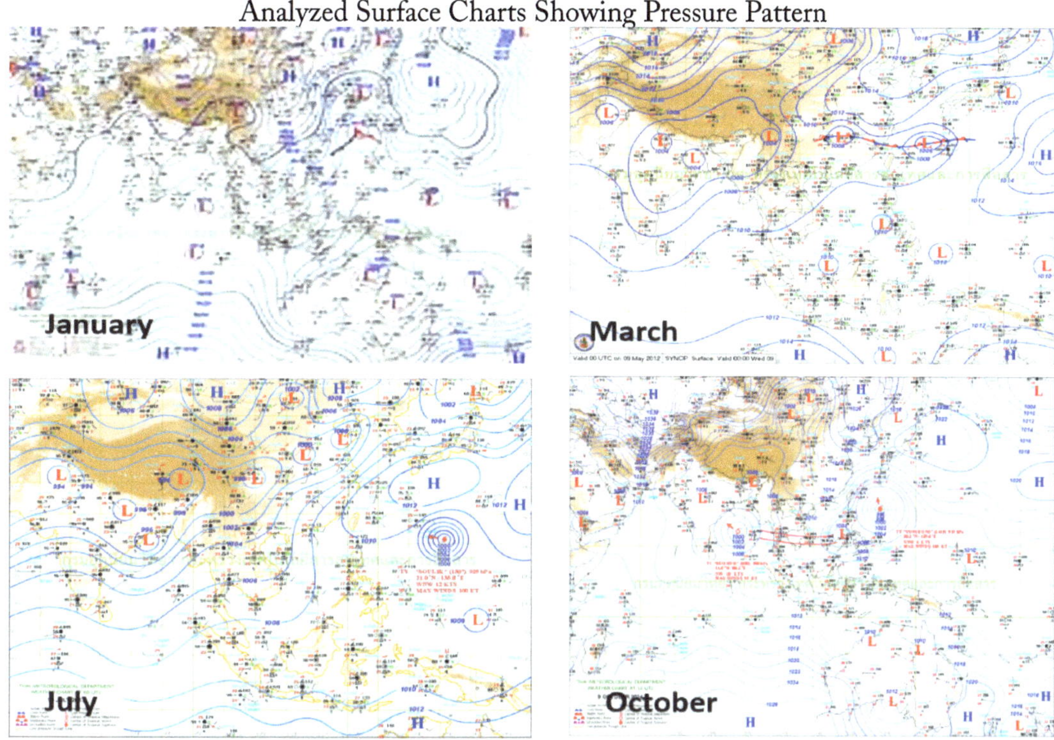

11. Study and analysis of the pressure system, their development, decay and movement is an important part of Meteorological services and is carried out on a day to day basis for every synoptic hours charts. The pressure systems, that is high pressure (H) and low pressure systems are of two types, Static and Migratory.

12. Static pressure systems are large in dimension. They may show some variation season to season but has little or no change within a season. Polar High, Equatorial Low are the examples

13. Migratory pressure systems affect a smaller area and remain for a short period from formation to decay. Their formation and subsequent movements are, however very important in terms of cross latitudinal transfer of energy. Tropical Cyclones and middle latitude Western Depressions (Extra Tropical Systems) are the examples of migratory systems.

14. **Diurnal Variation of Pressure**. Apart from change in atmospheric pressure due to movement of pressure system or changes due to change in season, pressure also undergoes changes on a daily basis. This is called diurnal variation of pressure. At a given station the pressure shows two high and two lows. On normal pressure day two maxima i.e. one at 10 a.m. and another at 10 p.m. and two minima i.e. one at 4 a.m. and another at 4 p.m. are observed.

15. Insolational heating and terrestrial radiation are mainly responsible for diurnal variations in pressure. The range between the maxima and minima is highest at the equator (about 3-4 mb) and negligible at the poles.

16. **Vertical Variation of Pressure**. Vertical pressure variation is the variation in pressure as a function of elevation. Depending on the fluid in question and the context being referred to, it may also vary significantly in dimensions perpendicular to elevation as well, and these variations have relevance in the context of pressure gradient force and its effects. However, the vertical variation is especially significant, as it results from the pull of gravity on the fluid; namely, for the same given fluid, a decrease in elevation within it corresponds to a taller column of fluid weighing down on that point.

17. A relatively simple version of the vertical fluid **pressure variation** is simply that the **pressure difference** between two elevations is the product of elevation change, gravity, and density. Greater the altitude, lesser is the air density and shorter is the air column on a particular point.

18. From the sea level to about 600 m the rate of fall of pressure is 4%. The rate of fall from 600 m to 1.5 km is 3% and that between 1.5 km to 3.0 km is 2.5%. At 6 km the atmospheric pressure is nearly half its value at the sea level. At a height of about 100 km the pressure is so low that it can be regarded as a vacuum.

19. A rough graph showing decrease in pressure with increase in altitude is shown below.

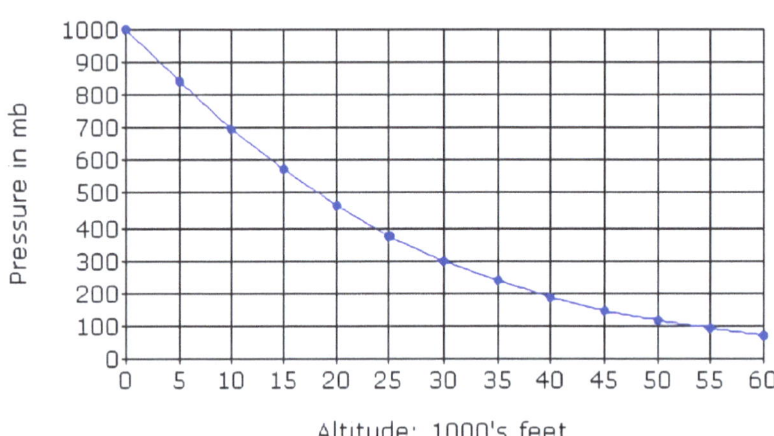

20. **Altimeters**. Altimeter is one of the many instruments the cockpit of an aircraft is equipped with. However, it is perhaps one of the most important instruments which is required to be set and monitored by the pilots before takeoff and landing and while in flying in cruising level. Altimeter indicates the height of an aircraft above sea level. When set in different pressure values, the height of the aircraft above the ground level of an airfield is also indicated.

21. The principle of 'decrease of air pressure with increase in altitude' is taken into consideration in the instrument. Basically, an altimeter reads the pressure of the atmosphere at a particular height or level of the aircraft. The pressure value is ultimately converted to the respective altitude of the aircraft.

22. As known to us, decrease in pressure with increase in height/altitude generally takes place at a certain rate. However, difference in the rate is obvious at different geographical position or different point in the atmosphere. Hence, setting of the altimeter with the pressure value of a certain airfield may not show the correct pressure (or height) at a different location. This is a drawback and is not suggestive. Now, what will happen if all the aircraft flying in air set their altimeters with a certain single pressure value? Will all the aircrafts not be indicated the correct proportionate reading? Is it possible? Yes, it is. Here comes the concept of Standard Atmosphere recommended by International Civil Aviation Organization (ICAO).

23. Before we talk about the standard atmosphere, we will discuss about the different pressure values. That is, how pressure values are observed or read at a certain observatory are given different corrections.

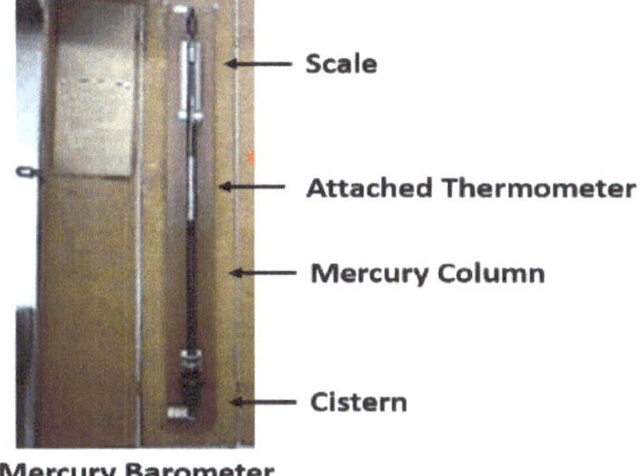

Mercury Barometer

24. **Corrections and Reduction of Pressure values**. Atmospheric pressure read at a place at a certain time is called 'Bar as read'. Immediately on taking the reading the following corrections are required to be applied to it: -

   a) **Instrument Correction**. There may be error in the instrument which is declared by the manufacturer or may be revealed during periodical calibration.

   b) **Temperature Correction**. The instrument is manufactured to give the correct reading at a certain standard temperature. Hence, with the change in the room temperature the reading will definitely show a deviated reading. Hence, temperature correction is required to be applied.

   c) **Cistern Height Correction**. A barometer indicates the pressure exerted at the level of the barometer cistern (Chamber for Mercury) or the sensor. In meteorology, the pressure value of a certain reference point is important and not the pressure of the level where the mercury is installed. For example, an airfield has a reference point which is referred for the geographical or topographical purposes. Actually, Airfield Reference Point (ARP) is a point which represent the average geographical or topographical position. Hence, the atmospheric pressure also has to be in respect of this point. The weather man may be recording pressure reading sitting in an office which may be located at any height. Hence, there is a requirement of reducing the pressure to the level of the ARP. The necessary correction required is generally applied once for all during installation of the instrument.

25. After applying the above corrections, the pressure value found is the pressure at the ARP level, or the aerodrome level pressure (When observatory is located in an airfield). This pressure in meteorology is given the abbreviated name as QFE as per ICAO guidelines. When an altimeter is set at QFE value, the reading in the instrument will show zero when the aircraft is on ground, and when it flies, the instrument will show the height of the aircraft above the ground level.

26. Setting the altimeter at QFE value cannot be a standard practice as the aircraft will be indicated the height above the airfield ARP. Therefore, there is a requirement to standardize the practice.

27. In standard practice, all the Aviation Met Observatories reduce QFE to Mean Sea Level (MSL) by applying some correction. This pressure is termed as MSL Pressure and is used in abbreviated form as QNH. In this method a vertical column of air extending from MSL to ARP is added to the QFE as per conditions of International Standard Atmosphere (ISA). That is, it is assumed that the column of atmosphere from the aerodrome level to the MSL will be as per ISA.

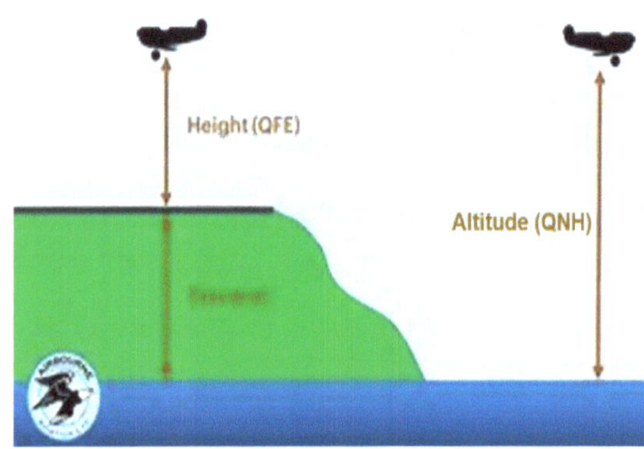

28. QNH value will certainly not be the same for the different airfields. Hence, there is a limitation in setting the altimeter in Local QNH. The local QNH is fine for taking off or while making a landing. However, as the aircraft ascends to the cruising level to cover a long distance, it will be difficult to maintain a given level in respect of the other aircrafts flying in that area and which are originated from different airfield setting their altimeter in different pressure values.

29. To overcome this issue, ICAO recommended setting of altimeters in the QNH value as per the ISA after takeoff and till the time the aircraft prepares for a landing. This facilitates all the aircrafts flying in a particular zone to have indicated level as per a single setting. In the image of an altimeter given at the beginning of this chapter, the altimeter was set in 1013.25 hPa or 29.92 which is the pressure at MSL as per ISA. The altimeter indicates the level of the aircraft as 25,000'. The pressure value at this level would be around 376 hPa.

Some of the specifications of ISA which are widely used in Aviation are as follows: -

Mean sea level temperature:	15°C(288.15 °K)
Mean sea level pressure:	1013.25 hPa
Surface density:	1225 g/m3
Acceleration due to gravity:	980.665 cm/sec2
Rate of fall of temperature with height upto 11 Km:	6.5°C/Km
Temperature:	Assumed constant at — 56.5°C (216.65°K) with an isothermal lower stratosphere above 11 Km upto 20 Km- From 20 to 30 Km there is a rise of temperature at the rate of 1° C/Km with a temperature of -44.5°C (228.65°K) at 32 Km.

30. The table below shows specifications of ISA in respect of Pressure, Temperature, Density and speed of sound at different levels.

$$\rho = \frac{p}{RT} \tag{9}$$

2. International Standard Atmosphere (ISA) Table [2]

The International Standard Atmosphere parameters (temperature, pressure, density) can be provided as a function of the altitude under a tabulated form, as given in Table 3:

Table 3 International Standard Atmosphere [2]

ALTITUDE (Feet)	TEMP (°C)	PRESSURE			PRESSURE RATIO $\delta = P/P_0$	DENSITY $\sigma = \rho/\rho_0$	Speed of sound (kt)	ALTITUDE (meters)
		hPa	PSI	In Hg				
40 000	- 56.5	188	2.72	5.54	0.1851	0.2462	573	12 192
39 000	- 56.5	197	2.58	5.81	0.1942	0.2583	573	11 887
38 000	- 56.5	206	2.99	6.10	0.2038	0.2710	573	11 582
37 000	- 56.5	217	3.14	6.40	0.2138	0.2844	573	11 278
36 000	- 56.3	227	3.30	6.71	0.2243	0.2981	573	10 973
35 000	- 54.3	238	3.46	7.04	0.2353	0.3099	576	10 668
34 000	- 52.4	250	3.63	7.38	0.2467	0.3220	579	10 363
33 000	- 50.4	262	3.80	7.74	0.2586	0.3345	581	10 058
32 000	- 48.4	274	3.98	8.11	0.2709	0.3473	584	9 754
31 000	- 46.4	287	4.17	8.49	0.2837	0.3605	586	9 449
30 000	- 44.4	301	4.36	8.89	0.2970	0.3741	589	9 144
29 000	- 42.5	315	4.57	9.30	0.3107	0.3881	591	8 839
28 000	- 40.5	329	4.78	9.73	0.3250	0.4025	594	8 534
27 000	- 38.5	344	4.99	10.17	0.3398	0.4173	597	8 230
26 000	- 36.5	360	5.22	10.63	0.3552	0.4325	599	7 925
25 000	- 34.5	376	5.45	11.10	0.3711	0.4481	602	7 620
24 000	- 32.5	393	5.70	11.60	0.3876	0.4642	604	7 315
23 000	- 30.6	410	5.95	12.11	0.4046	0.4806	607	7 010
22 000	- 28.6	428	6.21	12.64	0.4223	0.4976	609	6 706
21 000	- 26.6	446	6.47	13.18	0.4406	0.5150	611	6 401
20 000	- 24.6	466	6.75	13.75	0.4595	0.5328	614	6 096
19 000	- 22.6	485	7.04	14.34	0.4791	0.5511	616	5 791
18 000	- 20.7	506	7.34	14.94	0.4994	0.5699	619	5 406
17 000	- 18.7	527	7.65	15.57	0.5203	0.5892	621	5 182
16 000	- 16.7	549	7.97	16.22	0.5420	0.6090	624	4 877
15 000	- 14.7	572	8.29	16.89	0.5643	0.6292	626	4 572
14 000	- 12.7	595	8.63	17.58	0.5875	0.6500	628	4 267
13 000	- 10.8	619	8.99	18.29	0.6113	0.6713	631	3 962
12 000	- 8.8	644	9.35	19.03	0.6360	0.6932	633	3 658
11 000	- 6.8	670	9.72	19.79	0.6614	0.7156	636	3 353
10 000	- 4.8	697	10.10	20.58	0.6877	0.7385	638	3 048
9 000	- 2.8	724	10.51	21.39	0.7148	0.7620	640	2 743
8 000	- 0.8	753	10.92	22.22	0.7428	0.7860	643	2 438
7 000	+ 1.1	782	11.34	23.09	0.7716	0.8106	645	2 134
6 000	+ 3.1	812	11.78	23.98	0.8014	0.8359	647	1 829
5 000	+ 5.1	843	12.23	24.90	0.8320	0.8617	650	1 524
4 000	+ 7.1	875	12.69	25.84	0.8637	0.8881	652	1 219
3 000	+ 9.1	908	13.17	26.82	0.8962	0.9151	654	914
2 000	+ 11.0	942	13.67	27.82	0.9298	0.9428	656	610
1 000	+ 13.0	977	14.17	28.86	0.9644	0.9711	659	305
0	+ 15.0	1013	14.70	29.92	1.0000	1.0000	661	0
- 1 000	+ 17.0	1050	15.23	31.02	1.0366	1.0295	664	- 305

6

31. There is yet another method of reducing pressure. This is called QFF. It is the pressure at a place which is reduced to MSL as per standard meteorological practices. This value is used only in the preparation of surface charts as discussed previously.

32. **Altimeter Setting Procedure**. The pilot in flight has to know the height of his aircraft. On a circuit, the pilot has to know his height above the aerodrome level in order to adhere to the local approach procedures and to do a proper landing. On a route, the absolute height above the ground level is of little significance. The quadrantal height separation is more important from the point of view of flight safety.

33. It is, therefore, not necessary on a route flight for the aircraft to keep on re-setting the altimeter to QFE/QNH corresponding to the aerodromes directly below. If all the aircraft set their altimeters to any fixed base value, the desired height separation can be achieved. This fixed base value, which has been accepted as 1013.2 hPa (29.92 in), is the MSL pressure as per ISA and is termed as standard QNH. Thus, an aircraft leaving an airfield on a route flight will fly on local QNH setting initially, then at standard QNH at the cruising altitude, and finally change to local QNH on joining circuit at the destination airfield before landing. The necessary terms are defined below to understand the procedure better:

Altitude: The vertical distance of a level, a point or an object considered as a point, measured from mean sea level.

Height: The vertical distance of a level, point or an object considered as a point from a specified datum.

Elevation: The vertical distance of a point, or a level on or affixed to the surface of the earth measured from mean sea level.

Transition Altitude: This is the highest altitude below which the aircraft will always fly on local QNH. i.e. below which the vertical position of an aircraft is controlled by reference to the height above the aerodrome.

Flight Level: Surfaces of constant pressure at or above the transition level separated by pressure interval, corresponding to 500 ft. with MSL pressure as 1013.2 mb.

Transition Level: This is the lowest flight level above which the aircraft will always fly on standard QNH of 1013.2 hPa i.e. above which the vertical position of an aircraft is controlled by reference to the height above the pressure datum 1013.2 hPa. This is expressed in hundreds of feet.

Transition Layer: This is the airspace between the transition altitude and the transition level.

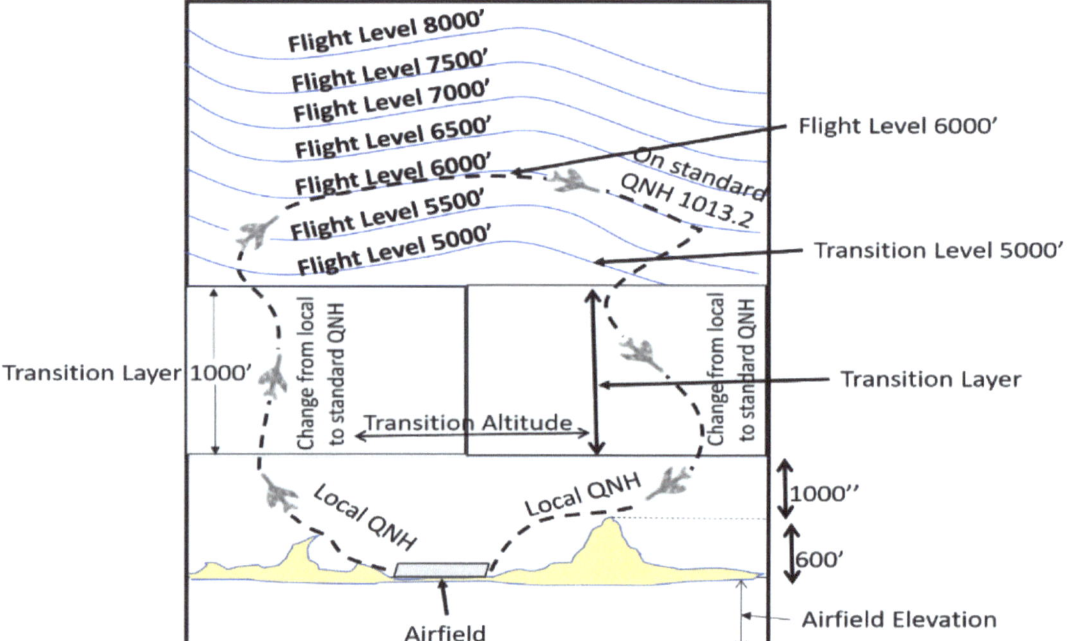

GENERAL METEOROLOGY

Chapter **10**

DENSITY OF AIR

1. The density of air or atmospheric density, denoted by , is the mass per unit volume of Earth's atmosphere. Air density, like air pressure, decreases with increasing altitude. It also changes with variation in atmospheric pressure, temperature and humidity. According to the International Standard Atmosphere (ISA), at 101.325 kPa (abs) and 20 °C (68 °F), air has a density of approximately 1.204 kg/m3 (0.0752 lb/cu ft), which is about 1/800 that of water. Pure liquid water is 1,000 kg/m3 (62 lb/cu ft).

2. Air density is a property used in many branches of science, engineering, and industry, including aeronautics, atmospheric research and meteorology.

3. Other things being constant, hotter air is less dense than cooler air and will thus rise through cooler air. This can be seen by using the ideal gas law as an approximation.

   Density of Dry and Moist Air. The density of dry air can be calculated using the ideal gas law, expressed as a function of temperature and pressure: -

$$\rho = P/RT$$

4. This formula gives a density of 1225 grams per cubic metre for dry air at ISA (sea level pressure of 1013.2 mb and temperature of 15°C or 288.15°A). For any other pressure, the equation can be modified as below for calculating .

$$\rho = 348.4\ P/T \text{ grams per cubic metre.}$$

Where p is in hPa and T in degrees Absolute

5. This equation applies only to perfectly dry air. Some water vapour is invariably present in the atmosphere. Water vapour also obeys the fundamental gas equation, but the gas constant R is different for water vapour from that for dry air. The total pressure p of moist air is the sum of the partial pressures exerted by the water vapour and the dry air, acting independent of each other. If the partial pressure of the water vapour is e, it can be shown that: -

Density of moist air = 348.4/T (p- 3e/8)

Thus, for a given pressure and temperature, moist air has lower density than dry air.

6. **Factors Affecting the Density of Air.**

   a) Air becomes denser as the air pressure increases. The pressure forces the air molecules together resulting in more mass in a given volume. Increasing altitude results in decreasing air pressure. The decrease in air density means that a mountain climber at high altitudes gets less oxygen when he breathes or the performance of an aircraft engine will be less.

   b) Temperature is another factor that affects the density of air. When the temperature is increased, the air molecules move faster and spread further apart. When air is denser, it creates a drag on objects moving through it. For example, a golf ball hit on a hot summer day will go further apart than one hit on a cold day. The higher temperature and lower air pressure found at high altitudes combine to lower the air density.

c)   Humidity or the amount of moisture in the atmosphere also changes the density of air. More moisture in the air, lower the density of air.

7.   **Density Altitude**. Density altitude is the altitude relative to standard atmospheric conditions at which the air density would be equal to the indicated air density at the place of observation. In other words, the density altitude is the air density given as a height above mean sea level. Density altitude can also be considered to be the pressure altitude adjusted for a non-standard temperature.

8.   Both, an increase in the temperature and a decrease in the atmospheric pressure, and, to a much lesser degree, an increase in the humidity, will cause an increase in the density altitude. In hot and humid conditions, the density altitude at a particular location may be significantly higher than the true altitude.

9.   In aviation, the density altitude is used to assess an aircraft's aerodynamic performance under certain weather conditions. The lift generated by the aircraft's airfoils, and the relation between its indicated airspeed (IAS) and its true airspeed (TAS), are also subject to air density changes. Furthermore, the power delivered by the aircraft's engine is affected by the density and composition of the atmosphere.

10.   Air density is perhaps the single most important factor affecting aircraft performance. It has a direct bearing on: -

a)   The efficiency of a propeller or rotor — which for a propeller (effectively an airfoil) behaves similarly to lift on a wing.

b)   The power output of a normally-aspirated engine depends on the oxygen intake, so the engine output is reduced as the equivalent dry air density decreases, and it produces even less power as moisture displaces oxygen in more humid conditions.

c)   Aircraft taking off from a 'hot and high' airfield, are at a significant aerodynamic disadvantage. The following effects result from a density altitude that is higher than the actual physical altitude: -

i.   An aircraft will accelerate more slowly on take-off as a result of its reduced power production.

ii.   An aircraft will climb more slowly as a result of its reduced power production.

11.   Due to these performance issues, an aircraft's take-off weight may need to be lowered, or take-offs to be re-scheduled for cooler times of the day. The wind direction and the runway slope may need to be considered.

12.   **Density Altitude in Sky Diving**.   Sky diving has been a common adventure for the adventure sports personnel as well as the tourist. However, with little knowledge of the atmospheric behavior in different parts of the year/season/month in respect of a particular geographical location, and not being guided by Meteorological professionals, there have been reports of some incidents and accidents. Sky Diving, being an exciting sport conducted in the atmosphere needs to respect the behavior and characteristics of the atmosphere at the time of operation to make the sport safe.

13.   In addition to the general change in wing efficiency that is common to all aviation, skydiving has additional considerations. There is an increased risk due to the high mobility of jumpers (who will often travel to a drop zone with a completely different density altitude than they are used to, without being made consciously aware of it by the routine of calibrating to *QNH/QFE*).

14.   Another factor is the higher susceptibility to hypoxia at high altitudes, which, combined especially with the unexpected higher free-fall rate, can create dangerous situations and accidents. Parachutes at higher altitudes fly more aggressively, making their effective area smaller, which is more demanding for a pilot's skill and can be especially dangerous for high-performance landings, which require accurate estimates and have a low margin of error.

15. To summarize, high altitude, high temperature, and high moisture content in the atmosphere reduces aircraft's performance due to increase in the density altitude. Hence, density altitude is required to be checked by the pilot of an aircraft, specially the cargo or passenger aircraft before takeoff on a summer humid day or while operating from a high-altitude airfield.

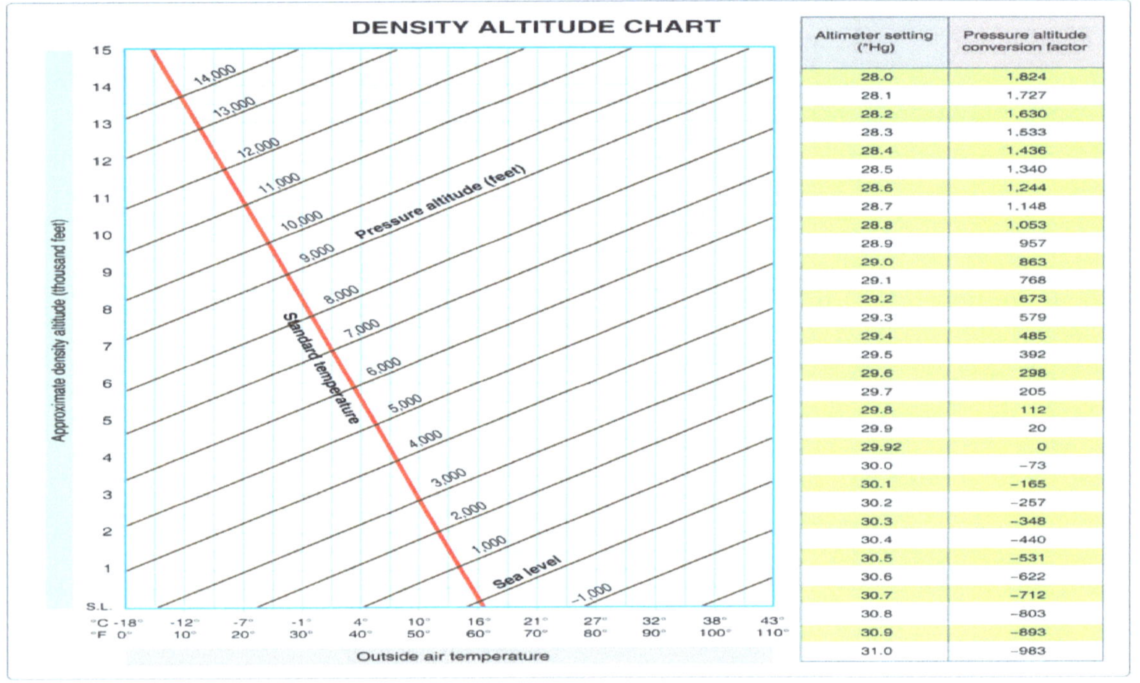

Altimeter setting ("Hg)	Pressure altitude conversion factor
28.0	1,824
28.1	1,727
28.2	1,630
28.3	1,533
28.4	1,436
28.5	1,340
28.6	1,244
28.7	1,148
28.8	1,053
28.9	957
29.0	863
29.1	768
29.2	673
29.3	579
29.4	485
29.5	392
29.6	298
29.7	205
29.8	112
29.9	20
29.92	0
30.0	−73
30.1	−165
30.2	−257
30.3	−348
30.4	−440
30.5	−531
30.6	−622
30.7	−712
30.8	−803
30.9	−893
31.0	−983

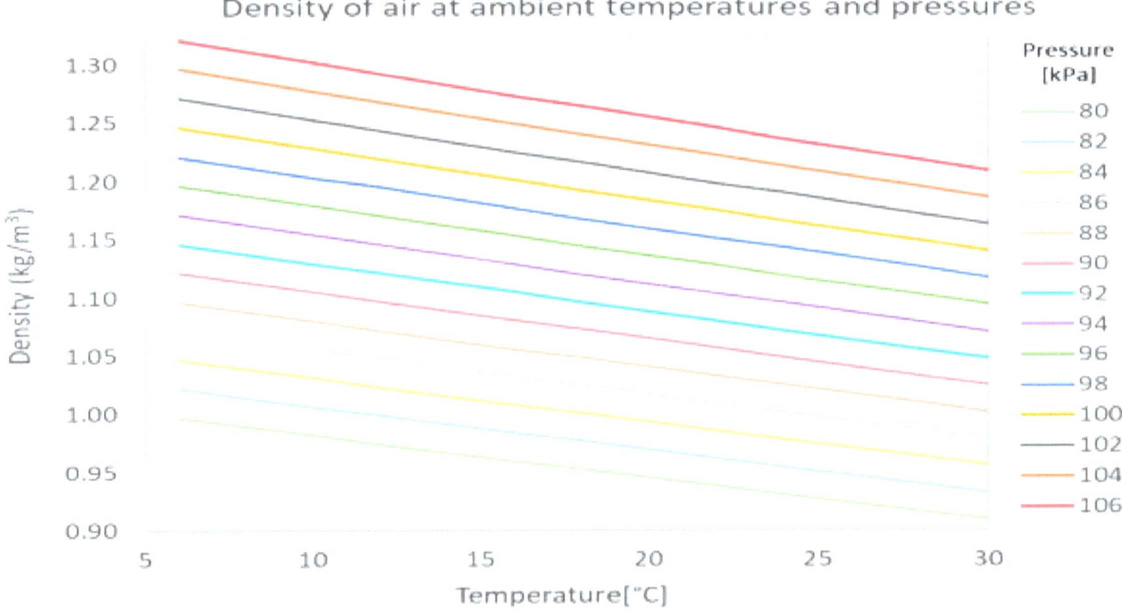

16. **Variation of Air Density with Altitude.** The density of air varies with pressure (the Ideal Gas Law) and the altitude above sea level. Mean absolute pressure at sea level is approximately 1013.2 hPa (760 mmHg) with a variation of about +/- 5 altitude and air pressure. Pressure decreases with height much more rapidly than temperature. Density decreases with height at all levels. Air density at about 6.0 Km is about 50%, 25% at 12.0 Km and at 20.0 Km it is just 10% of the standard density at sea level.

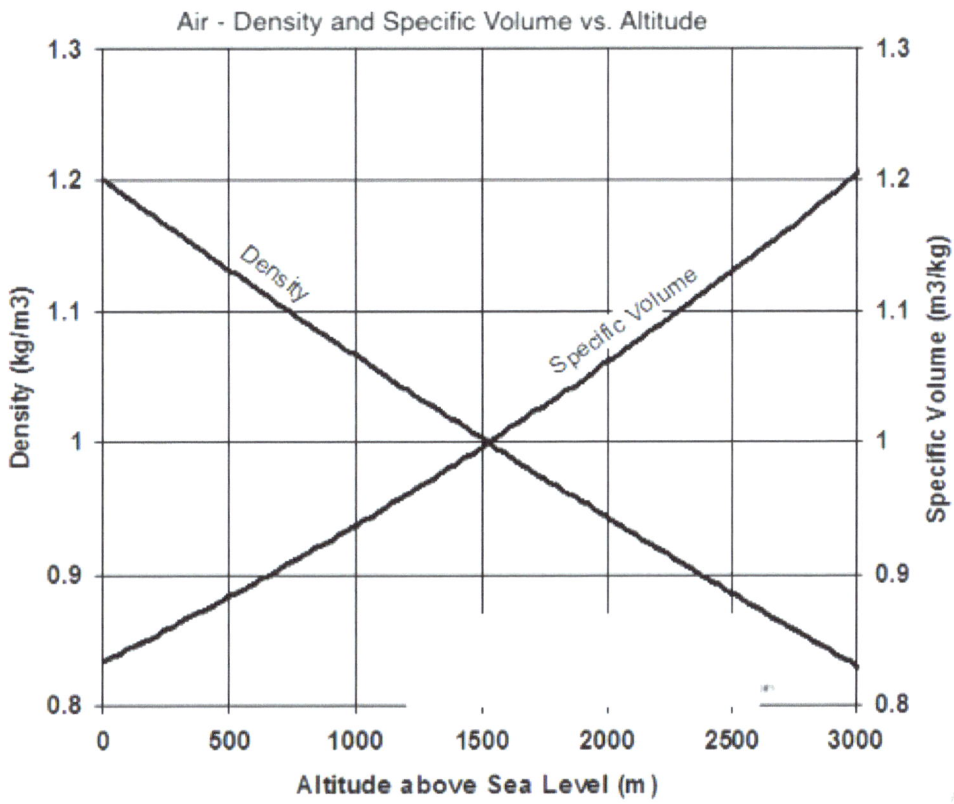

17. **Variation in Air Density with Latitude**.   At sea level, density is the lowest near the equator and greatest at the poles. This distribution is maintained up to about 8.0 Km. At about 8.0 Km, density is nearly uniform at all latitudes. Above 8.0 Km, a reversal of the distribution takes place, the density becoming more near the equator than at higher latitudes. Hence, we can conclude by stating that an aircraft flying below 8 Km will have better operational efficiency at higher latitude and an aircraft flying at higher altitude will have better efficiency at lower latitude.

CHAPTER 11

ADIABATIC CHANGES AND LAPSE RATES

1. Like pressure, temperature also decreases in the atmosphere with height. This decrease is not only due to the radiational effect. Convection is an important process by which heat transfer from the surface of the earth to the atmosphere takes place. This involves physical rise of air parcel. As a parcel of air rises, it expands. The expanded parcel cools from loss of heat energy because energy is dispersed over the larger volume. In the present chapter we shall examine the vertical distribution of temperature that results from such physical ascent of parcels of air.

2. **Lapse Rate.** The lapse rate is the rate at which an atmospheric variable, normally temperature in Earth's atmosphere, falls with altitude. Decrease in the temperature with height takes place within the troposphere. This decrease is a normal situation. At times temperature may remain the same within a certain height or it may increase. The shallow layer in which temperature remains constant is called *isothermal layer*. And rise in temperature with height is called *inversion*.

3. **The Adiabatic Process.** The temperature profile of the atmosphere is a result of an interaction between thermal conduction, thermal radiation, and natural convection. Sunlight hits the surface of the earth (land and sea) and heats them. The surface then heats up the air above the surface. If radiation were the only way to transfer energy from the ground to space, the greenhouse effect of gases in the atmosphere would keep the ground at roughly 333 K (60 °C; 140 °F).

4. However, when air is hot, it tends to expand, which lowers its density. Thus, hot air tends to rise and carry internal energy upward. This is the process of convection. Vertical convective motion stops when a parcel of air at a given altitude has the same density as the other air at the same elevation.

5. When a parcel of air expands, it pushes on the air around it, doing thermodynamic work. If we suppose that the parcel of air is insulated and transfer of heat from the parcel while expanding or contracting is not allowed. In such way it is possible to assess the change in temperature within the parcel while undergoing changes in volume. *An expansion or contraction of an air parcel without inward or outward heat transfer is called adiabatic process.*

6. Ascent and descent of parcels of air in the atmosphere occur under adiabatic conditions. The ascending or descending parcels of air undergo changes of pressure without appreciable loss or gain of heat with respect to the environment. This property gives us a useful tool in assessing the changes in temperature that an air parcel would undergo if made to ascend or descend in the atmosphere.

7. **Adiabatic Lapse Rate.** The rate at which the temperature of a parcel of air decreases with height when it is made to ascend adiabatically is known as adiabatic lapse rate. For unsaturated air it has been worked out from the gas equation as 9.8 °C per kilometre. This is known as the *dry adiabatic lapse rate* (DALR).

8. The presence of water within the atmosphere (usually the troposphere) complicates the process of convection. Water vapor contains latent heat of vaporization. As a parcel of air rises and cools, it eventually becomes saturated; that

is, the vapor pressure of water in equilibrium with liquid water has decreased (as temperature has decreased) to the point where it is equal to the actual vapor pressure of water. With further decrease in temperature the water vapor in excess of the equilibrium amount condenses, forming cloud, and releasing heat (latent heat of condensation). Before saturation, the rising air follows the dry adiabatic lapse rate.

9. Thus, the adiabatic fall of temperature is somewhat offset by the rise in temperature due to the latent heat of condensation. The residual lapse rate, known as the saturated adiabatic lapse rate (SALR) is, therefore, less than the DALR. At low temperatures very little water vapour is held by the air even when saturated and the latent heat released in condensation is small. Thus, at low temperatures, the SALR approaches the DALR. It follows that unlike the DALR, the SALR is not constant at all temperatures. It is about 5°C per kilo-metre at standard sea level temperature, and gradually approaches the DALR (9.8° C per kilo-metre) at temperatures below - 40° C. The SALR at any given temperature can be calculated from the gas equation and the value of saturated water vapour content at that temperature obtained.

10. While the dry adiabatic lapse rate is a constant 9.8 °C/km, the moist adiabatic lapse rate varies strongly with temperature. A typical value is around 5 °C/km.

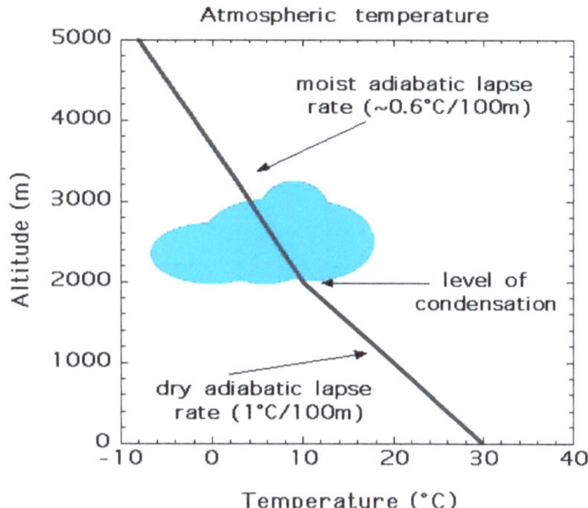

11. **Actual Lapse Rates in the Atmosphere**. Ascent and descent of air in the atmosphere is assumed to be adiabatic. However, actual lapse rates in the vary widely due to various causes. The actual lapse rate in the atmosphere, referred to as environmental lapse rate (ELR) is intermediate between the DALR and the SALR, being closer to the latter. The variance of the actual lapse rate from the adiabatic lapse rates is principally due to the following causes: -

a) Mixing of ascending or descending parcels of air with the surroundings.

b) Variation in quantity of water vapour held by air at different levels.

c) Horizontal transfer of heat by wind through advection.

12. Knowledge of adiabatic lapse rates is useful in predicting the changes in temperature which an air parcel would undergo if made to ascend or descend in the atmosphere. This knowledge can be usefully employed in the prediction of the behaviour of the atmosphere and the resulting weather processes.

CHAPTER 12

STABILITY AND INSTABILITY OF THE ATMOSPHERE

1. Air is said to be stable when it resists vertical ascent. It is unstable when it has tendency to rise up and cause weather or convective build up.

2. The simplest way to understand how stability works is to imagine a parcel of air having a thin, flexible cover that allows it to expand but prevents the air inside from mixing with the surrounding air. Next, imagine that we take the balloon and force it up into the atmosphere. Since air pressure decreases with altitude, the balloon will relax and expand, and its temperature will therefore decrease. If the parcel were cooler than the surrounding air, it would be heavier (since cool air is denser than warm air); and if allowed to do so, it would sink back down to the ground. Air of this type is said to be stable.

3. On the other hand, if we lifted our imaginary balloon and the air within it was warmer, and hence, less dense than its surrounding air, it would continue to rise until it reached a point where its temperature and that of its surroundings were equal. This type of air is classified as unstable.

4. If in the temperature of the air inside the balloon is cool and dense than the surrounding or than the layer of air immediately above, then the balloon even after given a force upward, will sink back. This stage is called absolute stability.

5. If after release, the balloon moves upward and continues till great height, it is called absolute instability.

6. It may happen that the atmosphere is stable to a certain height and aloft it is unstable. In this case if the balloon is brought to this height applying force, then the balloon will continue to move up even after removing the force. This condition of atmosphere is called Conditional Instability.

7. The concept is illustrated in a simple way in the following figure in respect of the current position of the object.

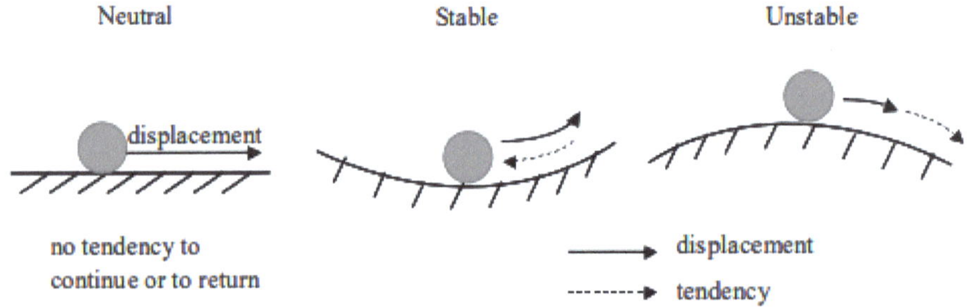

8. The second case in the figure above also illustrate that there is instability up to a certain position. After reaching this position the object attains stability. In the third case it is seen that, if the object can be brought to the top most position of the ramp applying force, then it will attain instability. That is, the case also represents conditional instability.

From the above, it can be stated that: —

a) A layer of air is said to be stable if a parcel of air within it, given a small initial push upwards, sinks back to the original level.

b) It is said to be neutral if a parcel of air, given a small initial push, remains at the level at which the force pushing it upwards is removed.

c) It is said to be unstable if the parcel of air continues to be displaced upwards on its own even when the initial force pushing it upwards is removed.

d) In case of conditional instability, a layer of air becomes unstable if it is raised to a certain height using some external energy.

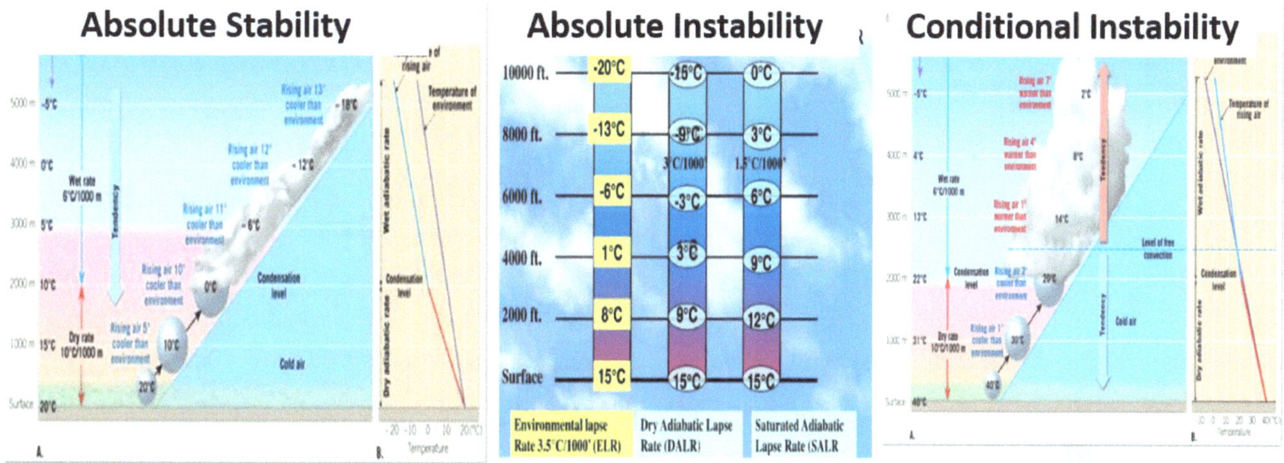

9. **Buoyancy (Archimedes' principle).** Physical law of buoyancy states that a body completely or partially submerged in a fluid (gas or liquid) at rest is acted upon by an upward, or buoyant force, the magnitude of which is equal to the weight of the fluid displaced by the body. The weight of the displaced portion of the fluid is equivalent to the magnitude of the buoyant force. The buoyant force on a body floating in a liquid or gas is also equivalent in magnitude to the weight of the floating object and is opposite in direction. The object neither rises nor sinks.

10. If the weight of an object is less than that of the displaced fluid, the object rises, as in the case of a block of wood that is released beneath the surface of water or a hydrogen filled balloon that is let loose in air. An object heavier than the amount of the fluid it displaces, though it sinks when released, has an apparent weight loss equal to the weight of the fluid displaced.

The principle of buoyancy may, therefore, be stated in terms of temperature as follows:-

a) A parcel of air whose temperature is higher than the surroundings, rises up to a level where the temperatures equalize.

b) A parcel of air whose temperature is lower than the surroundings, sinks to a level where the temperatures equalize.

11. In rising, the parcel of air cools at the adiabatic lapse rate (DALR or SALR as the case may be). By comparing the actual lapse rate with the adiabatic lapse rate, it is thus possible to determine whether the parcel of air will continue to rise or sink at any stage.

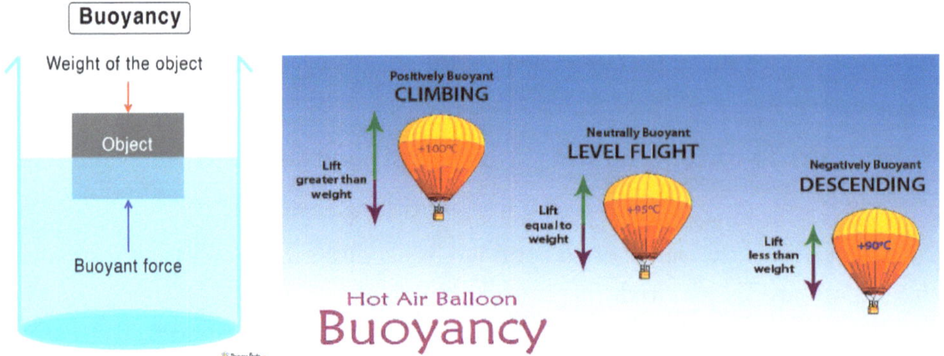

12. **Stability and Lapse Rate of Dry Air.** If an air parcel is dry, meaning unsaturated, stability is relatively straightforward. An atmosphere where the environmental lapse rate is the same as the dry adiabatic lapse rate, meaning that the temperature in the environment also drops by 9.8 K·km-1, will be considered neutrally stable. After some initial vertical displacement, the temperature of the air parcel will always be the same as the environment so no further change in position is expected.

13. If the environmental lapse rate is less than the dry adiabatic lapse rate, some initial vertical displacement of the air parcel will result in the air parcel either being colder than the environment (if lifted), or warmer than the environment (if pushed downward). This is because, the temperature of the air parcel if lifted, would drop more than the temperature of the environment. This is a stable situation for a dry air parcel and a typical scenario in the atmosphere. The global average tropospheric lapse rate is 6.5 K km-1, which is stable for dry lifting.

14. Finally, if the environmental lapse rate is greater than the dry adiabatic lapse rate, some initial vertical displacement of the air parcel will result in the air parcel either being warmer than the environment (if lifted), or colder than the environment (if pushed downward). This is because, the temperature of the air parcel if lifted, would drop less than the temperature of the environment. This is an unstable situation for a dry air parcel.

15. A simple illustration of the above principle is shown below. Consider a layer of dry air from the surface up to 3,000 ft, in which the vertical distribution of temperature is as shown in the figure below: -

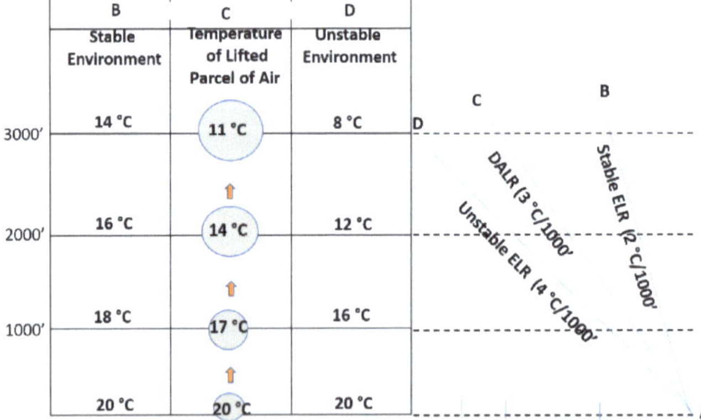

16. The lapse rate in this layer (ELR) is actually 2°C per 1,000 ft. Now let us test this air for stability. We take a sample of air at the surface where the temperature is 20°C and raise it to the 1,000 ft level; during this lifting it will cool adiabatically by 3°C (DALR) and will, therefore, have at 1,000 ft a temperature of (20 - 3) = 17°C as shown in the column C. This temperature is 1°C lower than that of the surrounding air at 1,000 ft and so the sample, being denser than the surroundings will tend to sink back to the surface. If the sample is lifted further to 2,000 ft. or 3,000 ft, it can be seen that at each level it will be colder and hence denser, than the environment. The layer of air is clearly stable because parcels of air, given an initial push, tend to sink back. A little reflection shows that the air is stable because the observed lapse rate in the layer is everywhere less than the DALR.

17. It is similarly easy to see that if the ELR is greater than the DALR, having a value of say 4°C per 1,000 ft as shown in column D, then a sample of lifted air is always warmer and hence less dense, than the surrounding air. In this case the layer is unstable.

18. **Stability of Saturated Air**. It can be shown in a similar way that saturated air is stable when the ELR is less than the SALR and unstable when the ELR is greater than the SALR.

19. The effects of moisture change the lapse rate of the air parcel and therefore, affects stability. However, the concepts are still the same and we still compare the air parcel temperature to the environmental temperature. We have just one added complication to worry about is that we need to know whether the air parcel is dry or moist. Some definitions are included below, which considered both dry and moist adiabatic lapse rates.

20. The atmosphere is said to be **absolutely stable** if the environmental lapse rate is less than the moist adiabatic lapse rate. This means that a rising air parcel will always cool at a faster rate than the environment, even after it reaches saturation.

21. The atmosphere is said to be **absolutely unstable** if the environmental lapse rate is greater than the dry adiabatic lapse rate. This means that a rising air parcel will always cool at a slower rate than the environment, even when it is unsaturated. This means that it will be warmer (and less dense) than the environment, and allowed to rise.

22. The atmosphere is said to be **conditionally unstable** if the environmental lapse rate is between the moist and dry adiabatic lapse rates. This means that the buoyancy of an air parcel depends on whether or not it is saturated. In a conditionally unstable atmosphere, an air parcel will resist vertical motion when it is unsaturated, because it will cool faster than the environment at the dry adiabatic lapse rate. If it is forced to rise and is able to become saturated, however, it will cool at the moist adiabatic lapse rate. In this case, it will cool slower than the environment, become warmer than the environment, and will rise.

23. **Latent Instability**. Consider a layer of air in which the lapse rate is not as expected but there is variation. There may be a case that the air is stable in the lower layer but is unstable at some higher layer aloft. In such situation the stable air in the lower layer, if applied some force to rise to the unstable layer, then the stable layer will also attain instability. Such situation of hidden instability is called Latent Instability.

    The above properties are summarized and illustrated below: –

    a)   Dry air is stable when the ELR < DALR.

    b)   Dry air is unstable when the ELR > DALR.

    c)   Saturated air is stable when the ELR < SALR

    d)   Saturated air is unstable when the ELR > the SALR.

    e)   Both dry and saturated air is stable when the ELR < SALR. This is known as absolute stability.

    f)   Both dry and saturated air is unstable when the ELR > DALR. This is known as absolute instability.

    g)   Air, in which the ELR is between the DALR and SALR has conditional stability. It is stable till it is dry but is unstable when the it gets saturated

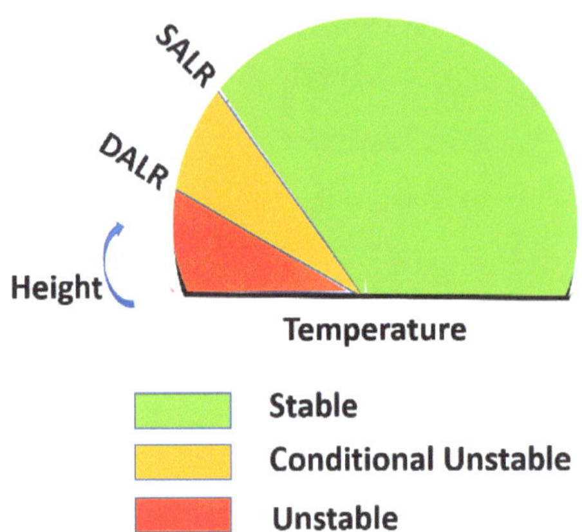

24. It follows from the above that a layer wherein the lapse rate is zero (isothermal) or negative (inversion), has absolute stability. On the other hand, a layer in which a superadiabatic lapse rate (lapse rate greater than DALR exists), has absolute instability.

25. **Uses of Stability Conditions**. We have seen how we can determine whether a sample of air, if given an initial lift, will continue to ascend or not. Such knowledge is useful to predict the extent of vertical development of clouds and the generation of convective weather phenomena like dust storms or thunderstorms.

CHAPTER 13

VERTICAL MOTION OF AIR

1. For occurrence of any weather activities like precipitation, thunderstorm, etc, there has to be upward rise of air. In other words, there has to be vertical motion of air. The vertical motion, though of very small order, is one of the most important factors for generation of weather.

2. The vertical motion of air is the **result of both thermal and dynamic action** and can be directly linked to precipitation. Heating of the underlying surface can be expressed directly as the vertical motion of air. The characteristics and intensity of vertical motion in the atmosphere are closely related to precipitation.

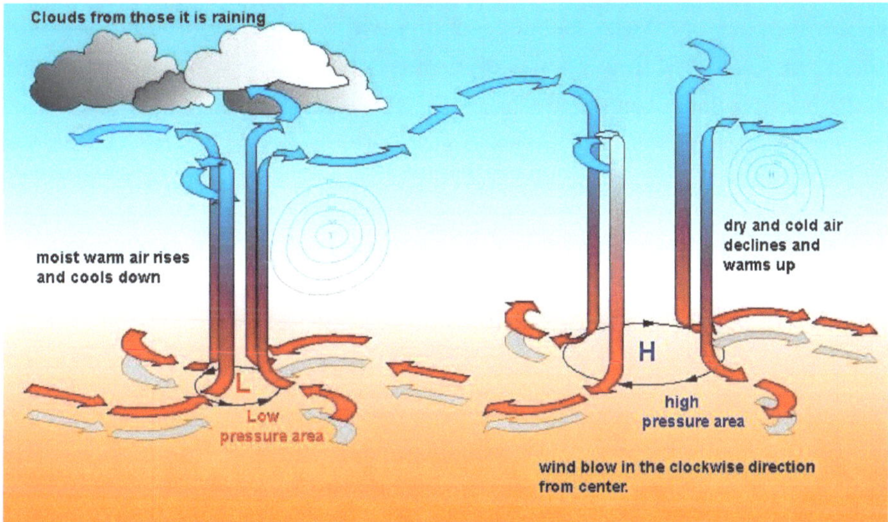

3. Wind, as we know, is the horizontal movement of air which happens due to the forces like pressure gradient, uneven heating of surface or development or movement of some significant pressure system. However, the causes of vertical movement, though effected by the above features, are little different.

4. For vertical movement to happen, the following atmospheric situations, and some other are required: -

   a) Heating of a parcel of air

   b) Lifting of a parcel of air by orographic lifting

   c) Lifting of air parcel due to frictional eddies

   d) Accumulation of air at a certain area or merging of different air masses due to: -

      i.   Cyclonic Circulation

      ii. Frontal Lifting

iii. Meeting of different air mass at a point or along a line due to the situations like confluence, wind discontinuity, significant decrease in wind speed over an area (speed convergence), etc.

It is important to note from the above that vertical rise of the air takes place when: -

a) There is a force like heating, friction or orographic lifting to lift a parcel of air.

b) When a warm air mass meets a cold air mass, it is raised (Frontal Lifting) and Vis -a -Vis, that is, when a cold air mass meets a warm one, it undercuts and lift up the warm air mass.

c) Convergence (accumulation) of air. What happens when air accumulates at a place for reasons like circulation, confluence, speed convergence, wind discontinuity, etc? Well, whenever there is accumulation, the following happen: -

   i. The accumulated air has to rise up as there is no scope for descending from surface.

   ii. Gross difference in large air mass meeting over a narrow geographical area (Front).

   ii. It is most likely that the accumulated air originating from different location will have different characteristics, mainly in respect of temperature and moisture content. The moist air mass or that with higher temperature will certainly rise up.

5. **Orographic Lifting**. Orographic lifting of air takes place when an air mass encounters a large and tall barrier, say a hill, on its way. The airmass, irrespective of its characteristics will rise up along the hill. In other words, Orographic lift occurs when an air mass is forced from a low elevation to a higher elevation as it moves over rising terrain. As the air mass gains altitude it quickly cools down adiabatically, which can raise the relative humidity to 100% and create clouds and, under the right conditions, precipitation and thunder.

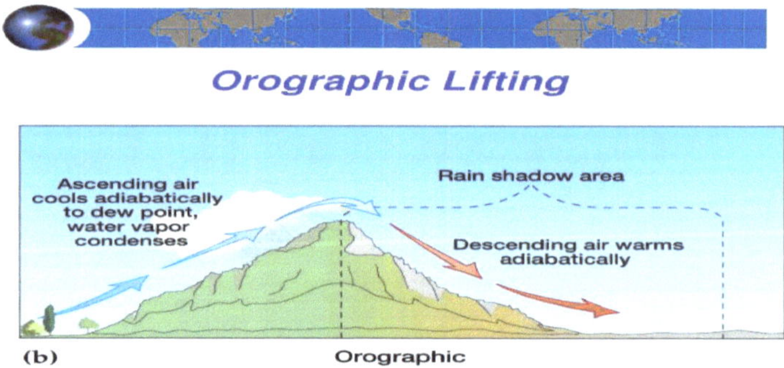

6. Orographic lifting may not be so effective in building visible weather if there is stability in the air, or there is inversion around the hill. In such situation vertical movement will be arrested and the air will tend to flow around the hill rather than surmount the top. Under such situation, the winds form eddies and often causes turbulence to the aircraft flying in such region.

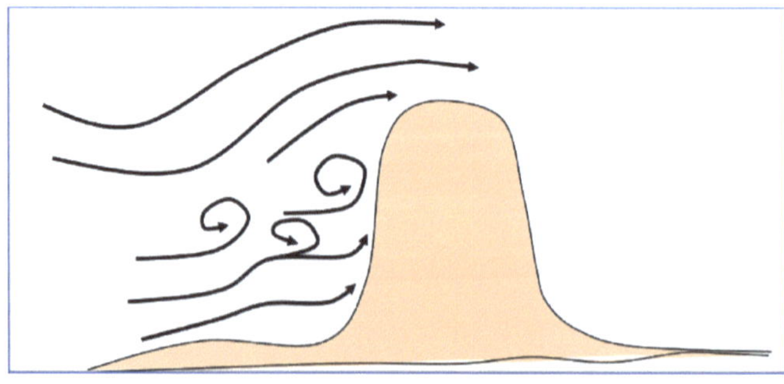

7. When the slope of the hill is steep and the wind speed is strong, the airflow ceases to be smooth. Formation of eddies become more prominent and form on either side of the hill. In such situation the eddies drift with the winds and create pronounced bumpiness in an aircraft flying in it.

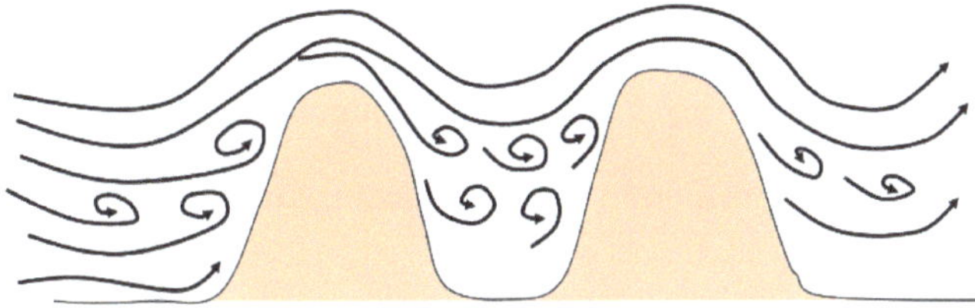

8. Another common type of disturbed airflow is that of standing waves formed on the leeward side of a range of hills. As the air stream ascends and passes over the top, a parcel of air at the crest of its displacement is subjected to buoyancy forces tending to restore it to its original level, so long as the atmosphere has stable lapse rate. The downdraft on the leeward side usually undershoots its original position and executes a few oscillations downwind before settling down to horizontal flow. Thus, standing waves are formed. These are also known as **mountain waves or lee waves**. Mountain waves are unlikely to be formed when the lapse rate at the mountain top level is steep. Lee waves are hazardous to an aircraft flying over mountainous terrain as they are associated with severe turbulence, strong vertical currents, and icing. The vertical currents in the waves can make it difficult for an aircraft to maintain altitude (level busts) and can cause loss of control. Loss of Control can also occur near to the ground prior to landing or after take-off. Aircraft can suffer structural damage as a result of encountering severe turbulence. In If caught unaware, passengers and crew walking around the aircraft cabin can be injured. There have been cases of fatal accident in such situation.

9. The wavelength and amplitude of the oscillations depends on many factors like the height of the hills, wind speed, instability, etc.

10. Mountain waves are generally associated with formation of Lenticular Clouds (lens shaped clouds). These clouds form in the crest of the mountain waves if the air is moist. Below the waves, Roll Clouds can also occur if the air is moist. These clouds are a good indication of the presence of mountain waves.

11. **Frictional eddies.** Surface wind is rarely steady, and exhibits gustiness. The reason for the gustiness is that the roughness of the ground does not permit smooth flow of air near the ground. The air stream breaks up into small irregular cells, known as frictional eddies which travel with the wind. In these eddies there is vertical motion.

12. The disturbance in the surface or low-level winds mentioned above generally persists up to about 3000-4000 Ft. Effect of frictional eddies are more pronounced during day time when there is instability due to solar heating.

13. Vertical motion induced by frictional turbulence is an important factor in the generation of fog and of thin layer clouds near ground level.

14. **Cyclonic Circulation (CYCIR).** Inward flow of wind around a low-pressure area in anti-clock wise direction in the northern hemisphere and in clockwise direction in the southern hemisphere is called cyclonic circulation. Convergence of great amount of air around the low-pressure area and subsequent rise are involved in a cyclonic circulation. As known to us, rising of the wind is associated with instability which results in formation of clouds and weather.

15. The intensity of clouding or weather depends on the vertical extent of the circulation, characteristics of the air masses that are converging and adequacy of divergence at higher levels.

16. In the figure below air column is shown extending from surface to 15,000 Ft.

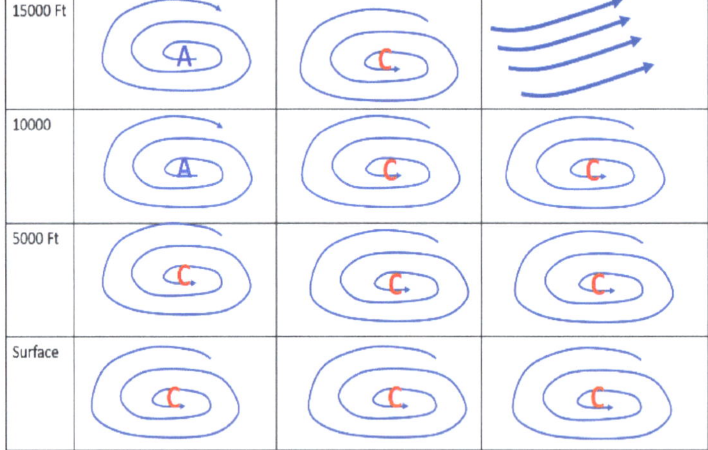

a) In the first case a CYCIR is seen extending up to 5,000 Ft indicating that, there is vertical ascent up to 5,000 Ft. However, aloft there is an anticyclone which is associated with stable air having tendency to sink. In such case, the CYCIR will not persist and will decay soon as further ascent of air is arrested by the overlying anticyclonic flow. Therefore, no significant cloud or weather will occur due to this CYCIR.

b) In the second case, the CYCIR extends up to great height. In this case, there has been no scope for the accumulated air to escape from the narrow area of the CYCIR. In other words, there is no divergence aloft. That is, the vertical movement of air will not increase and will remain low. In such case also the CYCIR will not be much effective as far as clouding and weather is concerned.

c) In the third case, the CYCIR extends up to 10,000 Ft and aloft at 15,000 Ft there is strong diverging wind. This will result in in enhancement of speed of vertical current which will result in formation of multi-layered clouds including embedded convective clouds which will, in turn cause weather activities like rain or thunder.

17. A three-dimensional model of a cyclonic circulation and anti-cyclone showing ascent, descent, convergence and divergence is given below: -

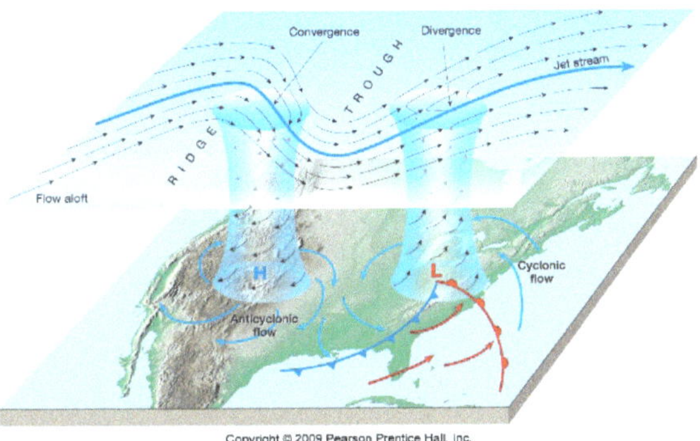

18. As said earlier, characteristics of the different air masses that have converged around a CYCIR also matter. If the air masses have vast difference in temperature and humidity profile, the severity of weather is more. In the numerically generated chart below, a CYCIR is seen over NW Bay of Bengal off Odisha and adjoining Gangetic West Bengal coast. Another CYCIR is marked over Saurashtra region. Note that different air masses have converged in these CYCIRs. One is the comparatively dryer and warmer continental (Originating from land) air mass and the other is moist and cool maritime (Originating from sea) air mass. Such features are prone to extensive clouding and weather.

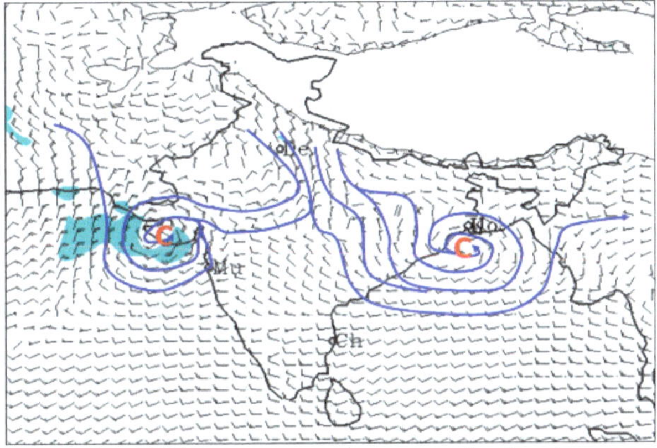

An NWP generated upper air chart

19. A manually analyzed upper air chart is shown below: -

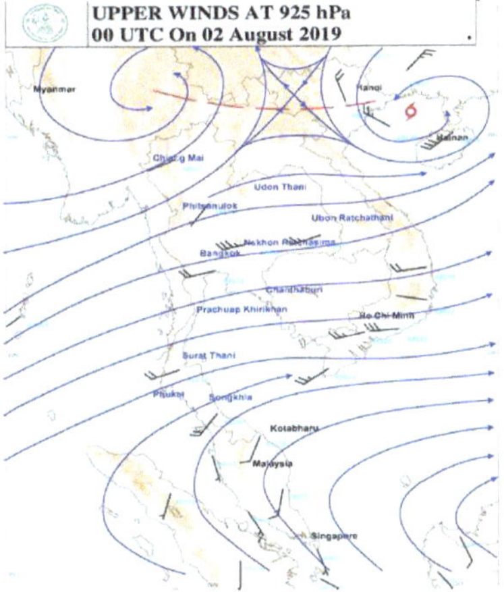

20. **Frontal Lifting**. Front in meteorology, is a transition where two air masses with different characteristics meet. Unlike in smaller synoptic feature like CYCIR, font covers a very large area. In general, one of the air masses is warm and moist and the other is dry and cold. When these air mass meet   along a line, the warm and the moist air mass gets lifted up along a slope resulting in formation of clouds and occurrence of weather. When a warm and moist air mass approaches a cold and dry air mass, it is called a warm front and when the case is other way round, it is called a cold front. Warm fronts are generally associated with gradual increase in clouding and light to moderate precipitation persisting for a longer period. Whereas, cold fronts are associated with rapid buildup of intense weather but lasting for short duration.

21. In both the cases, that is, warm and cold fronts, obviously the warm air mass gets lifted up. However, the lifting process is little different. The warm air mass in case of warm front gets gradually overlays the cold air mass. Whereas, in case of cold front, the cold air mass undercuts the warm air mass and eventually the warm air gets lifted abruptly resulting in change in weather in short spell of time.

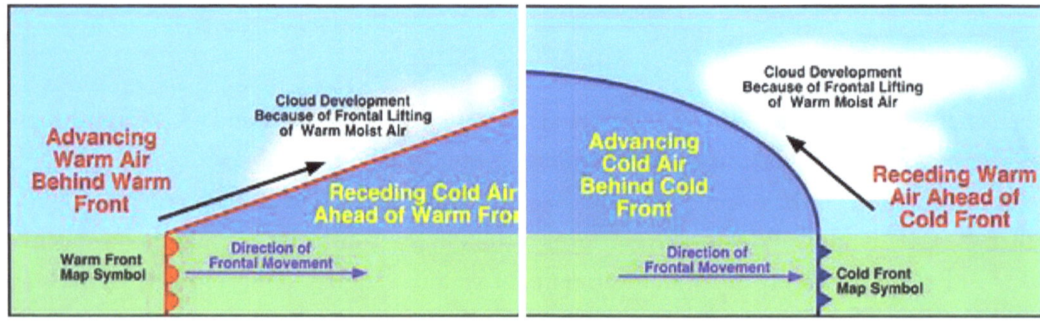

22. The western disturbances that affect the northern latitude including the areas of northwest India during winter season are the typical fronts. Following is a satellite image showing clouds associated with warm front in a Western Disturbance.

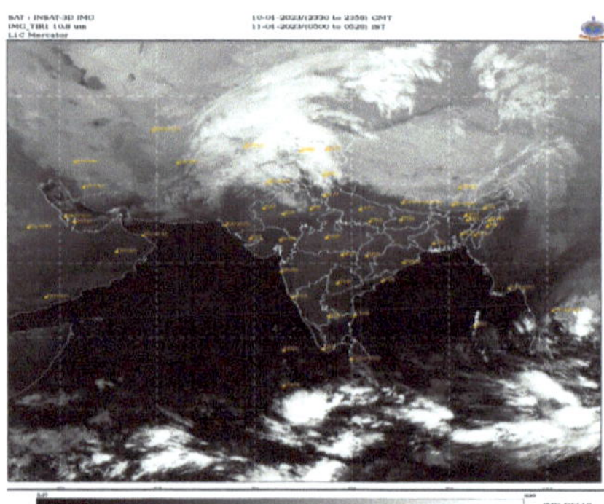

23. **Other Features**. Some of the other features which were discussed along with the 'Convective Clouds' are the following: -

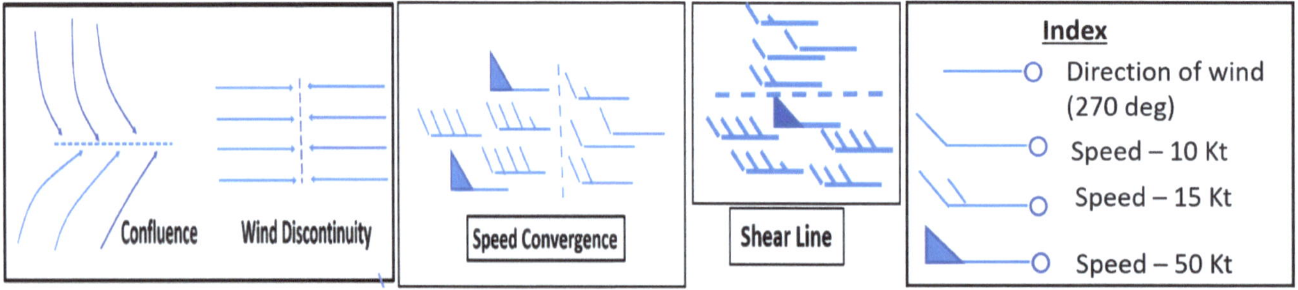

24. In the first two cases of the above figure (Confluence and Discontinuity), winds from different directions are seen converging along a line. Confluence can take place at any level. However, discontinuity is a lower level feature. In

both the case accumulated air gets lifted up resulting in cloud formation and weather subject to support at higher levels. Wind discontinuity is regular feature that can be marked almost on daily basis during pre-monsoon season aver Peninsular region causing scattered to widespread rain/thunderstorm. Following is an INSAT imagery (IR) of evening on a day during May 2020. The convective clouds marked extending from Kerala to Odisha is due to a north-south oriented wind discontinuity. The prevailing winds at lower levels over the region could be like the one shown at the right side.

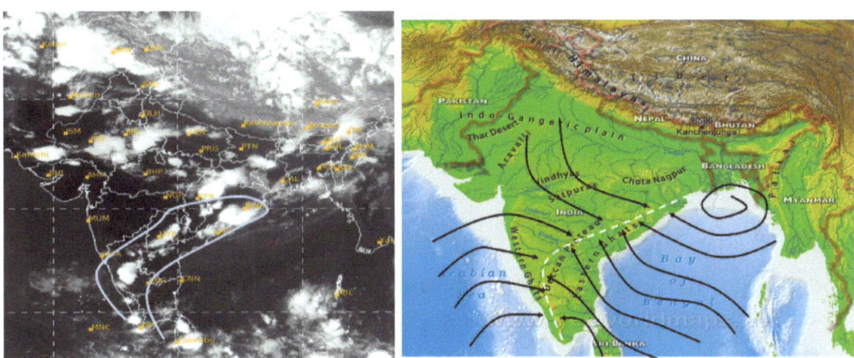

25. **Thermal Eddies**. The chapter will not be complete if Thermal Eddies are not discussed. Thermal Eddies forms due to heating of the earth surface during day time. However, this feature is pronounced only during hot season. Thermal Eddies form due to uneven heating of the surface. Uneven heating takes place due to varying nature of the surface in respect of absorption of heat. Some areas may become heated up earlier and more than the surrounding areas. Such uneven heating results in some pockets of air parcel to break off from the surrounding environment and rise upward. These eddies then, rises to a certain height and drift with the prevailing winds. In general, these eddies are effective only in lower levels and affect an aircraft with turbulence during approach for landing or after takeoff. In favourable lapse rate, that is, under very instable condition, these eddies show their effect up to greater height resulting in formation of convective clouds.

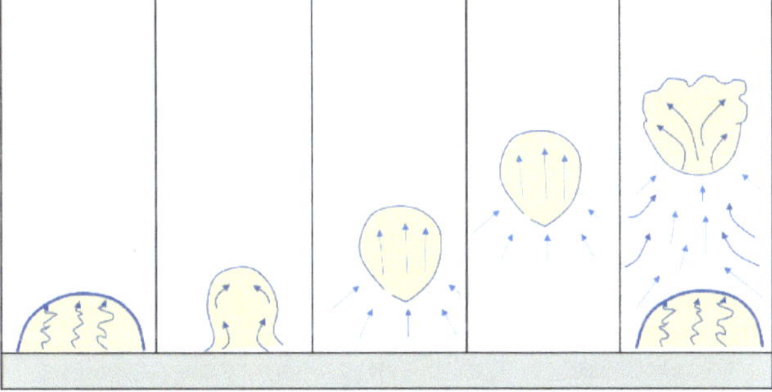

26. Thermal eddies are usually of larger dimensions than the Frictional Eddies and are more noticeable by occupants of aircraft through the pronounced bumpiness they create. Thermals are made use of by gliders for gaining height and remaining aloft by hopping from one thermal to another.

CHAPTER **14**

THUNDERSTORM

1. **Thunderstorm is** a violent short-lived weather phemenon generated by Cumulonimbus (Cb) cloud and is associated with lightning, thunder, other dense clouds, precipitation, and strong winds. Thunderstorm build up takes place in highly instable atmospheric condition. Up and down draft gets generated by the cloud and ultimately cause lightning and thunder. The precipitation caused by Cb cloud may be in form of hail as the grows beyond 0^0 C isotherm, where water is in its solid form.

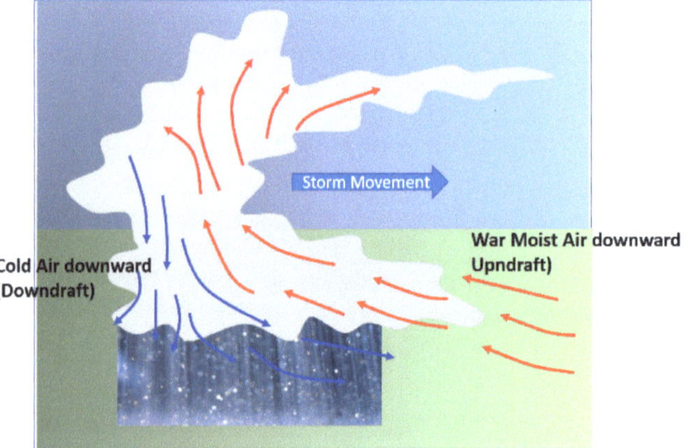

2. As far aviation is concerned, this cloud is considered the most hazardous one and all the aircraft, as far as possible, avoids this cloud. Due to prevailing up and down drafts, an aircraft experiences severe turbulence. Besides, due to presence of hail stone, the frame of the aircraft may get damaged. There are some other hazards associated with Cb clods which will be discussed little later.

3. **Lightning**. Lightning is a naturally occurring electrostatic discharge during which two electrically charged regions, both in the atmosphere or with one on the ground, temporarily neutralize themselves, causing the instantaneous release of an average of one gigajoule of energy. This discharge may produce a wide range of electromagnetic radiation, from heat created by the rapid movement of electrons, to brilliant flashes of visible light. Lightning

causes thunder, a sound from the shock wave which develops as gases in the vicinity of the discharge experience a sudden increase in pressure.

4. Lightning can be explained in terms of charges produced due to rubbing. During a thunderstorm, the air currents move upwards and the water droplets move downwards. And this is caused due to the separation of charge due to this vigorous motion. As a result of this process, the positive charges collect near the upper edge and the negative charges accumulate near the lower edge of the cloud and also near the ground.

5. As the charge gets accumulated, its magnitude becomes very large. Water droplets in the air act as a conductor of this charge. These charges flow to meet, thus producing strikes of lightning and thunder. For this phenomenon to occur, a sufficiently high electric potential between two regions and a high resistance medium must be present.

   The three main kinds of lightning are distinguished by where they occur: -

   a)   Inside a single thundercloud (intra-cloud)

   b)   Between two clouds (cloud-to-cloud)

   c)   Between a cloud and the ground (cloud-to-ground)

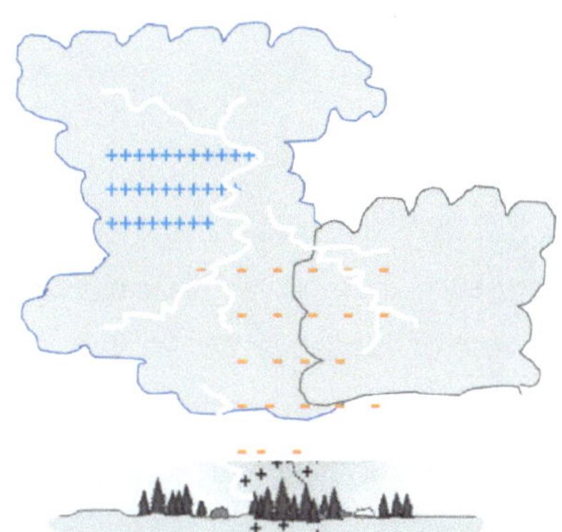

6. Many other observational variants are recognized, including heat lightning, which can be seen from a great distance but not heard; dry lightning, which can cause forest fires; and ball lightning, which is rarely observed scientifically.

7. The details of the charging process are still under study, but there is general agreement on some of the basic concepts of thunderstorm electrification. Electrification can be by the triboelectric effect as a result of ion transfer between colliding bodies. Uncharged, colliding water-drops can become charged because of charge transfer between them (as aqueous ions) in an electric field as would exist in a thunder cloud. The main charging area in a thunderstorm occurs in the central part of the storm where air is moving upward rapidly (updraft) and temperatures range from –15 to –25 °C. In that area, the combination of temperature and rapid upward air movement produces a mixture of super-cooled cloud droplets (small water droplets below freezing), small ice crystals, and graupel (soft hail).

The updraft carries the super-cooled cloud droplets and very small ice crystals upward. At the same time, the graupel, which is considerably larger and denser, tends to fall or be suspended in the rising air.

8. **Effect of Lightning in an aircraft**. The modern aircrafts are designed to withstand any ill effect of lightning. Aircrafts are made of advanced composite materials, which by themselves are significantly less conductive than aluminum. The composites contain an embedded layer of conductive fibers or screens designed to carry lightning currents. The composite material not only safeguard the outer frame but also protects the sensitive instruments and communication equipment inside the aircraft. However, an aircraft which flew through lightning strike should undergo thorough inspection by engineers after landing.

9. **General Protective Measures against Lightning Strike**.

a) **Stay indoors**. The best protective measure during lightning period is to be inside a house. However, even though your home is a safe shelter during a lightning storm, you might still be at risk. About one-third of lightning-strike injuries occur indoors. Following Protective measures to be adopted as suggested by Centers for Disease Control and Prevention (CDC), USA: -

   i.   Avoid water (bathe, shower, wash dishes, or have any other contact with water).

   ii. Don't touch electronic equipment.

   iii. Stay away from outer doors and windows.

   iv. Don't use corded phones.

b) **Outdoor Safety Tips.** In fact, no place outside is safe during a lightning. However, you can minimize your risk by assessing the lightning threat early and taking appropriate actions. The best defense is to avoid lightning. Here are some outdoor safety tips that can help you avoid being struck: -

   i.   Check the weather forecast before participating in outdoor activities. Either postpone your schedule or make sure suitable safe shelter is readily available.

   ii. When thunder roars, go indoors. Hard-top vehicles (with the windows rolled up) can also be chosen.

   If you are caught in an open area, act quickly to find shelter.

   If you are caught outside with no safe shelter nearby, take following actions: -

   (aa) Immediately get off elevated areas such as hills, mountain ridges, or peaks.

   (ab) Never lie flat on the ground. Crouch down in a ball-like position with your head tucked and hands over your ears so that you are down low with minimal contact with the ground.

   (ac) Never shelter under an isolated tree. If you are in a forest, shelter near lower trees.

   (ad) Never use a cliff or rocky overhang for shelter.

   (ae) Immediately get out of and away from ponds, lakes, and other bodies of water.

   (af) Stay away from objects that conduct electricity (such as barbed wire fences, power lines, or windmills).

10. **Installation of Earthing and Lightning Arrester Systems**. Earthing and Lightning protection is very important for everyone using electrical and electronics equipment. Nowadays there is no one in this world who is not using electrical equipment. We all have electrical and electronics equipment in one way or the other. Hence, Earthing and lightning arrester must be installed at all buildings. Installation of SPD (Surge Protection Device) **is also** important.

11. **Conditions Favourable for Formation of Cumulonimbus cloud**. In some of the chapters, namely Cloud, Atmospheric Instability, Vertical Motion of Air, etc, this topic has been discussed. A recap is as follows: -

    The Cumulonimbus cloud (Cb) forms when three conditions are met: -

    a)   There must be a deep layer of unstable air.

    b)   The air must be warm and moist.

    c)   A trigger mechanism must cause the warm moist air to rise: -

        i.   Heating of the layer of air close to the surface.

        ii.   Rising ground forcing the air upwards (Orographic Uplift).

        iii.   A front forcing the air upwards (Frontal Lifting).

        iv.   Convergence.

        v.   Radiational or Katabatic cooling.

12. **Structure of Thunderstorm**. There has been massive advance in this subject in recent years. Credit goes to enormous efforts put in by the government agencies and individual scientists. Modern technologies like satellite reports, weather radar outputs, etc. have also great contribution.

13. There are three types of thunderstorms, single-cell, multi-cell, and supercell.

    a)   Single cell thunderstorm or the isolated thunderstorm generally forms due to local instability and availability of adequate moisture up to middle or higher levels. These are short lived storms and generally are not associated weather of violent nature.

    b)   Multi-cell thunder storms are associated with instability over a larger area like fronts. Such storms are often observed as the squall line. A squall line, or more accurately a quasi-linear convective system (QLCS), is a line of thunderstorms, often forming along or ahead of a cold front. Linear thunderstorm structures often contain heavy precipitation, hail, frequent lightning, strong straight-line winds, and occasionally tornadoes or waterspouts.

    c)   Supercell thunderstorms are the strongest and most severe. Mesoscale convective systems formed by favorable vertical wind shear within the tropics and subtropics can be responsible for the development of such cells. These thunderstorms are characterized by the presence of a mesocyclone, that is, a deep, persistently rotating updraft. Supercells are the overall least common and have the potential to be the most severe. Supercells have the capability to deviate from the mean wind. Supercells can be any size – large or small, low or high topped. They usually produce copious amounts of hail, torrential rainfall, strong winds, and substantial downbursts. Supercells are one of the few types of clouds that typically spawn tornadoes within the mesocyclone.

14. **Life Cycle of a Cell**. As discussed earlier, warmer air rises upwards. The rising moist air cools and condenses. When the moisture condenses, it releases energy known as latent heat of condensation, which allows the rising packet of air to cool less than the cooler surrounding air continuing the cloud›s ascension. If enough instability is present in the atmosphere, this process will continue long enough for cumulonimbus (Cb) clouds to form and produce lightning and thunder. Meteorological indices such as convective available potential energy (CAPE) and the lifted index can be used to assist in determining potential upward vertical development of clouds. The three stages of development of Cb clouds are discussed below.

a) **Developing Stage**. The first stage of a thunderstorm is the cumulus stage or developing stage. During this stage, masses of moisture are lifted upwards into the atmosphere. The moisture carried upward cools into liquid drops of water due to lower temperatures at high altitude, which appear as cumulus clouds. As the water vapor condenses into liquid, latent heat is released, which warms the air, causing it to become less dense and less cool than the surrounding drier air. The air tends to rise in an updraft through the process of convection. This process creates a low-pressure zone within and beneath the forming thunderstorm.

To illustrate in simple words, cumulus stage occurs when one or more cumulus clouds begins to grow into a large cumulus. A general updraft prevails throughout the cell at this stage. A general up draft prevails throughout the cell at this stage, in which extreme velocities of 100 ft. per second have been reported. At the same time, inflow to the cell takes place through the sides at all levels as well as through the bottom of the cloud. The cloud might grow up to 20,000 Ft above ground level.

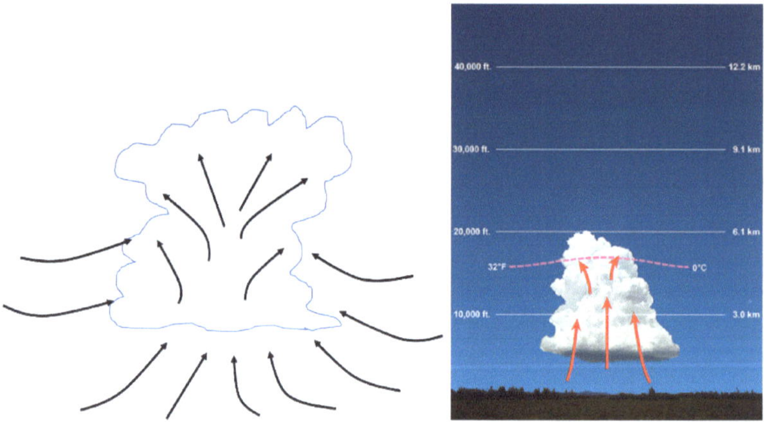

b) **Mature Stage**. In this stage the storm has considerable depth, often reaching 40,000 feet or above. Strong updrafts and downdrafts coexist. This is the most dangerous stage when severe weather phenomena can be generated by the cloud. This stage begins with the fall of precipitation. The release of the precipitation starts a downdraught in a part of the cloud where there was previously an updraft. Due to evaporation of the falling drops, the descending air is kept saturated with the result that it warms at the SALR during descent. Since the ELR is greater than the SALR, the descending air is colder at every stage than the environment and hence continues the downward motion on its own. Down drafts up to 40 ft per second have been observed, the maximum being reached a little after rain starts falling. The speed of up draft is about the same as in the cumulus stage. Thus, at this stage both up draft and down draft co-exist in the cell.

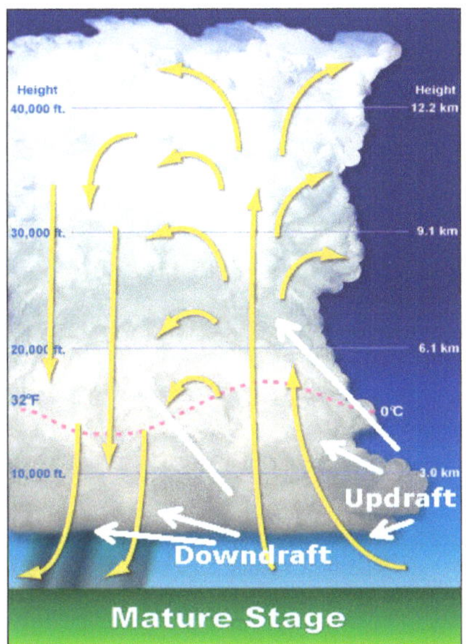

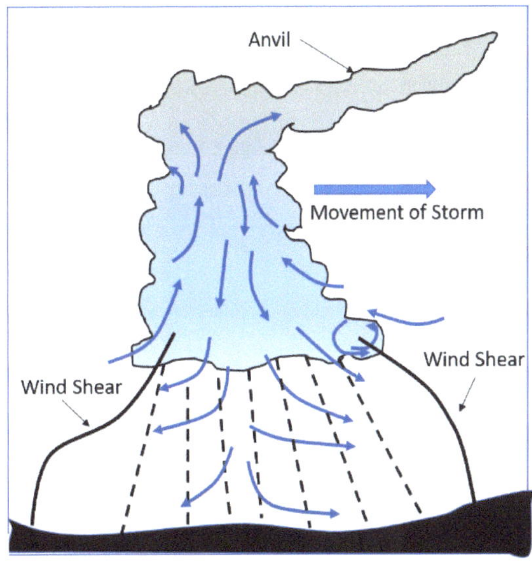

c) **Downburst**. Sometimes thunderstorms will produce intense downdrafts that create damaging winds on the ground. These downdrafts are referred to as macrobursts or microbursts, depending on their size. A macroburst is more than 4 km in diameter and can produce winds as high as 200 km per hour. A microburst is smaller in dimension but produces winds as high as 250 km per hour on the ground. When the parent storm forms in a wet, humid environment, the microburst will be accompanied by intense rainfall at the ground. If the storm forms in a dry environment, the precipitation may evaporate before it reaches the ground (such precipitation is referred to as virga), and the microburst will be dry.

Downbursts are a serious hazard to aircraft, especially during take-offs and landings, because they produce large and abrupt changes in the wind speed and direction near the ground leading to deviate the aircraft from the track.

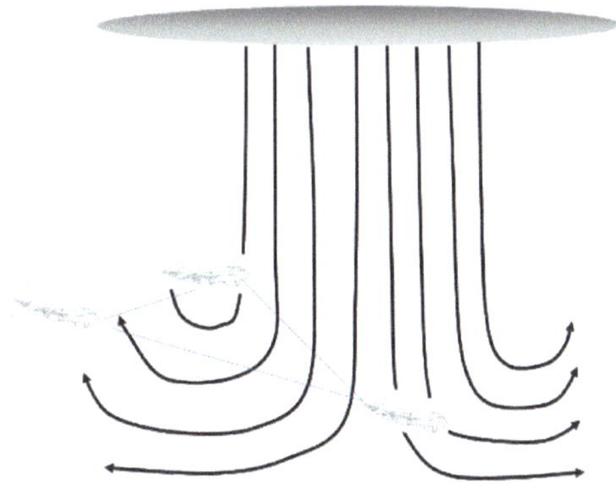

15. **Movement of Thunderstorm**. Maximum mass of a thunderstorm cloud is generally in the middle levels. Hence, in general, direction of the middle level winds determines the direction of movement of the cloud. The anvil, that is the top of the cloud tends to move with the higher-level winds and gradually dissociate from the mother cloud.

16. **Dissipating Stage**.   In this stage the cooler air outside of the cloud environment increasingly infiltrates the growing storm cloud, the storm's downdraft eventually overtakes its updraft. With no supply of warm, moist air to maintain its structure, the storm begins to weaken. The cloud begins to lose its bright, crisp outlines and instead appears more ragged and smudged.

17. The entire lower portion of the cloud exhibits downdraught, while only slight upward movement exists in the upper portion of the cloud. The lack of vigour in the upward movement results in the top portion spreading out laterally. Since this part of the cloud contains ice crystals, the spread-out portion has an appearance of cirrus cloud. It is therefore referred to as false cirrus or anvil cirrus. The lower part of the cloud cannot continue for long; it may be the first stage of dissipation leaving the anvil cirrus and other stratified remnants at higher levels. Figure below illustrates the circulation in the dissipating stage.

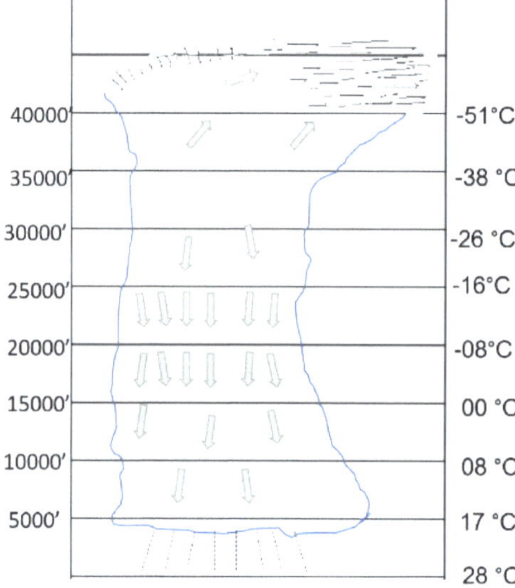

18. **Regeneration of Thunderstorms**. Due to the cold downdraught from a mature or dissipating Cb cloud the other cumulus clouds in the vicinity may get triggered with instability and moisture and subsequently build up to Cb cloud. This process is known as regeneration. Process of regeneration may repeat, giving the impression of movement of the same thunderstorm cloud. In reality it is a chain reaction by means of which thunderstorms occur in quick succession. This may take place over a large belt covering hundreds of kilometres. The Kalbaisakhis (Norwesters) over eastern part of India during pre-monsoon season is an ideal example of regeneration. In radar or satellite image it can be seen existing as a line squall for hours and moving eastward.

19. **Squall**. Squall is defined as "Sudden increase of wind speed by at least three stages on Beaufort Scale, speed rising to Force 6 (24 Kt or more) and lasting for at least one minute". Since it is a strong wind accompanied by sudden gust, the damage caused by squall could be of large scale. Squalls are very dangerous for an aircraft taking off or landing as well as parked on the ground.

20. The initial downdraft from the Cb cloud spreads out horizontally on reaching the ground and appears as a squall at its leading edge. The average velocity is of the order of 40 Kt, however, the peak velocities can be as high as 100 kt. Maximum speed is experienced in the direction of movement of the thunderstorm. When the terrain affected by a surface squall is uneven or rough, violent eddies may be generated near ground level rendering landing and take-off extremely unsafe. Squall associated with thunderstorm occurs generally during pre-monsoon season in India.

21. **Hail**. Hail is one of the most dangerous type of weather which is caused by Cb cloud. Hailstorm can be defined as any thunderstorm which produces hail. An ice crystal with a diameter of >5 mm is considered a hailstone. Hailstones can grow to 15 cm and weigh more than 0.5 kg.

22. Besides causing damage in aviation, hail can also destroy crops and structures and vehicles to large extent. There has been report of injury to men and animals at time being fatal if not moved to safe place when hailstorm is on.

23. Hail forms from supercooled water. Supercooled water is something unique It is water that is below its normal freezing point of 0°C and yet remains a liquid. Supercooled water will freeze when it comes in contact with something like an ice crystal, dust particle, or raindrop. Hail is associated with high, vertical Cb clouds, the kind of clouds that produce severe thunderstorms.

24. Within a cumulonimbus cloud, ice particles develop from supercooled water. The particles fall toward the bottom of the cloud from the pull of gravity, but they are forced back up by powerful updrafts of air within the clouds. In the upper part of the cloud, they encounter more supercooled water, which freezes on the ice particles, adding another layer of ice to them. This happens repeatedly. In this way, the small bits of ice grow larger and larger, becoming balls of ice or hailstones. The hailstones finally become too heavy to be lofted back up to the top of the cloud and they fall to the earth.

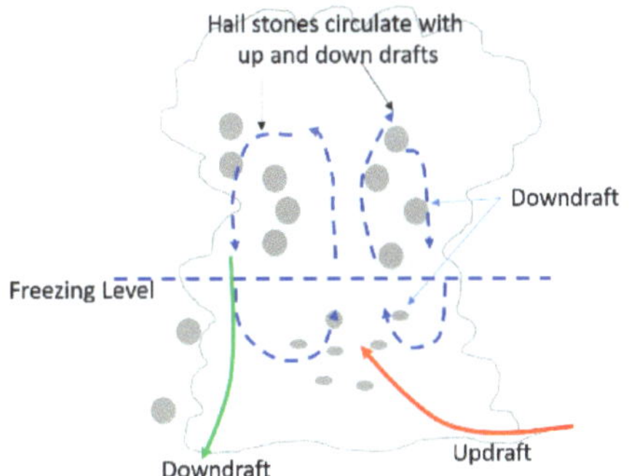

25. There are many reports of different intensities hailstorm and damages caused worldwide. Hailstorm at times may be so heavy that the area effected by hail give a look of snowfall. Below is a picture showing collection of hailstones on the ground over Moran, in the state of Assam, India on 22 Dec 22.

26. Hail is most common within continental interiors of the mid-latitudes, as hail formation is considerably more likely when the freezing level is below the altitude of 11,000 ft. Accordingly, hail is less common in the tropics despite a much higher frequency of thunderstorms than in the mid-latitudes because the atmosphere over the tropics tends to be warmer over a much greater altitude. Over India, hailstorm occurs over northern parts on approach of a cold front. Hailstorm also occur in India during pre-monsoon season when the freezing level is not yet so high and under the situation of high instability mainly over east India (associated with norwesters) and over close to the foothills over NE India (Associated with Katabatic cooling and cold front).

27. **Dust Storm**. A dust storm is a meteorological event that predominantly occurs in arid and semi-arid regions when large sections of fine loose dirt and sand are picked up by strong winds and blown into the atmosphere. It creates a dense wall of dust that can stretch for miles and be thousands of feet in height.

28. WMO defines dust storm as the result of surface winds raising large quantities of dust into the air and reducing visibility at eye level to less than 1000 m.

29. Dust storms can affect regions thousands of miles away. In fact, it can even cross oceans and affect countries on other continents. Dust from the Sahara desert in Africa can reach as far as the Amazon in South America and even parts of the United Kingdom. In India migration of dust from the site of dust storm occurring over Rajasthan or other parts of NW India has been seen up to the Brahmaputra valley.

30. During dry and hot season (Pre-monsoon season in India) when Cb cloud forms over desert or semi-arid region (Rajasthan and adjoining parts of NW India) under atmospheric instability conditions, the cloud cannot grow to greater heights due to poor humidity conditions aloft. However, the clouds, if grow above the freezing level, they can cause thunderstorm. These storms often raise dust and other small and loose articles up to a height of about 10,000 Ft reducing visibility to less than 1000 M. Dust storm is locally known as Andhi in India.

31. A dust storm is nothing but a thunderstorm over arid or semi-arid regions under low humidity condition. Hence, the mechanism of dust storm is the same as that of thunderstorm. Low humidity condition does not allow the Cb cloud to grow up to a greater height. However, there is quick down draft due to fall of the cold water drops from above the freezing level. These water drops can hardly reach the ground in form of rain as high temperature and low humidity conditions prevail at lower tropospheric levels and at surface. Hence, the down drafts become weaker. On the other hand, the updraft becomes vigorous and raise the loose dust particles to a great height. The raised dusts are then carried by the prevailing winds.

32. If for some reason, there is suitable amount of moisture at higher levels, the Cb cloud would grow higher. In such case, it will appear as a dust storm in the initial condition due to initiation of raising of dust due to prevailing up draft. However, the subsequent down draft causes rain and stops further raising of dust. Over NW India such weather events are seen when there is influx of moisture from the Arabian Sea due to the passage of extratropical system or before the onset of SW monsoon.

33. The vertical extent of the Cb cloud that causes dust storm is usually 25,000 to 30,000 Ft. It is seen that if the Cb cloud extends above 30,000 Ft, usually a thunderstorm occurs. The radar images give very good indication of the vertical extent of the clouds.

34. Prior advisory or forecast on occurrence of dust storm is very important for smooth and safe flying operation. Facing dust storm by the aircrafts suddenly can be highly dangerous as the visibility reduces abruptly at surface and at higher levels and it makes nearly impossible for the pilots to come into visual contact with the ground to make a landing or take off in poor horizontal visibility condition.

The picture below shows the regions that are prone to dust storm: -

Figure 1.1: Global Dust Potential Map. Source: DTF (2013).

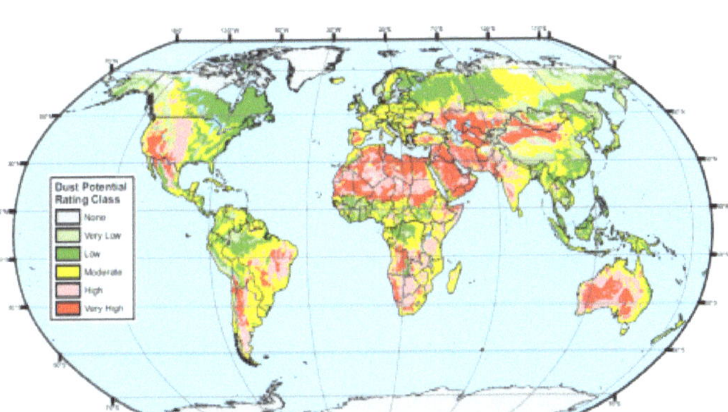

35. **Tornado.** Tornado is a small-diameter column of violently rotating air developed within a convective cloud and in contact with the ground. Tornadoes occur most often in association with thunderstorms over the region around the equator and mid-latitude regions, frequency being more around the equator. These whirling atmospheric vortices can generate the strongest winds known on Earth. Wind speed in tornado is very high. It is generally of the order of 200 – 300 Km per hour but highest speed of 500 Km per hour has been recorded. When winds of this magnitude strike a populated area, they can cause extremely heavy destruction and great loss of life mainly through injuries from flying debris and collapsing structures.

36. The rotating column of air that is in contact with both the surface of the Earth and a cumulonimbus cloud. It is often referred to as a twister. Tornadoes come in many shapes and sizes, and they are often visible in the form of a condensation funnel originating from the base of a cumulonimbus cloud, with a cloud of rotating debris and dust beneath it.

37. Technically, a type of thunderstorm known as supercells is what leads to the development of a tornado. Within the supercell, there is an air vortex known as a mesocyclone. As the mesocyclone moves down beneath the cloud, the air takes in cool and moist air from the lower sections of the storm. When the cool air comes together with warm air travelling upwards, the result is the creation of a low-pressure point at the surface that begins to pull the whole system down towards the surface of the earth. This puling action is what eventually leads to the visible funnel-shaped force that is known as a tornado. Supercells contain mesocyclones, an area of organized rotation a few kilometre up in the atmosphere, usually 1.6 to 9.7 km across.

38. The destructive nature of the tornado is primarily from something called the Rear Flank Downdraft (RFD). The RFD directs the base of the mesocyclone such that it assumes a funnel shape. When the funnel shape touches the earth, the RFD begins to move outwards leading to the creation of the powerful and destructive blasts of wind.

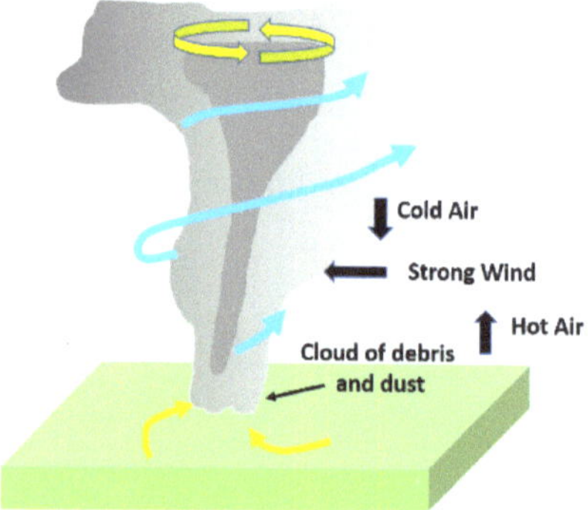

39. Occurrence of tornado is almost unknown in India. However, it doesn't mean that this phenomenon doesn't occur. Perhaps, the report with evidence has not been made. Life cycle of tornado being short spelled, and it's effects over a small area, remains unobserved by the meteorological observers. However, nature of some of the destructions seen during pre-monsoon season in India, particularly in east India (associated with Norwesters) and reports obtained from the locals of the areas about uplifting of roof top and other heavy objects reveal that the damage caused could be due to a tornado. With the present generation of instant recording of events through smart phones, it can be expected that evidence of occurrence of tornado in India will be available, which, in turn will help the weather men to undertake studies and come out with clue full conclusions. Following is a picture of damage caused by a storm in Purnea district of Bihar on 20 Apr 11. The report of a daily news paper is also shown alongside.

The Telegraph *online*

Tuesday, 14 February 2023 • E-paper

Home   Opinion   India   My Kolkata ▾   Edugraph ▾   States ▾   World   Business   Science & Tech   Health   Sports ▾   E

Home / Bihar / Storm leaves 5 dead in Purnea

Storm leaves 5 dead in Purnea

Read more below

JITENDRA KUMAR SHRIVASTAVA   Published 21.04.11, 12:00 AM

A woman cries in front of her wrecked home in Purnea. Picture by Mohan Mahato

Purnea, April 20: At least five persons, including a child of seven years, were killed in a storm that blew away the roofs of thatched houses in Dhamdaha sub-division of Purnea district last night.

Those killed in the storm were identified as Bindeshwari Mahato of Goriar village, Dinesh Chaudhari and his wife Nirmala Devi of Sreepur, Rupesh Uraon of Kukranwan and a seven-year child (name not known) of Latambari. The district administration today confirmed all five deaths and said all affected villages fall under the Dhamdaha sub division.

"The district administration has announced Rs 1 lakh for each family which has lost a member. The damage to standing crops is also being estimated and compensation would be given to the farmers soon," local legislator Lesie Singh told **The Telegraph**.

Most affected in the storm were people living in thatched-roof houses. The deaths were reported to be a result of roofs falling on persons living in the houses. The thatched roofs are generally pressed with heavy logs to save from being blown away by strong wind.

40. **Waterspout**. Waterspouts fall into two categories: Fair Weather Waterspouts and Tornadic Waterspouts.

41. Tornadic waterspouts are tornadoes that form over water, or move from land to water. They have the same characteristics as a land tornado. They are associated with severe thunderstorms, and are often accompanied by high winds and seas, large hail, and frequent dangerous lightning.

42. Fair weather waterspouts usually form along the dark flat base of a line of developing cumulus clouds. This type of waterspout is generally not associated with thunderstorms. While tornadic waterspouts develop downward in a thunderstorm, a fair weather waterspout develops on the surface of the water and works its way upward. By the time the funnel is visible, a fair weather waterspout is near maturity. Fair weather waterspouts form in light wind conditions so they normally move very little.

CHAPTER 15

GLOBAL CIRCULATION OF AIR

1. The sun is our main source of heat. Because of the tilt of the Earth, its curvature, our atmosphere, clouds, and polar ice & snow, different parts of the earth receive solar energy in different quantity. In other words, the heating of the Earth's surface is mot uniform. This leads to temperature difference between the poles and equator. However, due to natural circulation of air over the globe transfer of heat from hotter region (Equator) to the colder regions (Poles) in a manner that the temperature all over the globe remains more or less equal leaving apart the seasonal changes that every part of the globe undergoes due to the position and orientation of the Earth in respect of the Sun. In absence of the global circulation temperature over the equator would continue to rise and that over the poles would continue to fall making the earth uninhabitable for life.

2. Most of the physical and biotic phenomena of the earth are due to the incoming solar energy. The winds and ocean currents and the various weather phenomena, all ultimately owe their origin to solar energy. The high temperature radiation from the Sun is in the form of short waves while the radiation from the earth is in the form of long waves. The solar radiation or insolation in the short-wave radiation received from the Sun or the amount of solar energy received at any place on the Earth depends on the 'Angle of Incidence' of Sun's rays and the 'Length of the Day' which in turn determines the intensity and duration of solar radiation.

3. **Insolation**. Only approximately 52 per cent of this insolation reaches the earth's surface. The rest is absorbed by water vapour, dust and clouds, or is reflected by the Earth's surface and scattered by particles in the air.

4. Reflected heat, in the form of long-wave radiation, is trapped in our atmosphere and keeps our planet warm. This is known as the natural greenhouse effect.

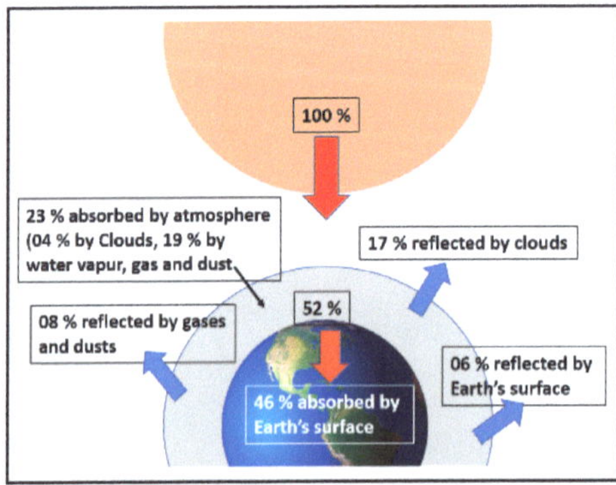

5. **The Heat Balance**. A balance is struck between the incoming radiation and the longwave radiation from the earth back to space. This balance is known as terrestrial heat balance. The mean annual temperature of the earth as a whole remains the same because the amount of incoming solar radiation absorbed is balanced by the amount of terrestrial energy radiated to space.

6. However, there is a latitudinal imbalance in the energy absorbed against energy radiated from the earth. In the low latitudes (equator to 37°) the incoming solar radiation exceeds the outgoing earth radiation while poleward of 37°, it is the opposite. In other words, there is a continued excess of gain over loss in the low latitudes and the reverse in the middle and high latitudes.

7. In order to maintain the overall heat balance, this situation requires a latitudinal transfer of energy from the excess to the deficit latitudes. This transfer is achieved by the earth's atmosphere and oceanic circulation i.e. the winds, ocean currents, storms and other weather phenomena.

8. **The Global Pressure Distribution**. The distribution of atmospheric pressure across the latitudes is termed global horizontal distribution of pressure. Its main feature is its zonal character known as pressure belts. On the earth's surface, there are seven pressure belts. They are the Equatorial Low, the two Subtropical highs, the two Subpolar lows, and the two Polar highs which make a pattern of alternate high and low-pressure belts over the earth. This is due to unequal heating of the Earth's surface by insolation. As discussed before, the Equatorial region receives the greatest amount of heat throughout the year.

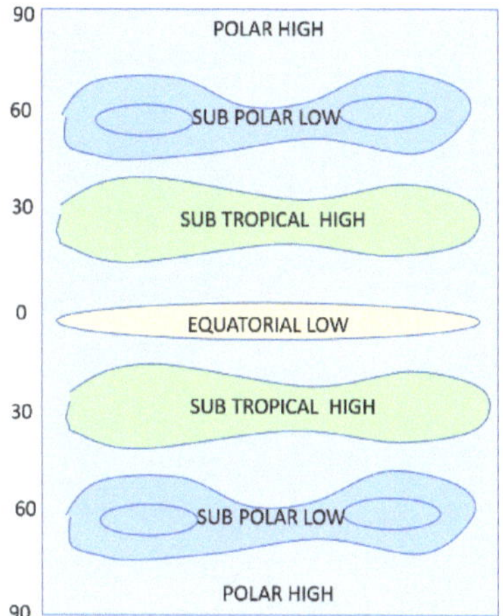

a) **Equatorial Low**. This low-pressure belt extends from 0 to 5° North and South of Equator. Due to the vertical incident of the sun rays, there is intense heating over this region. Therefore, the air expands and rises and a low pressure is developed. This low-pressure belt is also called as doldrums because it is a zone of total calm or light variable winds.

b) **Subtropical High**. At about 30° North and South of equator lies the area where the ascending equatorial air currents descend. This area is thus an area of high pressure. It is also called as the Horse Latitude. Winds always blow from high pressure to low pressure. So, the winds from subtropical region blow towards the equator (Trade Winds) and another wind blow towards Sub-Polar Low.

c) **Sub-Polar Low**. These belts are located between 60° and 70° in each hemisphere. The descending air in the Subtropical region gets divided into two parts, one part blows towards the Equatorial Low and the other part blows towards the Sub- Polar Low. This zone is marked by the ascent of warm subtropical air and descent over cold polar high regions.

d) **Polar High**. At the North and South Poles, between 70° to 90° North and South, the temperatures are always extremely low. The cold descending air gives rise to high pressures over the Poles.

9. **The Wind Cells**. Based on the above pressure distribution, three distinct wind cells prevail over the globe. These are Hadley Cell, Farrel Cell and Polar Cell. They exist in both the hemispheres.

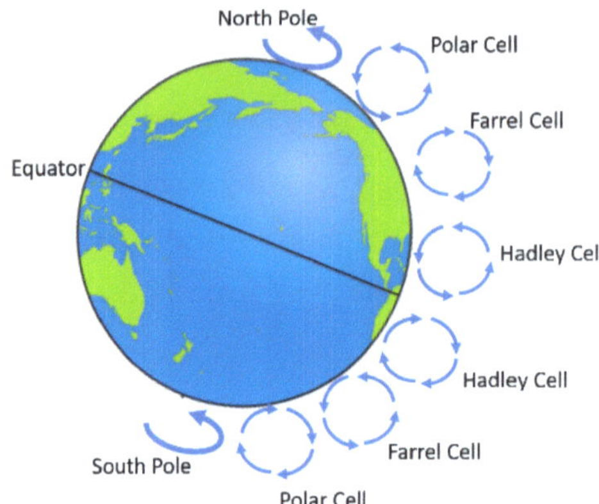

10. The cells on either side of the Equator are called Hadley cells and give rise to the Trade Winds at Earth's surface. Around the Earth's equator, the solar radiation is the highest throughout the year. Air near the equator is warm and hence, rises as it is less dense than the air around it.

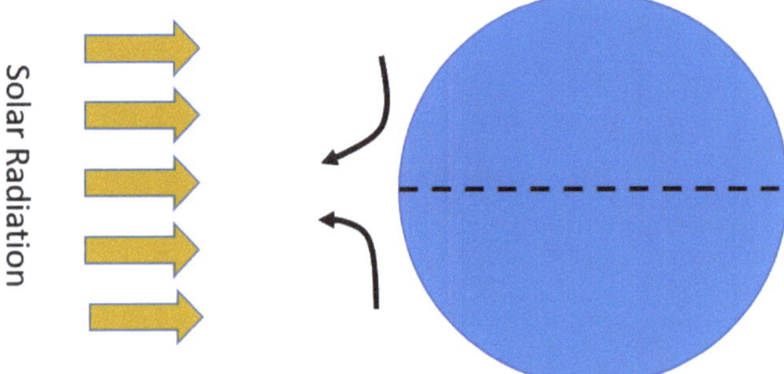

11. The rising air over the equatorial region creates a circulation cell, called the Hadley Cell in which the air rises and cools at higher altitudes and moves towards the poles. Eventually, this air descends back to the surface. The continual rise of air at the equator creates low pressure and causes wind to air to blow towards the equator to feel the area rising air. On the other hand, sinking air creates high pressure at the surface. A high to low gradient of pressure is formed that causes air to flow away from the high and towards the low pressure at the surface.

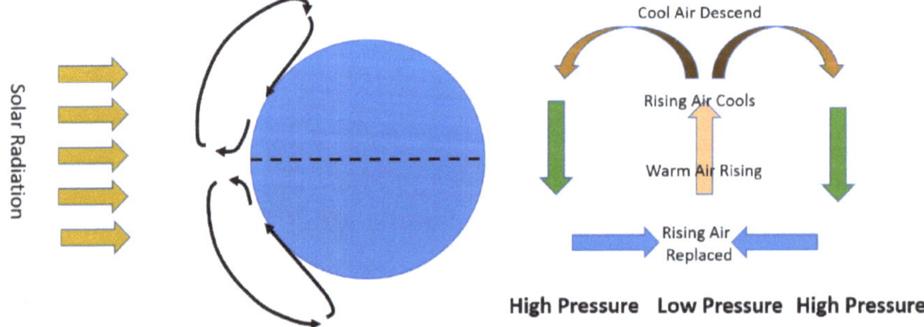

12. As illustrated above, the earth should have two large cells, one in each hemisphere. But actually same is not the case. Due to the Coriolis force created by rotation of the Earth, the Hadley cells break into three separate cells in each hemisphere. The Coriolis effect deflects the winds to the right in the northern hemisphere and to the left in the southern hemisphere resulting in breaking of the Hadley cells in each hemisphere into three cells, that is Hadley Cell, Farrel Cell and the Polar Cell.

13. **Normal Surface Wind Pattern**. The rising air at the equator produces a zone of surface low pressure known as the Equatorial Low. Air in both the hemispheres moves towards this Low where it converges and rises as part of the Hadley Circulation. The narrow zone where the air converges is called the Intertropical Convergence Zone (ITCZ). The region of Equatorial Low is also called the Doldrum as the winds here are light and variable. On the poleward

side of the Hadley Cell, there is descending of air. Hence, surface pressures are high over that region. This produces two subtropical high-pressure belts, each cantered at about 30° latitude. A large and stable anticyclones form within these belts. Air descends at the centres of these anticyclones and winds are weak.

14. Winds around the subtropical high-pressure centres move toward equatorial as well as middle latitudes. The winds moving equatorward are the trade winds. They blow from the northeast in northern hemisphere, and are called the northeast trade winds. To the south of the equator, they blow from the southeast, and are called the southeast trades. Also, towards the pole from the Sub-tropical highs, air blow outward towards Sub-polar Lows. In the northern hemisphere they blow from southwest and in the southern hemisphere they blow from northwest.

15. The pressure and wind patterns are complex between about 30° and 60° latitude. This is a zone where different air masses with different characteristics (dry & cool, warm & moist) meet. These winds generally move eastward. Hence, the region is also called the region of zonal westerlies.

16. At the poles there is high pressure and the air is extremely cold there. The winds are anti-cyclonic due to the prevailing high pressure. At the South Pole the winds are generally from east (easterlies) and are commonly known as polar easterlies.

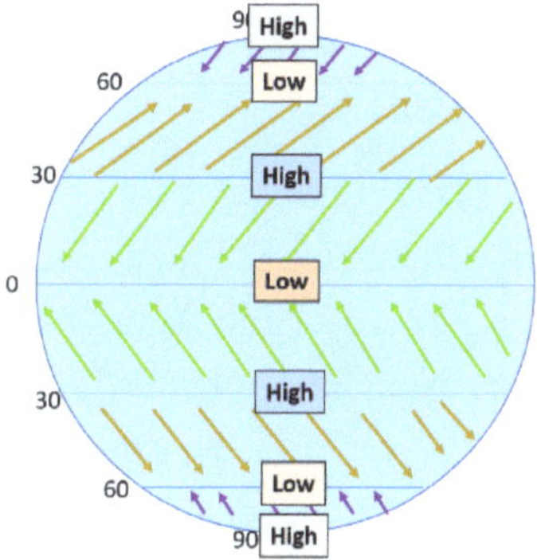

17. **The Seasons**. Different position of the earth in respect of the Sun causes the different seasons at different places. When we say position, we will refer the distance of a point from the Sun which determines the angle of incidence of Sun rays and length of the day at that point.

18. The earth rotates around an imaginary axis once in 24 hours. During this period most places on the earth are turned alternately towards the Sun and away from it, giving rise to day and night. The direction of rotation is towards the east.

19. The rotating earth revolves in a slightly elliptical orbit around the Sun. The time of one revolution is one year during which period the earth rotates around its axis approximately 365¼ times, thus determining the number of days in a year. The plane of the ecliptic is an imaginary plane passing through the Sun and extending outward through all points in the earth's orbit. The axis of rotation of the earth has a fixed inclination of about 66½° from the plane of the ecliptic (i.e. 23½° from the vertical). This position is constant and hence, the axis at any time during the yearly revolution is parallel to the position that is occupied at any previous time. This is called parallelism of axis. The rotation, revolution, inclination and parallelism of the axis act to produce the changing lengths of the day and varying angles of the Sun rays which in turn cause the seasons.

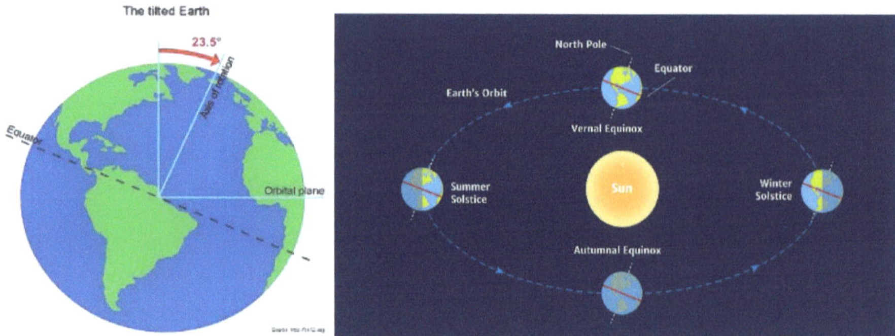

20. **Equinoxes**. Twice during the yearly period of revolution (21 March and 23 September) the Sun's noon rays are directly vertical at the equator with one half of the earth in light and the other half in darkness and days and nights are equal. These are known as spring and autumn equinoxes respectively.

21. **Solstices**. On 22 June and on 22 December, the earth is approximately midway in its orbit between the equinoxial positions and the north pole is inclined 23½° toward/away from the Sun. As a result of the axial inclination the Sun's rays are shifted northward/southward by the same inclination of 23½° so that the noon rays are vertical at the tropic of cancer/Capricorn (23½°) and the tangent rays in the northern hemisphere/southern hemisphere pass over the pole and reach the arctic/antarctic circle (66½°)

22. On the opposite hemisphere, the tangent rays do not reach the pole but terminate at the Antarctic/Arctic circle, 23½° short of the pole. Hence, during summer solstice (22 June) while all parts of the earth north of the arctic circle are experiencing constant daylight, areas poleward of the Antarctic circle in the southern hemisphere are entirely without daylight. This is the summer for northern hemisphere when days are longer and the period when maximum solar energy is received. In winter solstice (22 December), opposite conditions prevail i.e. it is constant daylight south of Antarctic circle, constant darkness north of arctic circle, summer in the southern hemisphere and winter in the northern hemisphere. Thus, the belt of maximum insolation swings back and forth across the equator during the course of a year.

23. **Seasonal Migration of Pressure and Wind Belts**. So far, we have assumed that the equator receives the maximum amount of radiation. This is not true throughout the year. Due to the inclination of the axis of rotation of the earth to the plane of its orbit around the Sun, the apparent elevation of the Sun is different in different parts of the year. Hence the radiation maxima execute oscillations between the tropic of cancer and the tropic of Capricorn on an annual scale. The pressure and wind belts, which are based mainly on thermal causes, also execute similar oscillations. Thus, in the northern summer we may find the equatorial low as far as latitude 25° north while in the northern winter it migrates to the south of the equator.

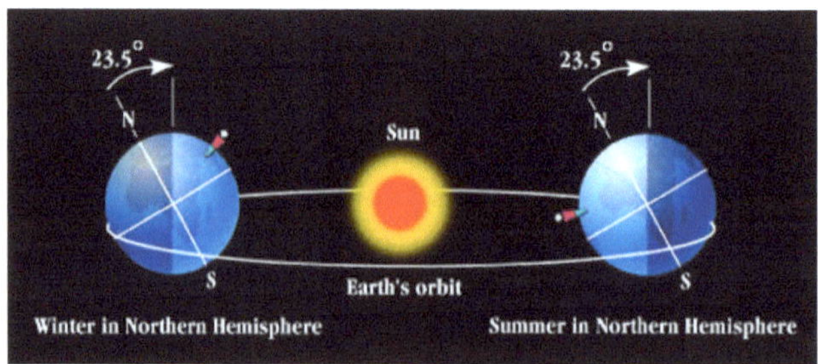

24. **Transient Disturbances**.   The patterns of pressure and wind described above, are actually the idealized pattern. The real pattern has a deviation. The deviation is mainly due to breaks in the pressure belts caused by various factors like distribution of land and sea; type of air mass (Maritime, continental, cold, warm, etc.) prevalent at a particular time over a zone; movement or shift of synoptic features, etc., often breaks the pattern. Such factors cause the transient disturbances in the ideal pattern.

CHAPTER 16

AIR MASS

1. An air mass is a large volume of air in the atmosphere with similar characteristics, that is, mostly uniform in temperature and moisture content. Air masses can extend thousands of kilometers across the surface of Earth with vertical extent of thousands of metre. The characteristics of the air mass depends on the source region. That is the region where the air remains for a long duration. In event of low wind speed, an air mass practically remains stationery over a region and it takes the features of that region. For example, an air mass originating from the Indian ocean around the equator will be moist and warm. Both these features will be in increased scale, that is, moisture content and temperature of the air mass will gradually increase if it travels through the region for a longer duration.

2. It is important to know the characteristics of an air mass to assess its influence on weather. With advancement in synoptic analysis and development in numerical models these days, it has been easy to understand the characteristics of an airmass. Following two images show temperature and humidity at a point at Indian Ocean at 1000 Ft above sea level on a certain day (Source: windy.com).

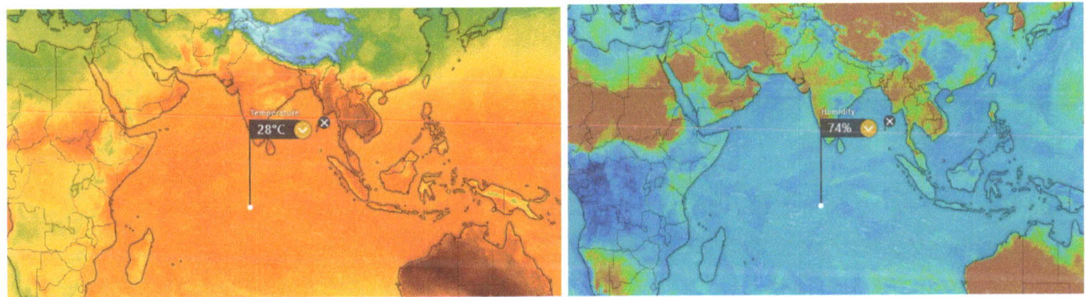

3. **Classification of Air Mass**. Air masses are identified and classified based on the source region. They are broadly classified into four types, namely, Arctic, Polar, Tropical and Equatorial. Further, based on the underlying surface (Sea or land) they are classified as maritime (m) and continental (c). Furthermore, the temperature of the airmass is compared with the temperature of the underlying surface. If the temperature of the air mass is warmer than the underlying surface, it is classified as warm (W) and if the temperature is less than the underlying surface, it is classified as cold (K). The classification is illustrated better in the following table.

Air mass		Symbol	Origin	Characteristics
Arctic		A	Polar Region	Low temp, low specific but high summer relative humidity, Coldest of winter air masses.
Polar	Continental	Pc	Sub-Polar Continental Region	Low temp, temp increasing with southward movement, Low humidity.
	Maritime	Pm	Sub-Polar Oceanic Region	Low temp, increasing temp with movement, Higher humidity

Tropical	Continental	Tc	Sub-tropical high-pressure land areas	High temp. Low moisture content
	Maritime	Tm	Southern border of Oceanic Sub-tropical Highs	Moderately high temp, High relative and specific humidity.
Equatorial Maritime		Em	Equatorial and Tropical seas.	High temp. High humidity.

4. **Change in Characteristics**. Once an air mass moves from the source region, it starts losing its original characteristics and undergoes modification due to the following reasons: -

a)   Due to passage over warm/cold or dry/moist areas.

b)   Due to mixing with other types of air masses.

5. While undergoing the modification, the air mass is called transitional and is often indicated by prefixing the letter n to the air mass designator. For example, nTm will mean Transitional Tropical Maritime air.

6. **Properties of Air Masses**. Property of an airmass is largely determined by the temperature and humidity of the underlying surface. The tropical air masses are generally warm and usually move towards higher latitudes. If they move over cooler surface, they get cooled. This generally results in formation of stratiform clouds causing light rain or drizzle and poor visibility due to fog formation. Air masses of polar origin are cold and move towards lower latitudes. In its passage over the warmer region, convection and turbulence develop and result in cumuliform clouds and thunderstorm activities.

7. It is important to track the movement of the air masses to study them in order to understand their properties. For identifying them correctly, temperature, relative humidity, pressure and associated clouds are not so reliable since these properties varies largely. Hence, we have to look for potential temperature, specific humidity and dew point which remain fairly constant irrespective of the modifications in travel.

Potential Temperature. The temperature that a sample of air attains if reduced to a pressure of 1000 millibars without receiving or losing heat from/to the environment.

Specific Humidity. Mass of water vapour in a unit mass of moist air, usually expressed as grams of vapour per kilogram of air.

8. **Air Masses of the Indian Region**. Over the Indian region, tropical and equatorial air masses are the most prevalent air masses and prevail in all the seasons. In winter season, sometimes there are intrusion of polar air. Often, two or three air masses may be present over an area and undergo transformation. The main features of air masses over the Indian region are illustrated in the following table.

Air Mass	Origin	Characteristics	Season/Period of arrival
Tm	Subtropical highs of the North Pacific Ocean	At source - high temp, high RH and high DP and largely retains the characteristics when it arrives over India.	SW Monsoon (Jun - Sep). Occasionally in other seasons in association with Tropical Storms

Tc	The great Siberian High	Cold and dry. Gathers some moisture while travelling through Bay of Bengal	Practically overruns the country in Winter. In monsoon season it is confined to NW India. In pre-monsoon season it undergoes heating and becomes unstable with little sea travel resulting thunderstorm.
Pc	Sub-Polar region	Cold and dry. Becomes nPc in the rear of strong depressions which move across Kashmir and the Punjab. Gives rise to more severe cold wave over north India	Mainly in winter. During monsoon season it can be found over NW India and in post monsoon it penetrates up to central and east India mainly in form of nPc
Em	Sub-tropical High soth of equator	Prevails over India south of 25^0 N during monsoon season. At its source region it is cool, humid and stable. As it crosses the equator into the northern hemisphere, it undergoes surface heating and becomes unstable.	Monsoon season. Sometimes seen over extreme south Peninsula during Winter.

9. Movement of Air Masses over India during the different seasons are shown below: -

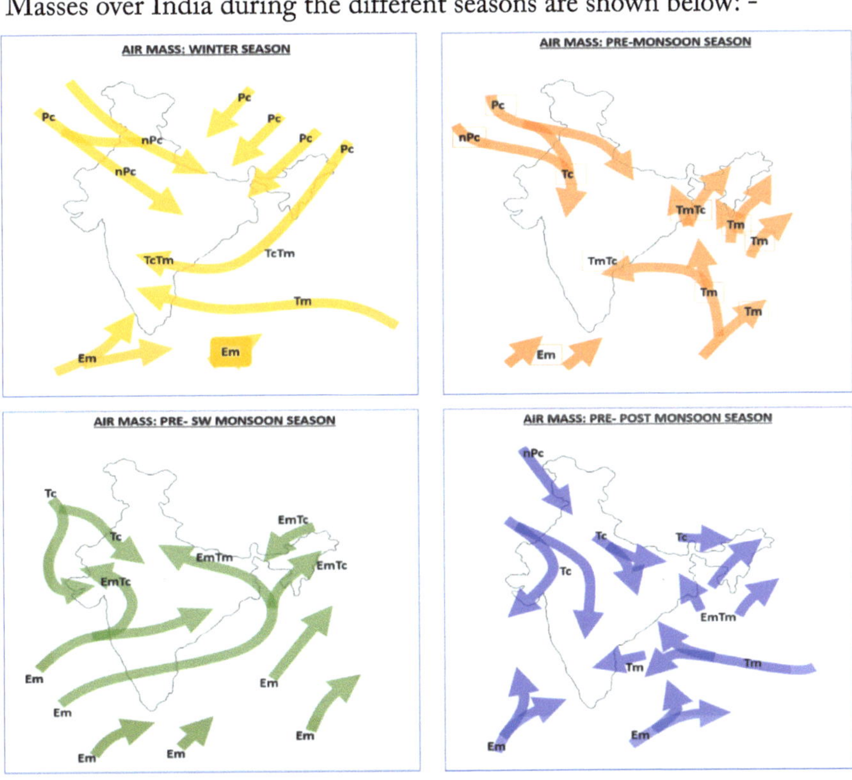

CHAPTER **17**

FRONTS

1. Having understood about the air masses and their characteristic features, it is now time to discuss on the results when two different air masses with different humidity and temperature profiles meet along a boundary. The subject is very interesting and is associated with weather phenomenon across the globe in the middle latitude regions. Air mass boundaries and disturbances in the middle latitudes are referred to as extra-tropical weather systems (cyclones or depressions). Some of the weather sequences encountered in the middle latitudes are met with in the disturbances affect the extreme northern parts of India in the winter season.

2. When two airmasses having different properties meet, there will certainly be a vertical wall like boundary along which the two meet. This boundary is called front in meteorology. It is defined as below: -

   A front is a weather system that is the boundary separating two different types of air. One type of air is usually denser than the other, with different temperatures and different levels of humidity. This clashing of air types causes weather like rain, snow, strong winds, cold wave conditions, etc.

3. Weather fronts mark the boundary or transition zone between two air masses and have an important impact upon the weather. The air around the globe has different properties. For example, air from the north is usually colder, and there is usually a sharp boundary where this meets the warmer air from the south. This air mass may also be dry when it travels through continent for a longer period. The air mass travelling across an ocean will certainly hold more moisture.

4. Across a front, there can be large variations in temperature, as warm air comes into contact with cooler air. The difference in temperature can indicate the 'strength' of a front. If very cold air comes into contact with warm tropical air, for example, the front can be 'strong'. However, if there is little difference in temperature between the two air masses, the front may be 'weak.'

5. Let us see what happens when two air masses, one being cold and the other being warm, meet along a front. When a warm air mass comes in contact with a cold air mass, it being lighter, will upglides over the cod air mass. On the other hand, when a cold air mass meets a warm air mass, it undercuts warm air mass beneath a sloping surface. The same can be demonstrated pictorially as below: -

6. The former, that is the warm air meeting a cold air is called Warm Front and the other one is called the Cold Front.

7. **Warm Front**. On coming in contact with a cold air mass, a warm air mass being lighter, upglides the cold air along a slant frontal surface. The ascent causes adiabatic cooling resulting in formation of clouds and subsequent weather. The uplift is gradual and hence, indication of its approach is visible in the sky in form of high clouds (Ci and Cs) at around 30,000 Ft ASL, about 500 miles ahead of the system at surface. There is a gradual change in the sky

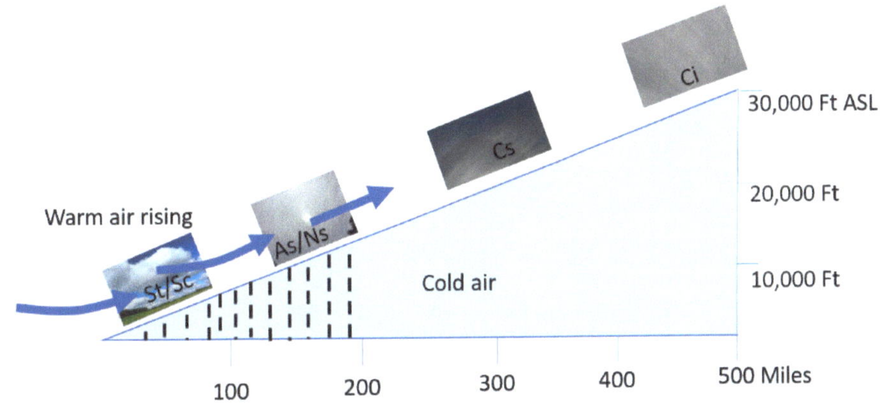

Condition. Amount of high clouds increases followed by rain bearing medium and low clouds respectively. Weather activities are not of intense nature. Light to moderate rainfall/snow fall of continuous or intermittent nature takes place and often without any thunderstorm spell.

8. Warm fronts are generally not associated with significant fall in atmospheric pressure at surface and can be mostly seen as upper air features like cyclonic circulation and/or middle/high level trough in the zonal wind field.

9. Some changes observed in various weather elements on approach, during passage and in the rear of a warm front are as follows: -

Elements	On Approach	During Passage	In the Rear
Pressure	Gradually falls	Fall ceases	Remains steady for some time and then rises
Wind	Direction changes anti clock wise (Backing)	Direction changes clock wise (Veering)	Direction become steady
Temperature	No change	Rises due to cloud cover	Becomes steady again
Clouds	Ci Cs Ac AS/Ns	As/Ns, Sc, St	Clouds passes away rapidly. Some remnants in form of Ac, Sc. St clouds may be there. St clouds form due to overnight radiational cooling
Weather	Light to moderate intermittent to continuous precipitation with formation of middle level sheet type cloud (As/Ns)	Precipitation stops abruptly	Fair/Clear sky condition. On occasion of adequate radiation cooling fog may occur or short spelled light drizzle may occur fron St clouds.

10. **Cold Front**. The frontal slope associated with cold front is comparatively steeper. The cold air undercuts the warm air abruptly and makes the warm air rise rapidly. This cause instability in short spell ofc time and thunderstorm accompanied by precipitation (sometime in form of hail) takes place and give the appearance of sudden change in weather.

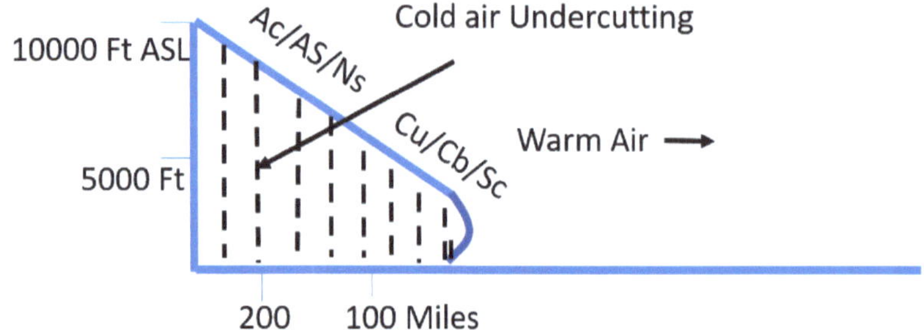

11. The weather associated with a cold front is generally confined to a smaller area and lasts for short duration. However, the intensity of weather is much more severe than that associated with warm front.

12. The behavior of various weather elements on approach, during passage and in the rear of a cold front are tabulated below: -

Elements	On Approach	During Passage	In the Rear
Pressure	Pressure is steady on approach but starts falling when the front is close by	Low pressure becomes steady but rises rapidly when the front is about to move away	As the front moves away filling up (rise in pressure) becomes slow and steady
Wind	Direction changes anti clock wise (Backing) and begins to become squally (Sudden rise in speed and direction)	Wind direction starts changing direction clock wise (Veering), Squally nature persists	Squally nature ceases and winds continue to veer steadily.
Temperature	No change	Falls suddenly	Becomes steady again
Clouds	Low clouds (St/Sc) are seen when the front is close by. Then medium clouds Ac/AS appears followed by convective clouds (Cu/Cb)	Cb clouds with associated rain bearing medium clouds are prominent. Stratus clouds are also seen.	Clouds disappear rapidly. Cb/Cu clouds may prevail for very short duration.
Weather	Drizzle from low clouds or fog may prevail followed by rain or thunderstorm	Moderate to heavy showers and Thunderstorm. May be with hail	Short spelled moderate to heavy showers, then clears up rapidly

13. **Movements**. The fronts, which are often in the form of a low-pressure system at the surface or disturbance in upper air, are the middle latitude or extratropical features. They move towards east or east-north-east ward. When the systems are stronger, another system may form at a lower latitude but within its wind field (circulation). Such baby systems at a lower latitude are called 'Secondaries'. These secondaries, like a satellite move in the direction of the parent system.

CHAPTER 18

JET STREAMS

1. Jet Streams are the comparatively stronger but narrow belt of wind flow within a zonal wind field. Jet Streams are generally located near the altitude of tropopause (Upper boundary of troposphere). The wind speed in a Jet Stream is 60 Kt or more. Any belt with speed less than 60 Kt is not considered as Jet Stream. The average speed in the core of the Jet Stream is around 100 Kt and maximum speed may be of the order of 150 – 200 Kt.

2. Jet Streams, which have great impact on the global weather systems are primarily associated with the speed of rotation of the Earth and its position in respect of the Sun. We will not go deep into this matter and will limit to the types of Jet Streams, their characteristics and their impact on weather and on a flying machine.

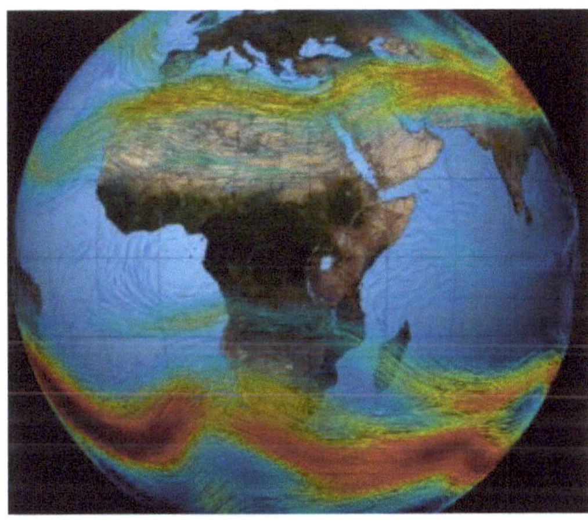

3. **WMO Definition of Jet Stream**. The WMO definition of a jet stream is: -

   "A strong narrow current concentrated along a quazi-horizontal axis in the upper troposphere characterized by strong vertical and lateral wind shears and featuring one or more velocity maxima. Normally a jet stream is thousands of Km in length, hundreds of Km in width and some Km in depth. The Vertical shear of the wind is of the order of 5m/sec per Km. An arbitrary lower limit of 30m/sec (60 Kt) is assigned to the speed of the wind along the axis of a jet stream".

4. In a jet stream, the path of the maximum speed is known as the axis and the tubular volume immediately surrounding it is known as the core. Since the speed of wind outside the core reduces rapidly as one goes away from the core, there is strong horizontal as well as vertical shear of wind near a jet stream.

5. The actual appearance of jet streams results from the complex interaction between many variables; such as the location of high and low pressure systems, warm and cold air, and seasonal changes. They meander around the globe, dipping and rising in altitude/latitude, splitting at times and forming eddies, and even disappearing altogether to appear somewhere else. In other words, Jet Streams are not a continuous stream of wind maintaining a particular speed but are often broken. In the same level, somewhere the Jet stream is visible with very high speed and on the other hand, it may not be seen at some area.

6. One way of visualizing this is to consider a river. The river's current is generally the strongest in the centre with decreasing strength as one approaches the river's bank. Therefore, it is said that jet streams are "rivers of air".

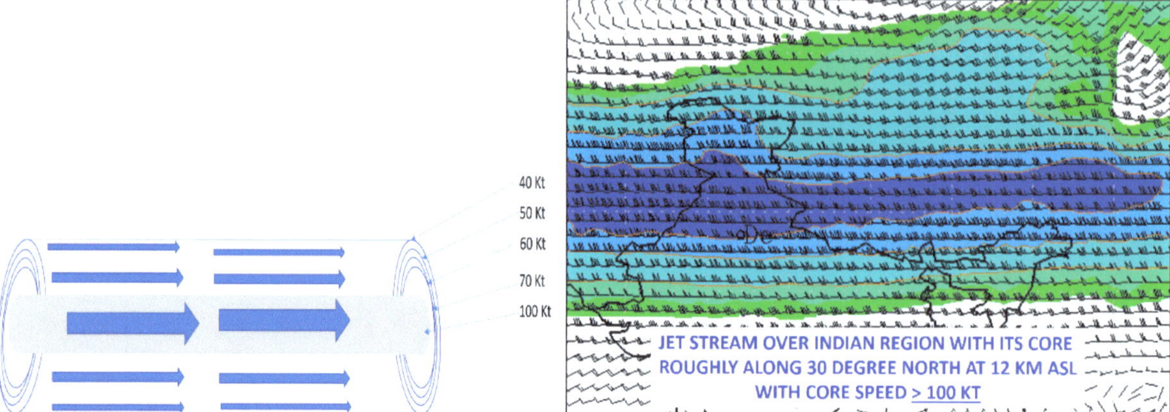

7. Jet streams are special features of the general circulation, i.e. they form part of the larger wind belts of the globe. Thus, they are also subject to seasonal oscillations.

8. **Types of Jet Stream**. There are four types of jet streams in the troposphere: -

    a) **Arctic Jet Stream (AJ)**. It is seen in the high-level westerlies of the Arctic regions. The mean height is 7-8 Km which is closer to the Polar Tropopause.

    b) **Polar Front Jet Stream (PFJ)**. It exists in the mid-latitude westerlies above the surface polar front. Mean height of PFJ is 9 Km. During winter season the PFJ shifts southward to as far as 30°N and in summer it shifts back to about 70°N. The core speeds increases from summer to winter. The mean direction is westerly and varies from south westerly to north westerly. Average speed is 80 to 100 kt during winter season.

    c) **Sub-tropical Jet Stream (STJ)**. It is found in the high-level westerlies above the sub-tropical high-pressure belts. In winter it shifts to about latitude 25°N and can be marked up to north of 35°N. Maximum speed is in the winter season. The STJ affects India in the non-monsoon months.

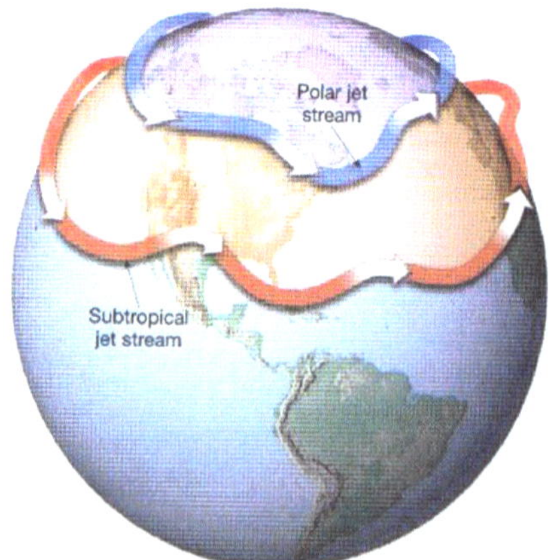

    d) **Tropical Jet Stream (TJ)**. TJ is also known as the Tropical Easterly Jet Stream (**TEJ**) as this is the only one flowing from east to west and is found within the zonal easterly winds over the tropical region. Tropical Easterly Jet Stream occurs near the tropopause over Southeast Asia, India, and Africa during summer between 5° and 20°N. The mean height is about 14 to 15 Km ASL. The speed is comparatively less than that of STJ. The average core speed is 60-80 Kt. However, on occasions winds up to 100 to 120 kt may be seen.

In July-August the axis of TEJ is around latitude 15 °N over the Indian Peninsular region. In other places there are large breaks. In winter this Jet Stream becomes very weak and almost disappears though, moderate easterly winds prevail just to the south of the equator.

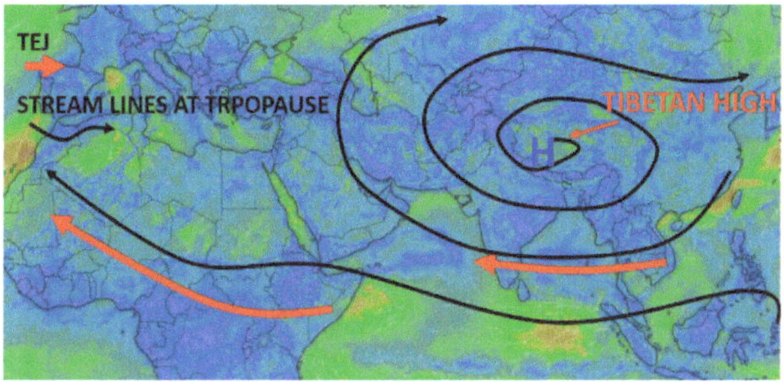

9. In addition to the above four types of Jet Stream which are found in the troposphere, another Jet Stream exists in the Stratosphere above 20 Km ASL. This Jet Stream is called the **Stratospheric Jet Stream (SJ)** and is found above Arctic and Antarctic Circles. These are westerlies in winter and easterlies in summer.

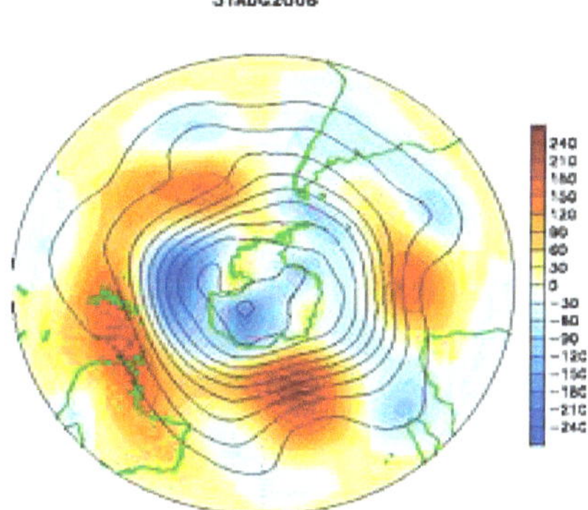

10. **The Low Level Jet (LLJ).** There is yet another type of Jet Stream known as the LLJ. This Jet Stream, unlike the above-mentioned ones are seen in the lower troposphere between 0.9 – 3.0 Km ASL. These Jet Stream, however, do not bear any of the characteristics of the actual Jet Stream that are discussed above. They are called Jet Stream because of the high speed of the order of 30-50 Kt but at very low levels. They are found within the prominent lower tropospheric wind flow and are mostly located over the sea or over the land area close to the sea. Over Indian region, such LLJ are seen over east Arabian Sea off west coast flowing in south-westerly or westerly direction during the southwest monsoon season. These LLJs origins near Mascarene High Pressure region around Madagascar Island and as monsoon current head towards western coast of India. The jet is not continuous and travels with multiple breaks. Arrival or prevalence of LLJ off west coast is associated with strong monsoon surge and often gives torrential monsoon showers over the places close to the west coast. The LLJ associated strong monsoon conditions extends further ahead if other factors are supportive.

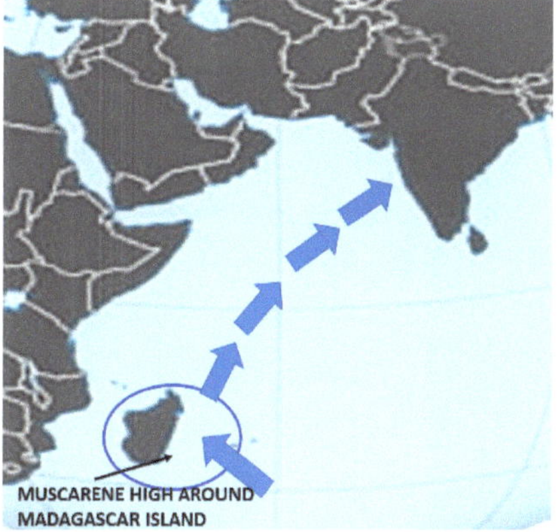

11. **Jet Streams Over India and Their Characteristics**. Two Jet Streams, namely the STJ and TEJ are seen over Indian regions.

a) The STJ is seen in the non-monsoon months.

    i. It is seen roughly along latitude 27 °N at a height of 12 Km.

    ii. The mean wind speed is 100 kt with maximum range reaching up to 150-200 Kt.

    iii. During winter season it follows the southern most track and can be marked as south as 22^0 N in the month of February.

    iv. The jet is not continuous and is often seen splitting. This is mainly due to the existence of the massive Himalaya. To the east of Afghanistan the jet splits into two branches, with one branch flowing along north of Himalaya while the other flowing along it's south. Later the two again form a single stream over China.

    v. The STJ can be seen over the Indian sub- continent from October to May. With the establishment of the southwest monsoon it shifts north of the Himalayas.

b) To the southern border of the Tibetan High, The TEJ appears over the South China Seas at upper levels between 150 and 100 hPa (13.5 to 16 Km ASL).

    i. The core of the jet is roughly along 15^0 N. It can be seen extending up to north Africa across SE Asia and India with distinct breaks.

    ii. The TEJ is present over the south Peninsula from June to August. During July-August it can be marked along the latitude 12°N to 15 °N. It disappears from the Indian area by September.

    iii. The maximum speed is over Peninsular India with average speeds of the order of 70 kt, though at times speeds of 150 kt can be seen.

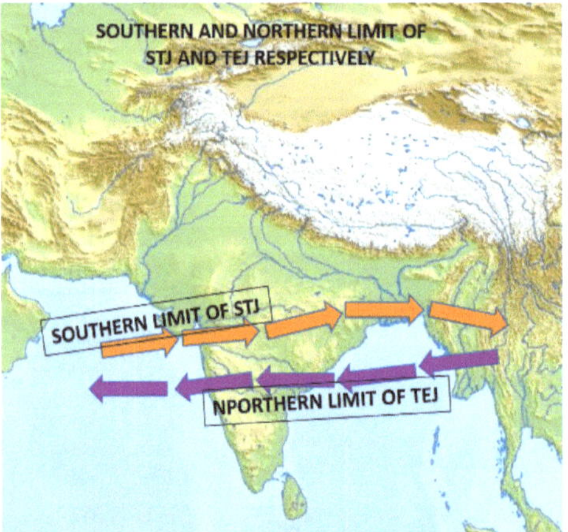

c) The Low-Level Jet Stream is associated with incursion of air mass with very high humidity. Thus, LLJ is characterized by continuous moderate to heavy rainfall, very low clouds and strong & gusty surface winds often associated with convective build up. Occasions arise when alerts are required to be issued to the people engaged in their activities on or off the coast. Besides, there have been occasions when traffic including air and railway traffics have been disrupted due to flash floods, low clouds, strong surface winds, poor visibility, etc.

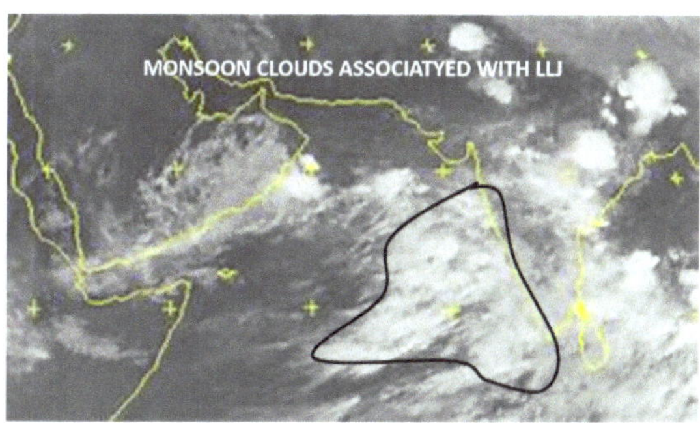

12. **Influence of Jet Atream on Weather**. One of the most important impacts of the Jet Stream is the weather it carries with it. It being a strong current of rapidly moving air, it has the ability to push weather patterns around the world. As a result, most weather systems do not just sit over an area, but they are instead, moved forward with the jet stream. The position and strength of the jet stream help meteorologists forecast future weather events more accurately.

13. Shifting of Jet Stream patterns can have a big impact on the weather. Jet streams are always changing. There is change in the latitude/altitude, breaking up, and shifting in flow, depending on the season and other variables. During winter, Jet Streams tend to follow the sun's elevation and move toward the equator, while they move back towards the poles in spring.

14. Air north of a jet stream is typically colder, while air to the south is usually warmer. As jet streams dip or break off, they move air masses around, creating shifts in global weather patterns. Jet Streams are the best media of transferring energy from tropics towards the pole. The best way of transferring energy is that Jet Streams play a significant role in pulling the tropical storms north ward. The best example is the movement of the fourth most devastating hurricane of in US, hurricane Sandy of October 2012. The storm originated over Caribbean Sea on 22 October, moved almost northward, travelled through the sea for seven days before it made landfall off New Jersey coast on 29 October.

15. The above said case of the path of Sandy is very interesting. It is associate with the deep amplitude trough in the STJ (discussed before). The area of the movement of hurricane Sandy falls on the sub-tropical region within the zonal westerly wind field. Normally, the storms move west to east with these westerlies. However, Sandy moved northward due to prevalance af a deep amplitude trough. Sometimes (on rare occasion) when the situation of deep trough remains longer, the Jet Stream may bend so much that it flows from east to west over short distances. When this happens, the jet stream may break off (close off) from the main branch and spin anti-clockwise in a large horizontal circle. It was a large buckle in the jet stream that developed and created an east to west flow in the case of Sandy during late October 2012.

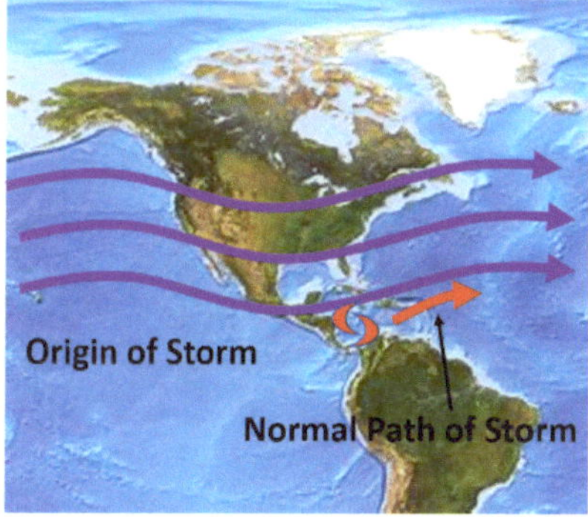

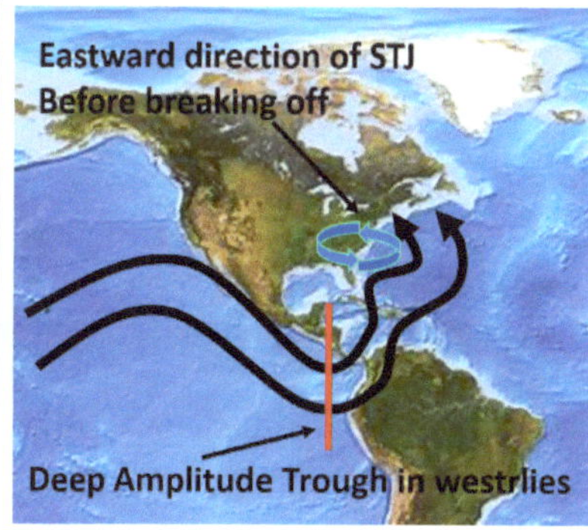

16. In the Indian region, the disappearance of the sub-tropical jet and the appearance of the tropical jet practically coincides with the onset of the southwest monsoon. Again, the withdrawal of the southwest monsoon is associated with the southward shifting of the sub-tropical jet into the northern parts of India. In the case of the TEJ, it has

been noticed that breaks in the monsoon are often associated with a more than normal northerly position of the TEJ. The large velocity gradients (both in the horizontal and in the vertical) associated with jet streams give rise to areas of convergence and subsidence. These naturally have an important bearing on the generation of dust storms and thunderstorms and also on the development, intensification and decay of depressions.

17. To the south of the jet stream, the air is usually warmer than average. When there is a large northward bulge (deep trough) in the jet stream during the winter, heavy coats and wool hats may not be needed. When this happens in the summer, millions of people (over middle and higher latitude region may face a sweltering heat wave.

18. To the north of the jet stream, the air is usually cooler than average. When the jet stream plunges southward (west of a deep trough), weather becomes cooler in summer. When this happens in the winter, severity of coldness increases

CHAPTER 19

CLEAR AIR TURBULENCE (CAT)

1. **Clear Air Turbulence (CAT)** is defined as sudden severe turbulence occurring in cloudless regions (Or outside convective clouds) that causes violent buffeting of aircraft. This term is commonly applied to higher altitude turbulence associated with wind shear. This includes turbulence in cirrus clouds, within and in the vicinity of standing lenticular clouds and, in some cases, in clear air in the vicinity of thunderstorms. It generally occurs in air between altitudes of 6,000 and 15,000 M (20,000 and 49,000 feet) and constitute a hazard to aircraft. This turbulence can also be caused by small-scale (i.e., hundreds of metres and less) wind velocity gradients around the jet stream, where rapidly moving air is close to much slower air. It is most severe over mountainous areas and also occurs in the vicinity of thunderstorms.

2. Apart from crew and passenger discomfort, CAT can give rise to difficulty in controlling an aircraft. For military aircrafts, it may lead to inaccuracy in bombing, aerial gun-fire or dropping operations. CAT associated with mountain waves is known to have caused disastrous accidents.

3. Clear-air turbulence is usually impossible to detected with the naked eye and very difficult to detect with a conventional radar, with the result that it is difficult for aircraft pilots to detect and avoid it. However, it can be remotely detected with instruments that can measure turbulence with optical techniques, such as scintillometers, Doppler LIDARs, or N-slit interferometers. Numerical Weather Prediction models have also extended technology to forecast of CAT at various levels. These days the aircrews are being briefed about the locations and levels at which probability of CAT exists.

4. **Factors responsible for CAT**.

   a) **Jet Stream**. Jet Streams are associated with significant vertical and horizontal wind shear on their edges. Such wind shears give rise to CAT. CAT is the strongest on the cold side of the jet stream where the wind shear is the greatest. CATs can prevail from 7,000 feet below the jet to a few thousands above the tropopause. CATs associated with jet streams are generally shallow in nature and can be avoided by an aircraft be descent or climb by a few thousand feet.

   Wind speed in a jet stream is not uniform. A significant change in the speed horizontally or vertically can causes CAT of high intensity. Also, a significant change in the direction in the high velocity jets also leads to CAT of severe nature.

   b) **Topography**. High ground feature like hills or mountain terrains disturb the horizontal flow of air over it and cause turbulence. The intensity of the turbulence depends on the prevailing speed of wind, the degree of unevenness of the terrains, shape of the terrain (the rate of change and curvature of contours), and the elevation of the terrains above surrounding features and above sea level.

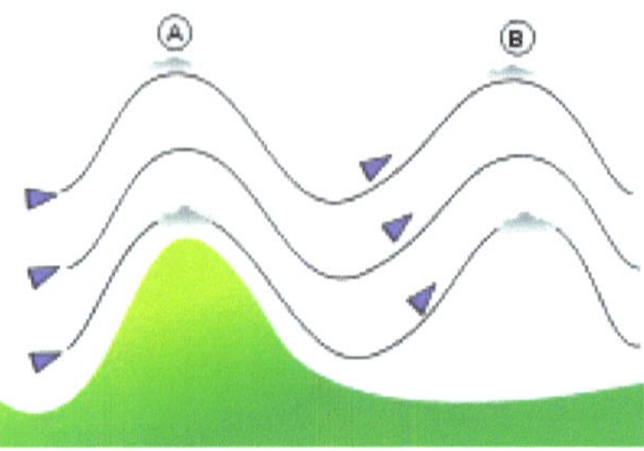

The intensity and frequency of CAT is maximum in the lee side of the mountainous terrains. Numerous equally spaced lee waves are often seen where they are not interfered with by other mountains, such as over the sea. They may produce clouds, called wave clouds (Lenticular clouds, rotor clouds, cap clouds), when the air becomes saturated with water vapour at the top of the wave. Lee waves occur most often when a airstream with strong winds in the higher levels and stable air in the lower levels flows across a long ridge having steep slopes. The strongest up current then occurs not over the wind-facing slope but at the front of the first lee wave.

c)   **Prevalence of Thunderstorm in the Vicinity**. The turbulence generated by the Cb clouds can affect as far as 35 to 40 Km horizontally and about 1.5 Km above it. Hence, avoiding Cb clouds maintaining adequate distance and height from it is always advisable.

d)   **Temperature Gradient**. Temperature gradient is the change of temperature over a distance in some given direction. Significant change in temperature along horizon as well as vertical also causes change in the air density in the same manner, which ultimately can cause CAT. Such CATs, however, are of feeble nature in the middle and upper troposphere where the change in temperature is not rapid but more or less uniform.

e)   Significant synoptic features like fronts, depression/cyclones are associated with moderate to severe CAT.

f)   CAT is also encountered in the regions of upper air features like cyclonic circulation, deep amplitude trough/ridge, confluence (Convergence of two jets or two streams of same jet) of high-speed winds from different directions, etc. Trough/ridge and confluence, when associated with jet stream are known to have caused very severe CAT.

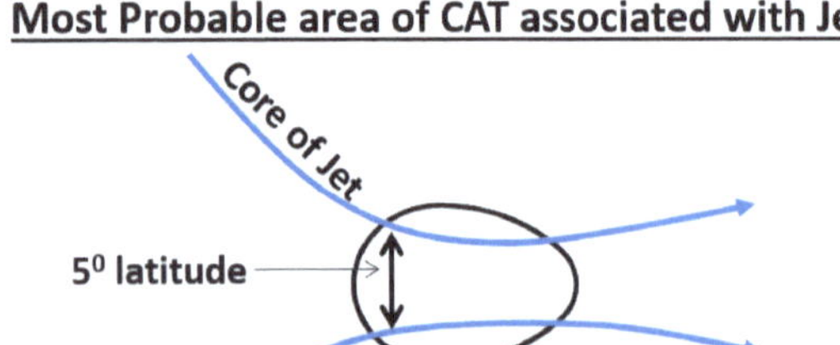

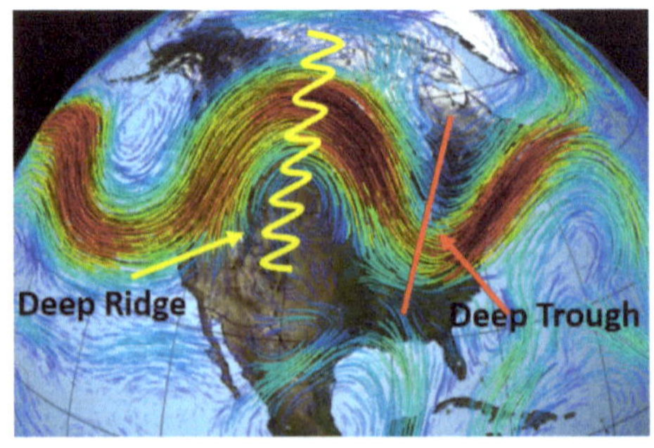

On 18 December 2022 a Hawaiian Airlines flight HA35 from Phoenix to Honolulu which was maintaining flight level 36,000 Ft, encountered severe CAT about 30 minutes prior to landing. The turbulence was so severe that it caused several people to fly out of their seats and hit the ceiling of the plane. This happened right around the time the plane was starting

*its descent. 36 passengers including three attendants suffered injury. 11 were seriously injured. **This episode of CAT was associated with a deep amplitude trough in sub-tropical jet stream at upper troposphere west of Hawaii.***

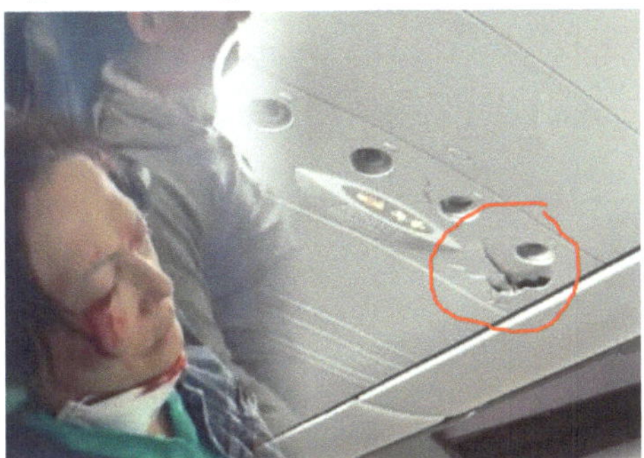

5. **Vulnerable Regions of CAT**. Based on long studies carried out so far and the available information on encountering CAT by aircrafts, it has been worked out that the following regions are vulnerable for CAT.

    a) The vertical wind shear is greater than 4 Kt per 300m

    b) Horizontal temperature shears of 5° C per 100 NM.

    c) Horizontal wind shears greater than 25 Kt per 90 NM (moderate) and 50 Kt per 90 NM (severe).

    d) Left or polar side of jet stream at all altitudes around jet stream and just below tropical tropopause.

    e) In the convergence zone of two jet streams.

    f) Inversions (Increase in temperature with height in troposphere) especially near and downwind of mountain ranges.

    g) Flying over sharp trough and out of lows.

6. **Effects of CAT**. The incidence of CAT over India has been studied from a large number of reports from aircrafts. Some of the more important results of the study are given below: -

    a) **On encountering CAT of severe nature, an a**ircraft may suffer structural damage. In extreme cases this can lead to the break-up of the aircraft. In moderate turbulence, damage can occur to fittings within the aircraft, especially as a result of collision with unrestrained items of cargo or passenger luggage.

    b) **P**assengers and crews in the aircraft cabin can be injured if caught unaware. Hence, it is mandatory that all crews and passengers are informed about the prevailing or expected CAT so that adequate precautions are taken by all.

    c) Prolonged CAT may make it difficult for the pilots to monitor and control various instruments. This may reduce the performing efficiency of the pilots.

7. **CAT over Indian Region**.

    a) Major CATs in India are linked to Jet Streams. In association with Sub-tropical Jet, CAT frequency is highest from October to May over central and northern India and in association with the Easterly Jet, it is highest in July-August over southern India.

    b) The period of peak activity of Sub-tropical Jet is during the period from December to Feb and during this period maximum incidence of CAT takes place.

    c) Like, Jet sStreams, CAT zones are also not uniform but occurs in patches. Usually, the dimensions of CAT zones are 150 Km in width (north-south) and about 300 Km in length (east-west).

    d)   Vertical extent of a CAT zone is usually less than 1 Km but may extend through a deeper layer.

    e)   CAT over Indian region are of light or moderate nature. However, there has been a few cases of CAT of severe nature and is restricted to December to February period.

    f)   Most prone area prevails about 300 Km to the south of the Sub-tropical Jet axis.

    g)   During October, when Sub-tropical Jet makes it's appearance over western Himalayas, frequency of CAT is maximum over that region. And in eastern Himalayas maximum frequency is noticed during mid-winter months.

8.   **Avoiding CAT**. CAT is meso-scale in nature. It is encountered in a small region with few hundred Km in length, few Km in breadth and few thousand Ft in height. Best way to get control on CAT is not to get panic and take all required safety measures. Following are suggested by Meteorologists to the aircrews: -

    a)   The vertical and horizontal extension of CAT is not much. Its better to escape the CAT area by deviating the route to south or north or climb or descent by a few thousands of feet.

    b)   Avoid flying just below tropopause. A level about 3000 Ft above the tropopause is known to be free from CAT.

    c)   Try to maintain a route along the windward side of hilly terrain (If possible), when flying over mountains.

9.   When strong wind deviates, the change of wind direction implies a change in the wind speed. A stream of wind can change its direction by differences of pressure. CAT appears more frequently when the wind is surrounding a low-pressure region, especially with sharp troughs that change the wind direction more than 100°. Extreme CAT has been reported without any other factor than this.

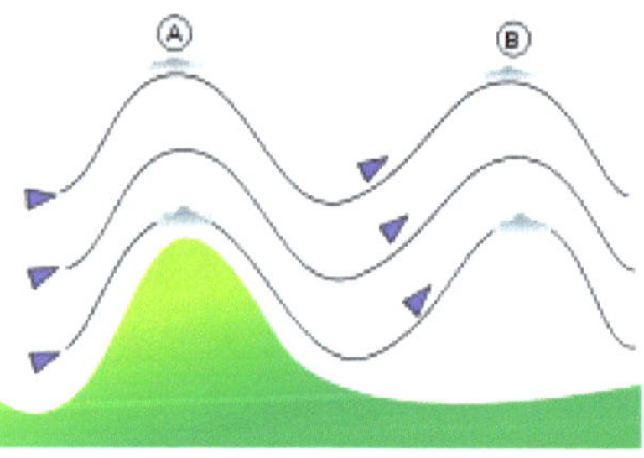

CHAPTER 20

WESTERN DISTURBANCE

1. **Wester Disturbances**. As far as the weather over India is concerned, the extra tropical disturbances or middle latitude disturbances (Frontal Disturbances) play a great role. Its influence over the weather over India prevails through out the year, that is, in all the four seasons. The middle latitude disturbances that form over West Atlantic or Mediterranean regions and move east-north-east wards are commonly known as the **Western Disturbance (WD)** in India.

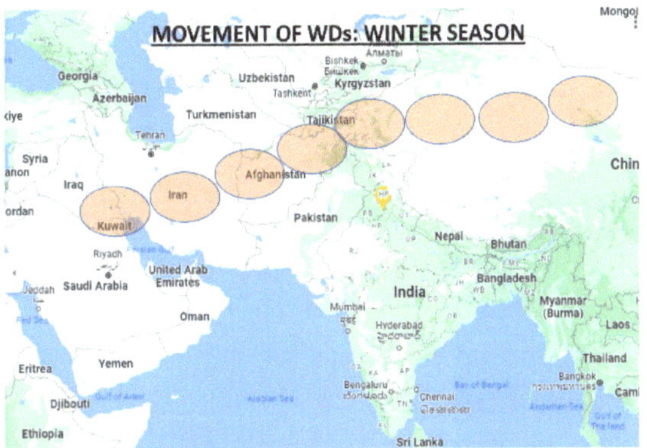

2. These disturbances are mostly seen as upper air features in form of trough or cyclonic circulation at lower/middle level zonal westerlies. The trough that are marked at middle levels may extend to higher levels (Upto 30,000 Ft or above). Sometimes these disturbances can be marked on surface as low pressure area or depression. While moving east-northt-east ward, these systems mostly move across north of India except for the winter months when they pass across extreme north India, that is, J&K region. Hence, these systems have maximum direct influence on weather over Indian region during the winter season. The snowfall activities over J&K and Himachal region during the winter season or accumulation of huge amount of snow over Siachen can rightly be attributed to the Western Disturbances.

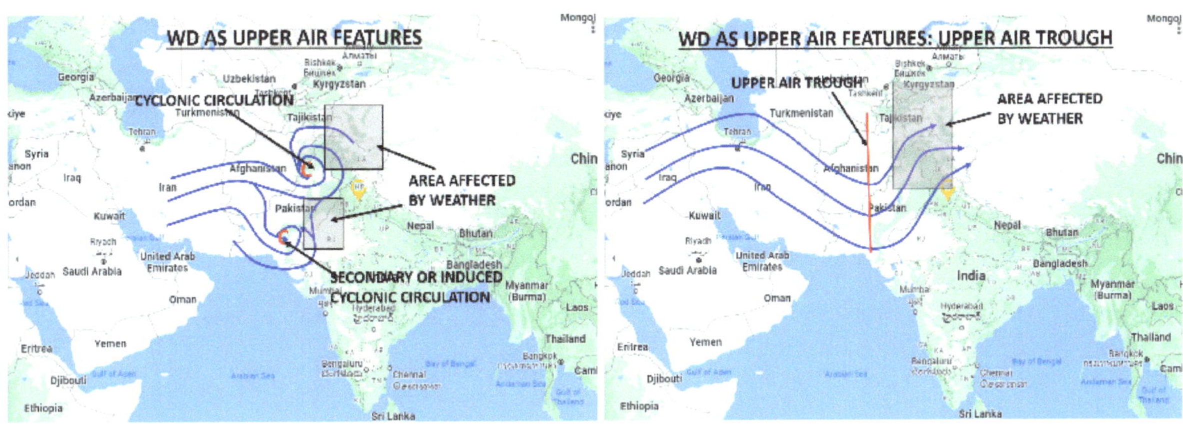

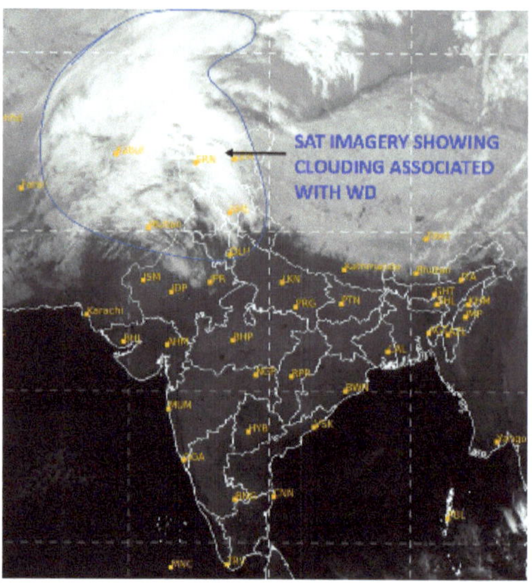

3. However, the indirect impact of western disturbances often becomes more influential to the weather pattern over India in all the four seasons. In fact, most of the significant weather events or the synoptic situations are directly or indirectly influenced by the east ward movements of the western disturbances across middle or higher latitudes. The effects are of varied nature and are co-related to the position, intensity and track of movement of the WDs and other weather features. Influence of WDs in some of the features during different seasons are discussed below: -

a) **Winter Season**. During winter season the track of WDs shift much to the south and often the features directly affect the extreme north India. Formation of the secondary or the induced system over central Pakistan and moving across Rajasthan and UP is also common during the season. Under the influence of these systems the north Indian regions experiences precipitation and relief from cold wave condition. After the systems move away, cold condition returns and widespread dense fog occurs due to availability of moisture and radiational cooling. The WDs, while moving eastward mark their effect up to NE India. Sometimes the induced system follows more southward track and even affect the weather over central and east India.

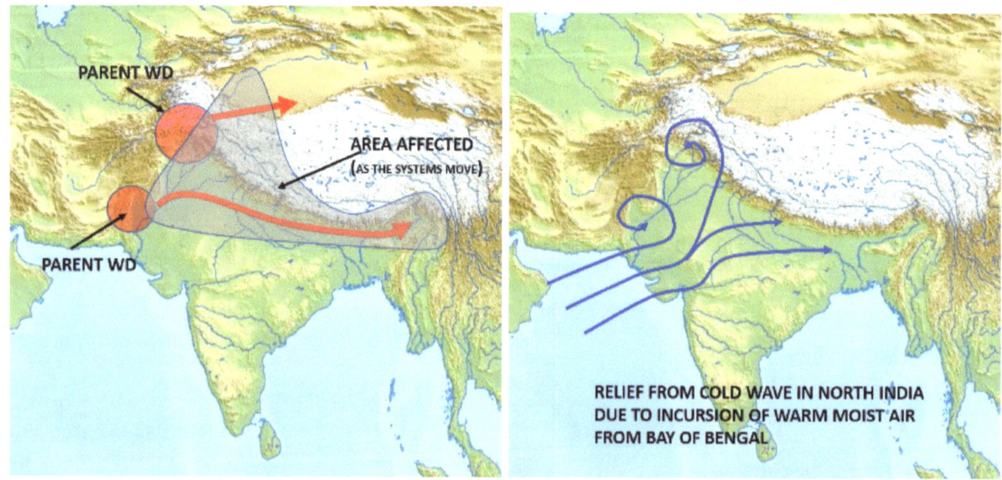

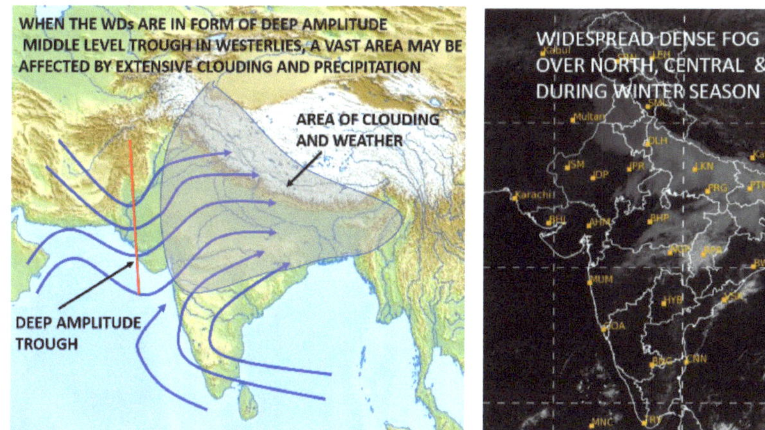

b) **Pre-Monsoon Season**. During this season, mainly the secondary systems influence the weather pattern. The most visible and important weather phenomena are abating of heat wave condition and occurrence of dust storm/thunderstorm over north, east and NE regions.

When a secondary system moves across north India, moisture incursion takes place at lower tropospheric levels. The moist cool air replaces the hot and dry continental air which leads to decrease in air temperature and subsides the heat wave condition. On the other hand, moisture incursion enhances convective buildup due to existing instability that is existing due to solar insolation and results in dust storm/thunderstorm activities.

Due to movement of the eastward moving systems, the seasonal low-level anti-cyclone is pushed eastward to north Bay of Bengal. Location of the anti-cyclone over north Bay results in moisture incursion into Gangetic West Bengal and its adjoining regions. This results in occurrence of thunder squall or nor wester (Kal Baisakhi) over the regions.

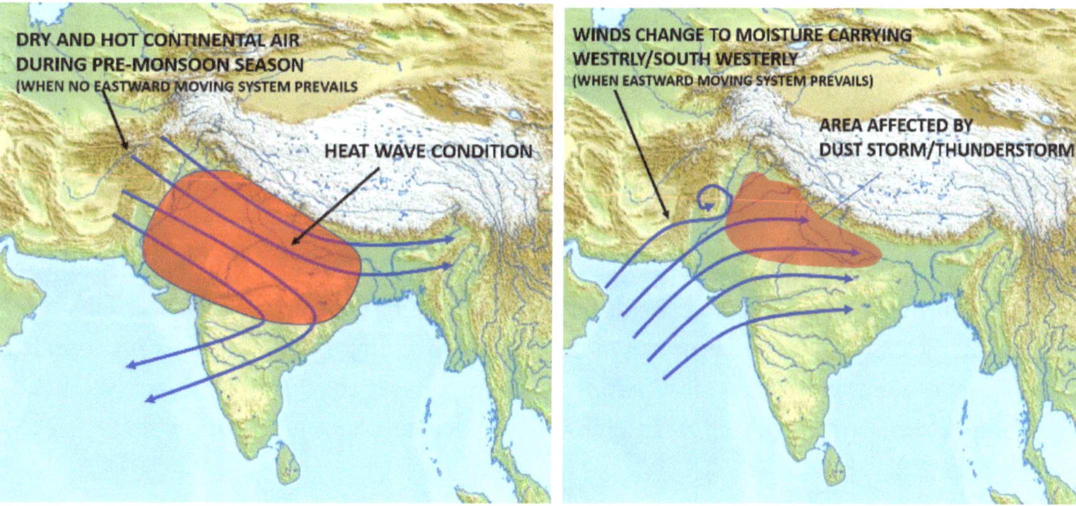

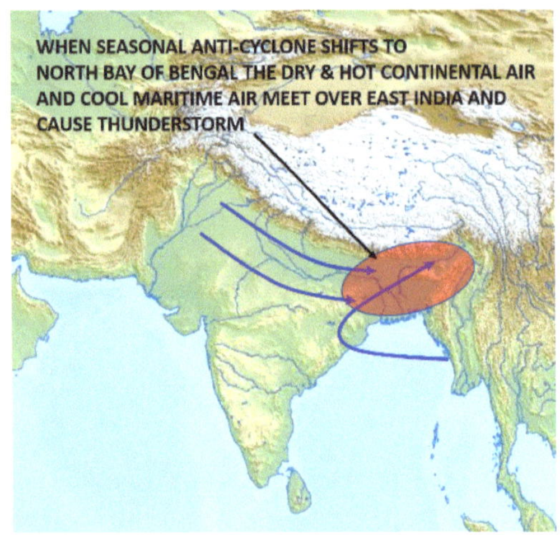

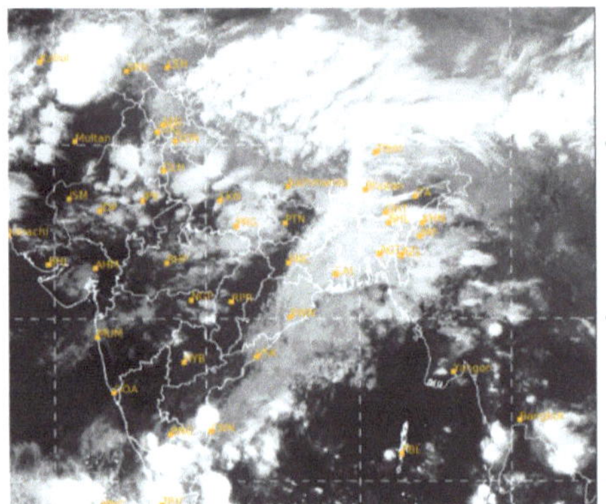

Movements of WDs also influence the track of the tropical cyclones that form over Bay of Bengal during this season. This will be discussed separately in 'Tropical Cyclone' chapter.

c) **Southwest Monsoon Season**. Intensity and distribution of rainfall depends to large extent on the west ward movement of Monsoon Lows from Bay of Bengal and position & orientation of the Axis of Monsoon Trough (AMT) during the south west monsoon season. Movement of WDs has great role to play in both these features.

Normally, the Monsoon Lows move west-north-west ward along the AMT and causes strong monsoon condition over the Southwest sector of the system. The normal position of AMT is from NW Rajasthan to south east Bengal across Gangetic plains, Gangetic West Bengal and north Bay. In normal situation the Monsoon Lows should move from North Bay towards NW Rajasthan. However, on many occasions it has been observed that the Monsoon Lows recurve from the normal path towards north or northeast. Such re-curving of the monsoon lows takes place due to its interaction with a WD that is moving across middle or higher latitudes.

The strength of monsoon to a large extent depends on the position and orientation of the AMT. So long the AMT prevails around it's normal position, the strength of monsoon remain normal when other factors do not change significantly. When an active WD moves along middle or higher latitude the AMT shifts northward and passes along the foot hills of Himalaya and extents up to NE India. In this situation, the easterly and south westerly components of monsoon flow are replaced by dry continental westerlies/north-westerlies. This results in drastic decrease in rainfall over the country outside the foothills areas including NE India. This situation is called 'Break Monsoon' when monsoon weakens and most part of the country experiences sultry and hot weather. However, the rivers originating in Himalaya may overflow and cause flood over the plains under sunny sky.

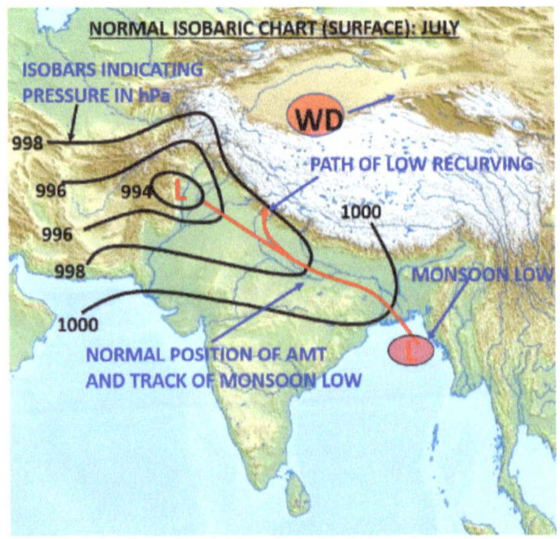

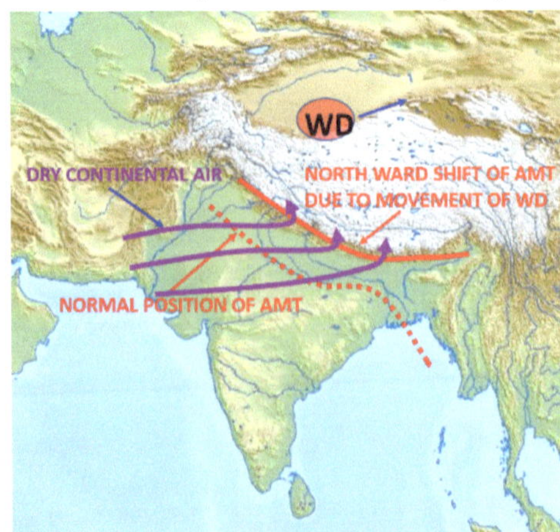

d) **Post Monsoon Season**. During this season (Oct – Nov) monsoon withdraws from most parts of north, east and central parts of the country. However, most parts of Deccan plateau may still be under monsoon coverage during the retreating period. During such stage a deep amplitude trough extending southward from a middle latitude upper air system at middle level may enhance rainfall activity over the Deccan region, specially over Maharashtra and adjoining areas of Karnataka and Telangana.

e) Another important upper air feature that plays an important role in monsoon activity over the north Deccan region is the east-west oriented middle level trough which is commonly known as the east-west shear zone. Enhanced rainfall activity takes place about 02 degree latitude to the either side of this shear zone. The shear zone line shifts north or south from around 14^0 north to about 21^0 north. The shifting of the line is controlled by the movement of the middle latitude disturbances. When the disturbance is close to the Indian region, the shear zone shifts northward and after the disturbance passes away, the shear zone shifts back south ward.

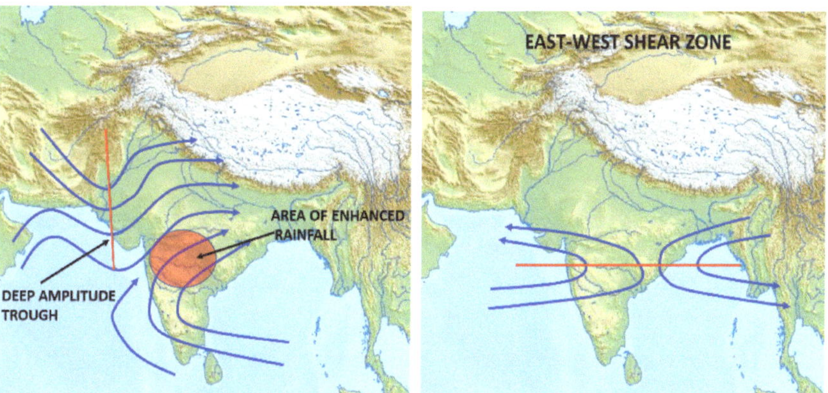

4. **Occlusion**. As the western disturbances move across large area comprising land, lakes and sea, the system starts losing their original characteristics of warm or cold front. Most of these systems when reaches close to India are in an occluded state. A stage is reached when the surface warm and cold fronts merge and the warm sector completely disappears from the surface and lower levels. This process of the merging of the warm and cold fronts is known as occlusion. The composite front is known as an occluded front.

CHAPTER 21

INTER TROPICAL CONVERGENCE ZONE (ITCZ)

1. The Inter Tropical Convergence Zone, or ITCZ, is the region near the equator, where the trade winds (north-easterlies and south easterles from north and south tropics respectively) meet. Incoming solar radiation is the maximum over this region making the air and the sea surface above it warmer. The warm air mass contains adequate moisture and remain boyant. Due to convergence, the buoyant air rises. As the air rises it expands and cools, releasing the accumulated moisture in and causes long lasting rain and thunder storm almost over the entire width of convergence.

2. With the shifting of the solar equator from south to north in the summer and back to south in the winter, ITCZ also shifts accordingly. Seasonal shifts in the location of the ITCZ drastically affects rainfall in many equatorial and tropical regions, resulting in the wet and dry seasons of the tropics rather than the cold and warm seasons of higher latitudes.

3. Basically, ITCZ is a zone of discontinuity and can be called the tropical discontinuity. The discontinuity is between the tropical air masses of the two hemispheres. It is a zone with tropical maritime or continental air to the north and invariably tropical or equatorial maritime air to the south. The former originates in the sub-tropical highs of the northern hemisphere which are partly over land and partly over oceans. The latter originates over the sub-tropical high of the southern hemisphere which lies mainly over ocean.

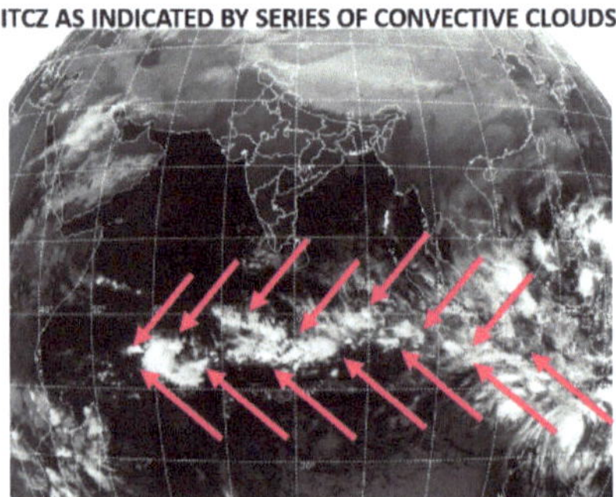

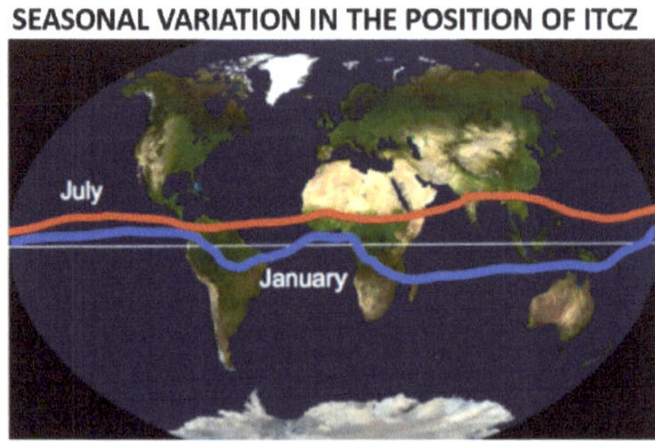

4. Monsoon, a major wind system that seasonally reverses its direction, such as one that blows for approximately six months from the northeast and six months from the southwest. The most prominent monsoons occur in South Asia, Africa, Australia, and the Pacific coast of Central America. As seen above, the ITCZ runs across north India in the month of July. This portion of the ITCZ over India is commonly known as the Axis of monsoon trough (AMT) along which the south-westerly current from the Arabian Sea and the eastery current from the Bay of Bengal converge.

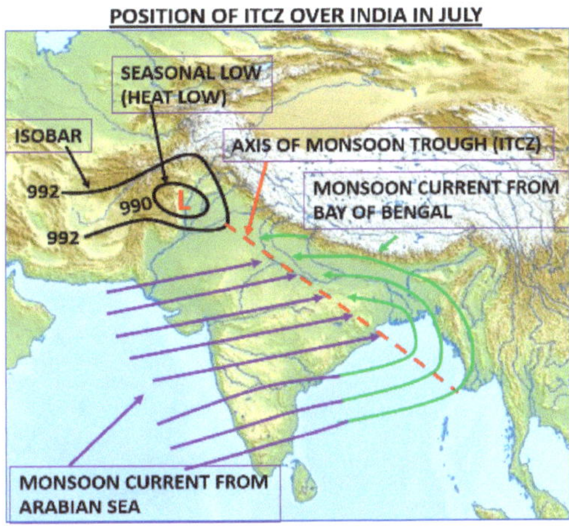

5. The weather or the clouding associated with ITCZ do not exist in a continuous pattern. The cloud imageries show distinct breaks in the ITCZ. On the other hand, the ITCZ is not active on all the days and at all the places. The intensity of weather or clouding depends on the difference in characteristics between the two air masses meeting. Both the air masses, except for the northern hemisphere summer months, converge close to the equator. Hence, both the air masses are mostly maritime and has less difference in terms of temperature and humidity. Hence, on all the days or at all the places, the ITCZ is not associated with significant clouding or strong convection. In other words, the activity of ITCZ in terms of active and non-active state, is not uniform. When the ITCZ is active, its width may extend from 75 to 150 Km. In this zone the density of intense convection (Cb cloud) is high and are associated with heavy showers. When the ITCZ is close to the equator, it sometimes splits into two and encloses a belt of doldrums (fair weather and light winds). This happens because of the existence of what are known as equatorial westerlies in a narrow zone. These westerlies act as a buffer between the two split portions of the ITCZ.

6. ITCZ undergoes diurnal variation. Its effect is maximum during the afternoon/evening hours and minimum during late night/early morning hours. Hence, in judging the state of ITCZ (active or non-active) it is important to keep in mind the time of the charts or satellite imageries.

7. **Formation of Low Pressure System on ITCZ**. The ITCZ, being itself an area of instability and disturbance, is also a hub for the formation of stronger tropical systems. The more intense systems that get the breeding bed in ITCZ respectively are waves, depressions and cyclonic storms.

8. As mentioned earlier, Tm and Em air masses converge along the ITCZ. The Tm air mass, which is stronger lies to the north of the ITCZ and the weaker Em lies to the south. This is a situation of an undisturbed ITCZ. If a situation arises when there is a surge in the Em air mass to the south, the ITCZ gets disturbed and a wave forms in the ITCZ. The winds in the area of wave tends to become cyclonic and the wave may turn into a cyclonic circulation. Under favourable conditions, the circulation may grow higher and under it's influence a low-pressure area may form at surface. The low-pressure area, which was in a weak stage move along the ITCZ from east to west in the nature of a wave without much intensification. Due to their direction of movement these waves are also referred as the easterly waves and are steered by the lower or middle level tropospheric winds.

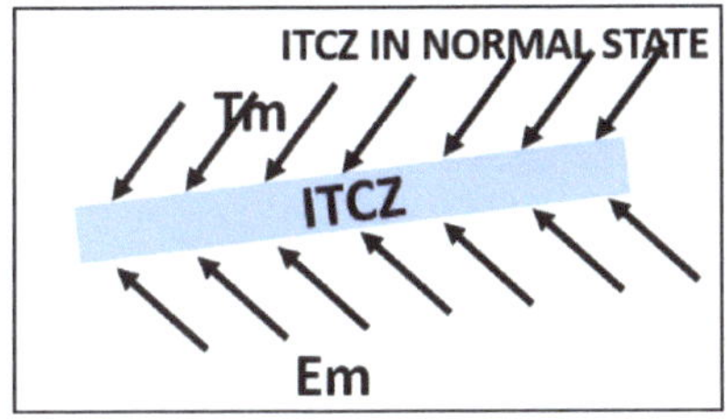

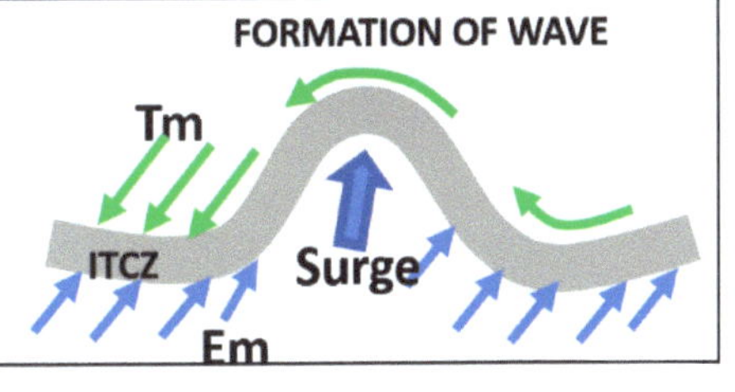

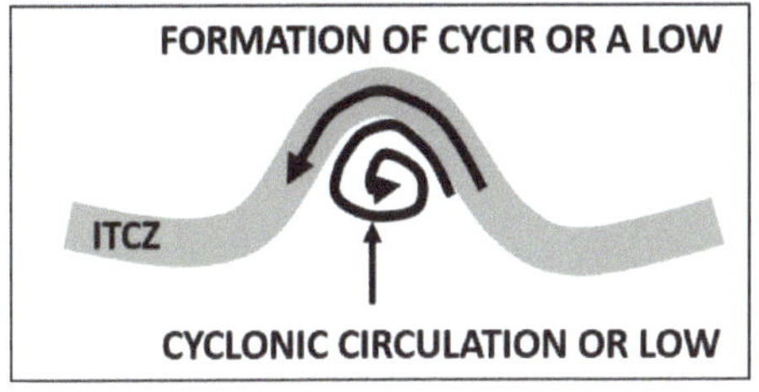

9. Easterly wave is a perturbation in the waves of easterlies close to ITCZ propagating from east to west. It is normally seen in the lower & middle troposphere. Over Indian region it is seen in the middle & upper troposphere propagating from Bay of Bengal to Arabian Sea through Southern Peninsula.

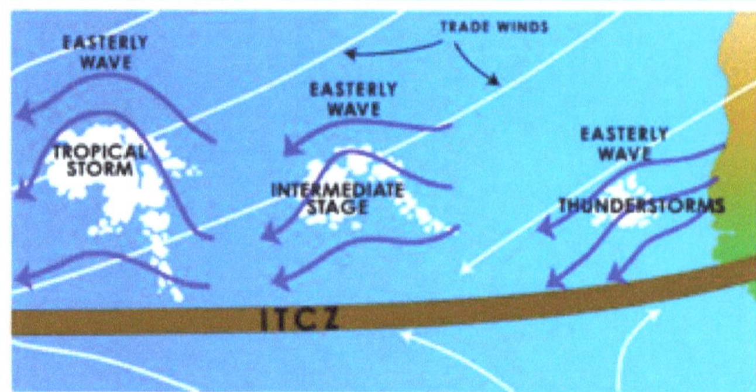

10. **Troughs and Ridges in Easterly Waves**. The formation of a curve in the geostrophic wind, concave toward lower pressure, is termed a trough in the wave. As the wind crosses a reference latitude toward the south, we observe a curve in the wind concave toward high pressure which is called a ridge in the wave.

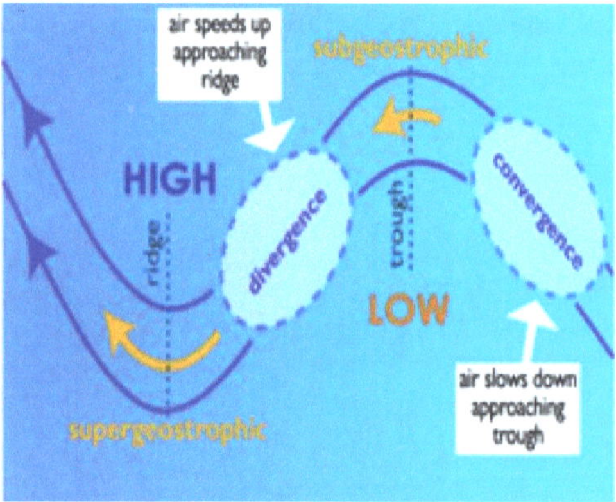

11. **Regions of convergence and divergence in an easterly wave**. Ahead of a trough, where the air in the wave is slowing down and converging, some air gets 'pushed up' away from the surface, producing lower pressure near the surface. Conversely, ahead of an upper level ridge, where the air is speeding up and diverging, air gets 'sucked down' into the long wave, producing subsidence and higher pressure near the surface. Lower layer divergence, subsidence, and fair weather are found ahead (upwind, or to the west) of the trough axis. Convergence, ascending motion and heavy weather are concentrated to its rear (to the east).

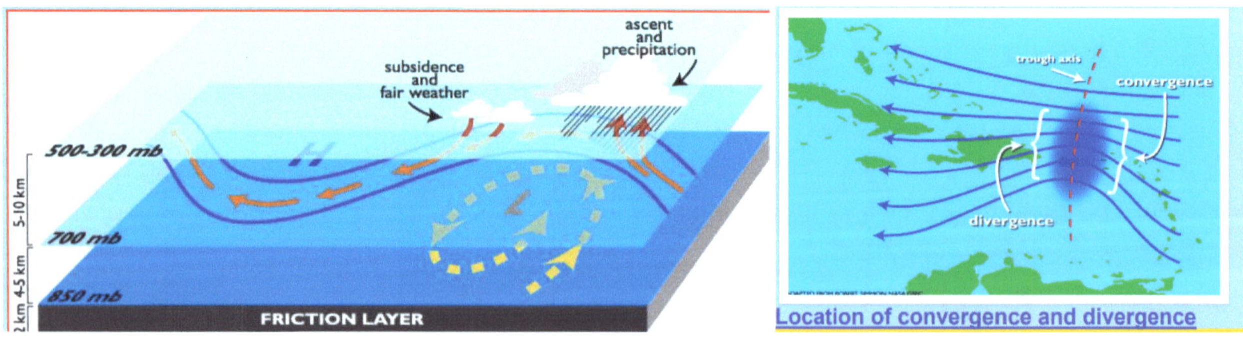

Weather Pattern Associated with Easterly Wave.

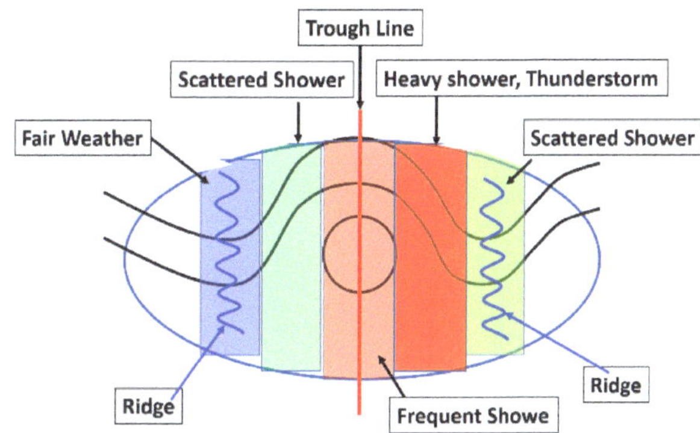

a) The area the ridge ahead of the trough experience fair weather. Some weak convective clouds may form but they do not cause rain.

b) Just ahead of the trough large convective clouds are seen but without precipitation.

c) Scattered shower occurs over the area close to the trough.

d) At the trough line there is frequent showers.

e) At the rear of the trough frequent moderate to heavy rain, shower and thunderstorm are experienced.

f) The area over the ridge at the rear of the trough experiences moderate showers which gradually decreases.

g) Further to the eastward there is dry weather with fair sky condition.

12. When the disturbance moves into the land area, the wave characteristics get disturbed due to friction and surface heating. Also, the sequence of weather also differs. The disturbances tend to weaken after making land fall. However, if the land area is smaller in size and a sea surface lies ahead, the wave may regain strength and intensify.

13. **Easterly Waves over Indian region**. During the period when ITCZ is close to India, easterly wave forms over Bay of Bengal and move west ward towards India. The waves that form over south Bay of Bengal heads towards south Peninsula. On arrival over the east coast the waves tend to weaken for the existing topography. However, often the systems do not die down. They continue to move west ward in the weak stage and on crossing the west coast they again revive and move away further with the prevailing wind field.

14. During winter season the ITCZ is far to the south. Hence, the easterly waves do not affect Indian region. However, on some occasion, when a surge pushes the ITCZ northward to the extreme south Bay of Bengal, the south Peninsular region and the Bay islands get affected with rain for a few days. As the hot season sets the ITCZ advances northward and the frequency of incidence of easterly waves over India increases. In the southwest monsoon season, the waves are seen moving eastward across the SE Asian countries to Bay of Bengal and further to the main land of India. Some of these waves travel further westward along the ITCZ.

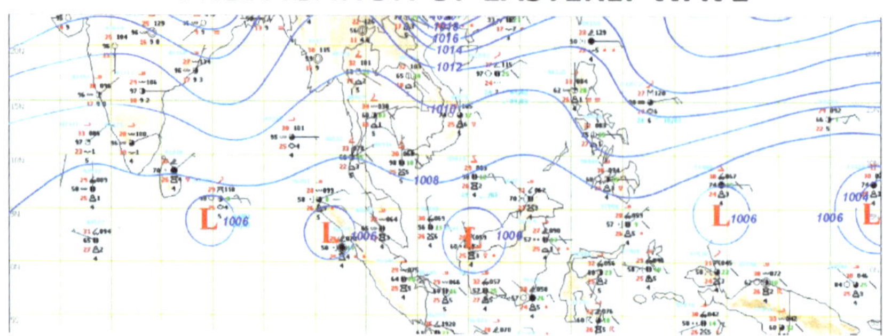

15. The greatest disturbance that get produced from the initial stage of a wave is a Tropical Cyclone. Formation, growth and movement of this type of disturbance will be discussed separately.

CHAPTER **22**

TROPICAL CYCLONE

1. Tropical Cyclone is an intense storm that forms over warm sea surface. It is characterized by low atmospheric pressure, high winds, and heavy rain. It gradually grow from a wave in upper air or a trough or low pressure area over sea surface. Under favourable conditions it intensify into a depression and then into a storm. The storm moves in the direction of the steering wind current (wind at the topmost level up to which the circulation of the storm extends). The storm maintains its energy or grow further as long as it is over the warm sea surface. On making a landfall or on reaching a cooler sea surface the system weakens.

2. Winds generated by a cyclonic storm are very strong and exceeds 119 Kmph and in extreme cases may exceed 320 kmph. The devastating wind is accompanied by very intense rain and thunderstorm and elevation of sea surface up to about 20 Ft above normal level due to storm surge. Such a combination of intense weather elements make a cyclonic storm the most hazardous weather phenomenon in the tropical and sub-tropical regions. Every year during the summer months in both the hemispheres cyclones strike most of the countries which are located in the tropical or sub-tropical regions. India and neighbouring countries adjacent to sea also experience tropical cyclone every year.

3. Tropical Cyclones are known in different local names in different parts of the world. In the North Atlantic Ocean and the eastern North Pacific they are called **hurricanes**, and in the western North Pacific around the Philippines, Japan, and China the storms are referred to as **typhoons**. In the western South Pacific and Indian Ocean they are referred to as **cyclones** and in Australia they are also known as **Willy Willy**. All these different names refer to the same type of storm.

4. It has been observed that the tropical cyclones origin in certain areas of ocean and follow pre-dominant tracks. The track, no doubt depends on the seasonal wind pattern of the area. Following image shows the normal position of origin of the storms and the paths they usually follow.

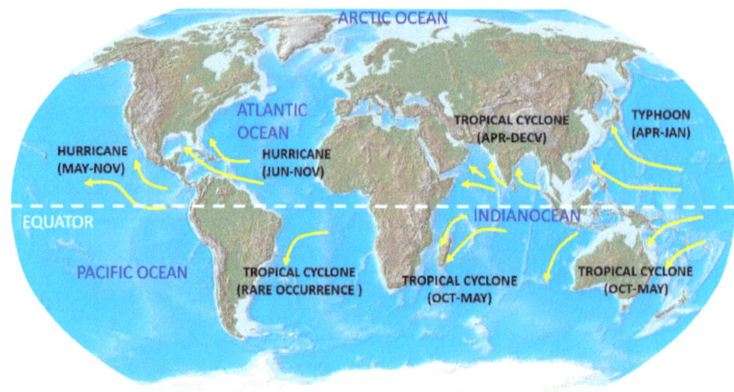

5. **Life cycle of Tropical Cyclone**. The life cycle of a tropical cyclone can be boroadly divided into Formative Stage, Mature Stage and Dissipating or Decaying Stage.

6. **Formative Stage**. The energy for a tropical cyclone is provided by the transfer of heat from the warm sea surface to the overlying air. Primarily, the transfer of energy takes place by evaporation from the sea surface. As the warm and moist air rises, it expands and cools and then becomes saturated and releases latent heat due to condensation. The air above the disturbance is warmed and moistened by this process. This results in higher temperature in the column of disturbed air as compared to the surroundings which makes the rising air more boyant and enhancement in vertical movement.

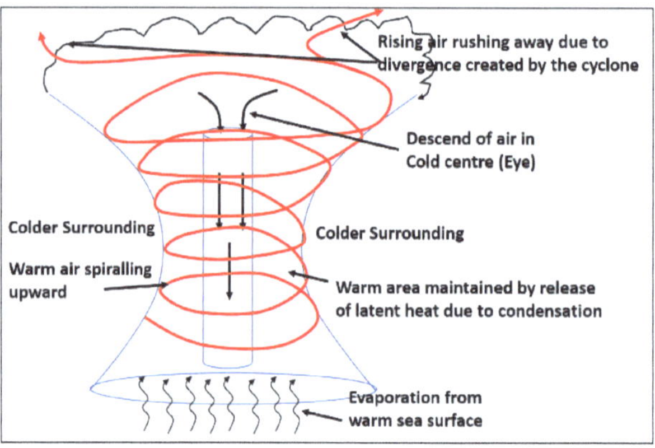

7. The above said situation is required to be sustained for the energy supply to continue and the disturbance to grow further and further. The supply chain may stop if the sea surface is not sufficiently warm and depth of warm water is not adequate. This is because, the moderate disturbance may cause convective build up and rain which will cool the sea surface if it is not sufficiently warm and if the warm water layer is shallow.

8. Existence of only the vertical motion of warm air is not adequate to initiate the formation of a tropical system. It is important that there is flow of moist and warm air into a pre-existing disturbance for further development to occur. The rising air increases the temperature at the core of the disturbance through release of latent heat and direct heat transfer from the sea surface. This results in decrease in the atmospheric pressure in the centre of the disturbance. The decreasing pressure causes flow of wind with increased speed around the centre (anti-clock wise over northern hemisphere and clockwise over southern hemisphere). This increases the process of evaporation and heat transfer thereby contributing to further rising of air. The warming of the core and the increased surface winds thus reinforce each other in a positive feedback mechanism.

9. **Anatomy**. The cyclones are circular storms, generally 250-320 km in diameter, whose winds spin around a central region where atmospheric pressure is very low. The winds are driven by this low-pressure centre and the Coriolis force of the earth. As a result, tropical cyclones rotate in a counter-clockwise direction in the Northern Hemisphere and in a clockwise direction in the Southern Hemisphere.

10. A tropical cyclone can be divided into three regions of wind fields.

    a) A ring-shaped outer region with outer radius of, about 160 km and an inner radius of about 30 to 50 km. There is uniform increase in wind speed towards the centre. This region is also known as the **rain band**. This is a band of secondary cells arranged around the centre and spiral into the centre of the storm. Sometimes the band is seen stationary while in other cases they rotate around the centre. In some cases the band is seen shifting side to side unsteadily. In such case, the actual place of land fall on the coast may largely vary from the forecast location.

    As a tropical cyclone hit the land mass, there is increase in frictional force due to prevalent land topography. Increase in friction causes increase in convergence into the eye wall which subsequently increases the vertical motion of air leading to extreme and rapid convective build up. The situation causes torrential rain. The rainfall is so heavy that it may be close to 300 mm in 24 hours.

    b) The second region is the **eye wall** which is about 15 to 30 Km from the centre. In this region wind speed is the maximum. This region is characterised by extremely strong winds and is responsible for all types of destructions and calamitic factors of a cyclone. Maximum wind speed is observed around 1000 Ft above surface. The region is also characterised by heavy to extremely heavy rainfall and very intense convective clouds. Clouds may vertically grow to the height of more than 45,000 Ft. There is upward movement of air in the eye wall with great vertical velocity. The updraft becomes stable when the air reaches tropopause. Here the air flows outward and due to Coriolis force the outward flowing air creates anticyclonic circulation aloft.

    c) Surrounded by the eye wall is the **eye** in the centre. Winds in the eye are light or calm and is also characterised by clear sky, warm temperature and very low atmospheric pressure which may be as low as 960 hPa against average surface pressure of 1000 hPa. In case of a super cyclone the pressure may further deep to less than 900 hPa. There is some amount of subsidence of air in the eye. Subsidence results in compression and increase in temperature. Hence, temperature at the eye is about 5^0 C more than the surrounding areas under cyclone. As the warmer air can hold more water before condensation, this region generally remains free of any significant cloud.

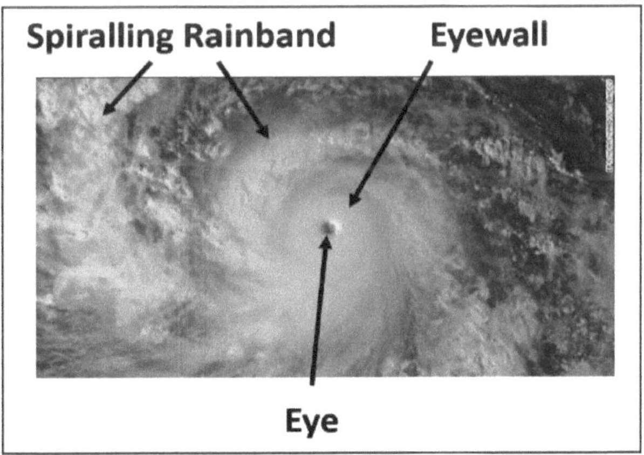

10. **Intensification and Mature Stage**. For a tropical cyclone to continue developing, the surrounding air in the vertical column must be cooler so that the instability is maintained to great height. The vertical movement of air results in deep convective clouds to develop. The rising air at centre, on the other hand, also draws air from the surrounding atmosphere from 1.5 Km to about 5 Km. If this air is sufficiently humid, the circulation and subsequent vertical movement of air will continue. However, if the surrounding air is dry, then evaporation of some amount of moisture in the centre will take place. This will decrease the temperature in the centre. This situation will arrest further intensification of the system.

11. As a tropical system grows, the rapidity of the circulation of air increases or the increase is required to be sustained. For this, it is important that the low-pressure area is located at least about 500 Km away from the equator. If it is too close to the equator, then the effect of the coriolis force will be too small to provide the necessary spin. The air that is drawn towards the centre is deflected by the Coriolis force. As mentioned earlier, the direction of the resulting circulation around the low in the northern hemisphere is anti-clock wise and in the southern hemisphere it is clockwise.

12. Another important requirement for intensification is that, there must be a little change (increase) in the wind speed with height. Too much increase in wind speed with height disturbs the vertical alignment of the system over the warm surface which is providing the energy. When verticality is not maintained, the energy feeding mechanism gets disturbed. The ideal condition for vertical orientation of tropical cyclone is minor north to south.

13. **Dissipating stage**. The conditions required to weaken a storm are as follows: -

    a)   Decrease in moisture feed from warm sea surface.

    b)   Mechanism to counter the prevailing wind speed.

14. Both of the above requirements are met when a cyclone makes landfall. Hence, there is rapid weakening of the storm after it arrives over land. In fact, the frictional force caused by the land topography starts acting soon after the eye wall enters land. The dissipation is so rapid that the system becomes a depression or a low-pressure area within a day or so.

15. It is seen that a tropical storm weakens over the sea itself. This happens when the system remains over the sea for a long time and travels across a great distance. When the cyclone crosses the latitudes northward, it encounters decrease in sea surface temperature. As the system moves to higher latitude, it becomes an extra tropical system and the looses its original characteristics of tropical cyclone. The core temperature decreases, the central pressure increases and subsequently, the speed of spinning winds and vertical movement also decreases. As a result, the diameter of the storm increases and it gradually defuses in a few days after moving with the zonal winds.

16. **Classification of Cyclonic Disturbances over North Indian Ocean**. Based on the associated maximum sustained wind speed, the tropical disturbances over north Indian Ocean region are classified as follows: -

Type of Disturbance	Associated Maximum Sustained Wind Speed
Low Pressure Area	Not exceeding 17 Kt (< 31 Kt)
Depression	17 to 27 Kt (31 to 49 Kmph)
Deep Depression	28 to 33 Kt (50 to 61 Kmph)
Cyclonic Storm	34 to 47 Kt (62 to 88 Kmph)
Severe Cyclonic Storm	48 to 63 Kt (89 to 117 Kmph)
Very Severe Cyclonic Storm	64 to 90 Kt (118 to 167 Kmph)
Extremely Severe Cyclonic Storm	91 to 119 Kt (168 to 221 Kmph)
Super Cyclonic Storm	120 Kt and above ($\geq$ 222 Kmph)

17. **Life Period**. The average life period of cyclonic disturbances over North Indian Ocean region are as follows: -

Disturbance	Average Life
Depression (D)	2 days
Deep Depression (DD)	3 days
Cyclonic Storm (CS)	3.5 days
Severe Cyclonic Storm (SCS)	4 days
Very Severe Cyclonic Storm (VSCS)	5 days
Extremely Severe Cyclonic Storm (ESCS)	5.75 days
Super Cyclonic Storm (SuSc)	

18. VSCS have higher mean life period over both the Arabian Sea and the Bay of Bengal in pre-monsoon, post-monsoon and year as a whole. While the VSCS stage has significantly higher duration over the Arabian Sea than over the Bay of Bengal in pre-monsoon and the year as a whole, it is significantly higher over the Bay of Bengal than over the Arabian Sea during post-monsoon season. During the monsoon season, the duration D, DD and CS stages are significantly higher over BOB than they are over the ARB.

19. The track of longest ever recorded cyclone over the North Indian Ocean is shown in below. It originated over the South China Sea, moved west-northwestwards across Vietnam, Bay of Bengal, South India and Arabian Sea to Oman during Oct. 1924.

20. **Vertical Structure**. The vertical structure of tropical cyclones is divided into **three sections.**

   a) The lowest layer, known as the inflow layer, extends up to 3 km and is crucial for generating the storm. The wind flow is towards the centre of the storm. Most of this inflow layer occurs in the planetary boundary layer where friction plays a great role.

   b) The primary cyclonic storm occurs in the middle layer, which extends from 3 km to 7 km.

   The outflow layer is located above 7 kilometers and extends up to the top of the storm. The wind is anticyclonic (clockwise). Outflow is most pronounced around 12 Km level.

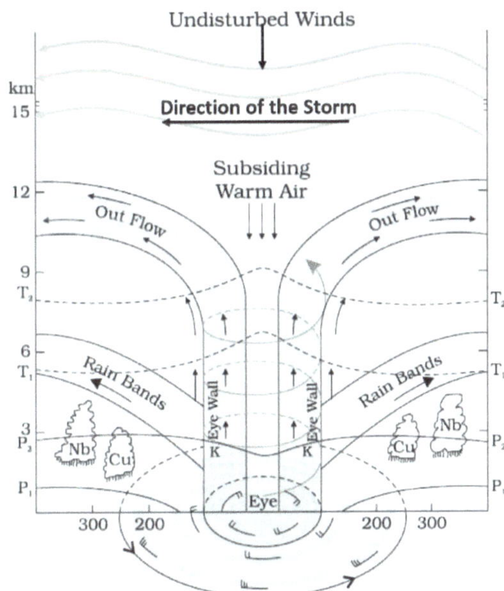

21. **Size of Cyclone**. Tropical cyclones are about 300 miles (483 km) wide although they can vary considerably. Size is not necessarily an indication of cyclone intensity. Hurricane Andrew (1992), the second most devastating hurricane to hit the United States, next to Katrina in 2005, was a relatively small hurricane. On record, Typhoon Tip (1979) was the largest storms with gale force winds (39 mph/63 km/h) that extended out for 675 miles (1087 km) in radius

in the Northwest Pacific on 12 October, 1979. The smallest storm was Tropical Storm Marco with gale force winds that only extended 11.5 miles (18.5 km) radius when it struck Misantla, Mexico, on October 7, 2008.

22. **Movement**. Tropical cyclones initially move west to east in the direction of the trade winds. However, due to the prevailing Coriolis force cyclones turn to right in the northern hemisphere and to left in the southern hemisphere as shown in a picture above while discussing the origins of the tropical cyclones.

23. In general, the average direction of the winds at various layers in portion of the atmosphere where the system prevails, determine the direction of the movement of the cyclones. The week systems like depressions or deep depression, the lower and middle level winds tend to move the storm. However, the stronger system extends up to upper tropospheric levels. In such case the middle and upper level winds play greater role. In general, it has been found that the direction of the winds just above the layer up to which the associated circulation extends, give a good indication of the movement of the tropical cyclones.

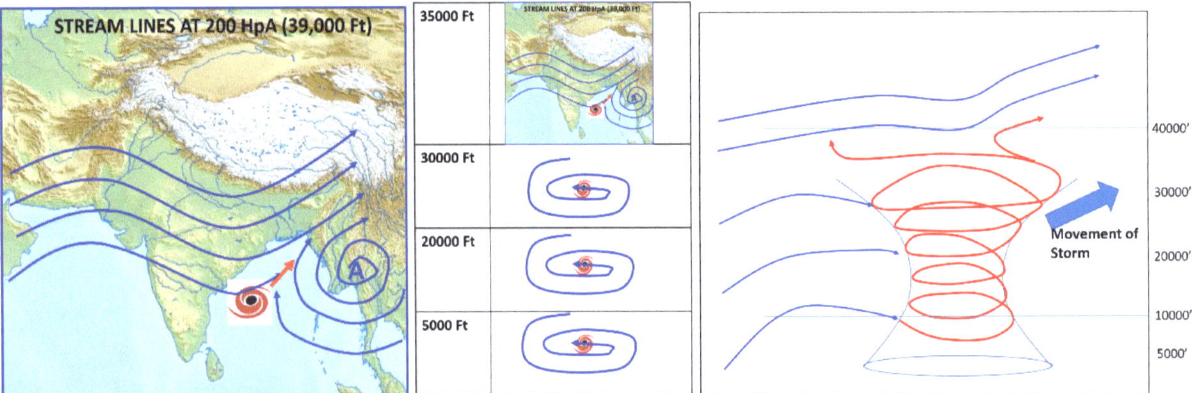

24. **Hazards Caused by Cyclone**. Cyclones are the low pressure systems. The severity of weather increases with the intensity of the low pressure. Intense low pressure systems like depressions and cyclones originate in the equatorial trough zone over warm ocean surface under certain favourable atmospheric conditions. The cyclonic storms cause heavy rains, strong winds and also high seas and devastate coastal areas at the time of landfall, leading to loss of life and property. Types of damages associated with a tropical cyclone are also show in the following figure.

25. The expected damage associated with the cyclonic disturbances of different intensities along with action suggested by IMD to disaster managers is given in the following table.

Intensity	Damage Expected	Action Suggested
Deep Depression (DD)	Minor damage to unsecured structures	Fishermen advised not to venture into the open seas.
Cyclonic Storm (CS)	Damage to thatched huts. Breaking of tree branches. Minor damage to power and communication.	Total suspension of fishing.
Severe Cyclonic Storm (SCS)	Extensive damage to thatched roofs and huts. Minor damage to power and communication. Flooding.	Total suspension of fishing. Coastal hutment dwellers to be moved to safer places. People to remain indoors.
Very Severe Cyclonic Storm (VSCS)	Extensive damage to kutcha houses. Partial disruption of power and communication line. Minor disruption of rail and road traffic. Threat from flying debris. Flood.	Total suspension of fishing. Mobilise evacuation. Regulation of rail and road traffic. People to remain indoors.
Extremely Severe Cyclonic Storm (ESCS) 168-221 kmph (91-119 Kt))	Extensive damage to kutcha houses. Damage to old buildings. Large-scale disruption of power and communication. Disruption of rail and road traffic due to flood. Threat from flying debris.	Total suspension of fishing. Extensive evacuation. Diversion or suspension of rail and road traffic. People to remain indoors
Super Cyclonic Storm (SuSc)	Extensive structural damage. Total disruption of communication and power supply. Extensive damage to bridges. Large-scale disruption of rail and road traffic. Large-scale flooding and inundation of sea water. Air full of flying debris	Total suspension of fishing. Large-scale evacuation. Total suspension of rail and road traffic. People to remain indoors inside strong structures.

26. **Strong Winds**. Strong winds cause extensive damage in the cyclone affected areas. Sometime the damages produced by strong wind are very extensive and cover areas greater than the areas of heavy rains and storm surges. The impact on the areas through which the eye passes, is further more. While the areas away from the eye are generally affected by unidirectional winds, an eye passage brings with it rapid changes in wind direction, which imposes torques and can twist the vegetation or even structures. Parts of structures that were loosened or weakened by the winds from one direction are subsequently severely damaged or blown down when hit upon by the strong winds from the opposite direction.

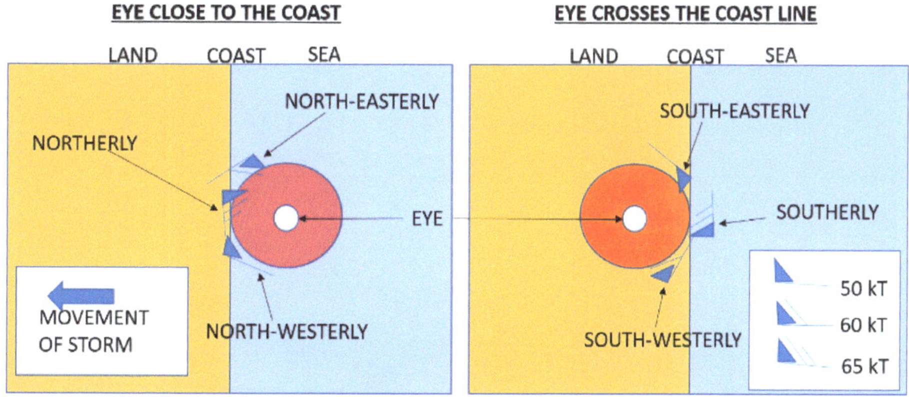

27. Many of the overhead communication networks are susceptible to damage when the winds reach 85 kts (158 kmph), This is especially the case for secondary telephone lines. Microwave towers are susceptible to misalignment when winds reach 85 kts (158 kmph). This affects local telephone, cellular service and long distance service. Microwave and radio towers are susceptible to destruction when winds reach 100 kts (186 kmph). At higher wind speed even larger antennas are also vulnerable and are blown off. Even large satellite communication dishes can be damaged in cyclones with sustained wind speeds of 135 kts (251 kmph). Coastal roads/locations are vulnerable to damage from

inundation/waves runup. The most detrimental hazards to roadways are uprooted trees, power poles and lines, and debris falling on roads and blocking them. This becomes a serious problem when winds reach 80 kts (149 kmph) or more.

28. **Rainfall**. Generally heavy to extremely heavy intensity takes place over the affected area thus leading to excessive amount of water. The amount of fall in 24 hours may be more than 30 Cm. Persistent rains give rise to unprecedented floods. Rainwater on the top of storm surge worsens the situation. Rain is an annoying problem for the people who become shelter less due to a cyclone. It creates problems in post cyclone relief operations also.

29. **Storm Surge**. Though, the deaths and destruction are caused directly by the winds in a tropical cyclone, storm surge is the major cause of devastation from tropical storms. The winds lead to massive piling of sea water in the form storm surge that leads to sudden flooding of coastal regions due to rise of sea level. In India such rise in sea level is maximum over West Bengal and Odisha coast where the storm surges get amplified due to the prevailing topography and depth of sea. The northward converging shape of the Bay of Bengal provides another reason for the enhanced storm surge in these areas. The other cause of enhanced surge is the astronomical tide. The rise due to high tide may be as high as 4.5 m above the mean sea level at some parts of Indian coast. The worst devastation is caused when the peak surge occurs at the time of high tide. The sand and gravel carried by the moving currents at the bottom of the surge can cause sand papering action of the foundations of the structures. The huge volume of water can cause such pressure difference that the house "floats" and once the house is lifted from the foundations, water enters the structure that eventually collapses.

30. **Marine Impact ('T' Number of Tropical Disturbance)**. This is another method for estimating intensity of tropical cyclones over North Indian Ocean Region. In this method, maximum wind speeds associated with tropical cyclones are estimated operationally by a very widely used tool, the Satellite-based Dvorak technique (SDT) of cloud pattern.

Satellite-based Dvorak Technique (SDT). A statistical method for estimating the intensity of tropical cyclones from interpretation of satellite imagery. It uses regular Infrared and Visible imageries. It is based on a "measurement" of the cyclone's convective cloud pattern and a set of rules.

31. In this technique, the current intensity in terms of T numbers of TCs is estimated by analysing satellite image patterns (e.g. eye, shear, banded, central dense overcast). It is used to estimate intensity by assigning a T number which ranges from 1 to 8 with increments of 0.5. A unit T number corresponds to the climatological rate of TC intensity. The T number is mainly used to issue advisory to those operating in the Sea. The wind speed, condition of Sea and wave height associated with 'T' numbers of various categories of cyclonic disturbances, are given in the following table.

Intensity	Strength of wind in Kt/Kmph	Satellite 'T' No.	Condition of Sea	Wave height (m)	Action suggested
Depression	17-21 / 31- 40 22-27 / 41- 49	1.5	Moderate Rough	1.25 – 2.5 2.5 – 4.0	
Deep Depression	28-33 / 50–61	2.0	Very Rough	4.0 – 6.0	Fishermen advised not to venture into the open seas.
Cyclone	34-47 / 62–87	2.5 - 3.0	High	6.0 – 9.0	Total suspension of fishing operations
Severe Cyclone	48-63 / 88 -117	3.5	Very High	9.0 – 14.0	
Very Severe Cyclone	64-90 / 118-167	4.0 - 4.5	Phenomenal	> 14.0	
Extremely Severe Cyclone	91-119 / 168-221	5.0 – 6.0		> 14.0	
Super Cyclonic Storm	120 or more /and more	> 6.0		> 14.0	

32. **Cyclone hazard prone districts of India.** Based on frequency of cyclones, total number of severe cyclones, maximum wind speed, probable maximum surge, and precipitation for all districts is proneness to cyclones has been worked out by IMD. Ninety-six districts including 72 districts touching the coast and 24 districts not touching the coast, but lying within 100 km from the coast have been classified based on their proneness. Out of 96 districts, 12 are very highly prone, 41 are highly prone, 30 are moderately prone, and the remaining 13 are less prone. Twelve very highly prone districts include South and North 24 Parganas, Medinipur, and Kolkata of West Bengal, Balasore, Bhadrak, Kendrapara, and Jagatsinghpur districts of Odisha, Nellore, Krishna, and east Godavari districts of Andhra Pradesh and Yanam of Puducherry. The remaining districts of Odisha and Andhra Pradesh, which touch the coast are highly prone districts. The north Tamil Nadu coastal districts are more prone than the south Tamil Nadu districts (south of about 10°N latitude). Most of the coastal districts of Gujarat and north Konkan are also highly prone districts. The remaining districts in the west coast and south Tamil Nadu are either moderately prone or less prone districts.

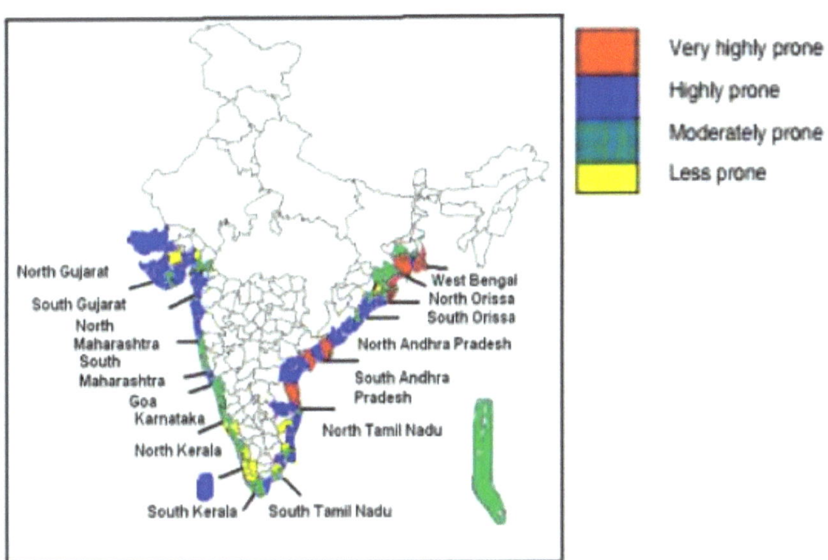

33. **Records of Tropical Cyclone of North Indian Ocean.** Some of the data pertaining to cyclones that originated over North Indian Ocean region since 1890 are given below.

Year	D	CS	SCS
1890-1899	112	60	28
1900-1909	105	69	29
1910-1919	95	54	21
1920-1929	140	62	18
1930-1939	151	63	21
1940-1949	162	55	23
1950-1959	134	39	10
1960-1969	153	61	38
1970-1979	153	66	44

Year	D	DD	SC	SCS	VSCS	ESCS	SuSC
1980-1989	110	84	44	21	15	9	1
1990-1999	93	69	41	25	18	9	3

The SuCS that hit Odisha in 1999 is the most intense Tropical Storm in North Indian Ocean so far. The 1999 Odisha cyclone organized into a tropical depression in the Andaman Sea on 25 October. The disturbance gradually strengthened as it took a west-northwes ward path, reaching cyclonic storm strength the next day. Aided by highly favorable conditions, the storm *rapidly intensified, attaining super cyclonic storm intensity on 28 October, before peaking on the next day with winds of 260 km/h (160 mph) and a record-low pressure of 912 hPa. The storm devastated Odisha in which population of 1,25,71,000 were affected, 18,28,532 houses were damaged and 9,887 people were killed.*

Year	D	DD	SC	SCS	VSCS	ESCS	SuSC	Strongest Cyclone
2000-2009	110	84	44	21	15	9	1	SuSC Gonu, ESCS Nargis

Super Cyclonic Storm Gonu was an extremely powerful tropical cyclone that became the strongest cyclone on record in the *Arabian Sea. Gonu developed from a persistent area of convection* in the eastern Arabian Sea on June 1, 2007. It *rapidly intensified* to attain peak winds of 240 km/h on June 4. Gonu made *landfall* on the easternmost tip of *Oman, becoming the strongest tropical cyclone to hit the Arabian Peninsula. It then turned northward into the Gulf of Oman, and dissipated on June 7, after making landfall in southern Iran, the first landfall in the country since 1898. The cyclone caused 50 deaths in Oman, where the cyclone was considered the nation's worst natural disaster. Gonu dropped heavy rainfall near the eastern coastline, reaching up to 610 mm, which caused flooding and heavy damage. In Iran, the cyclone caused 28 deaths.*

Extremely Severe Cyclonic Storm Nargis was an extremely destructive and deadly *tropical cyclone* that caused the worst *natural disaster* in the *recorded history* of *Myanmar* during early May 2008. The cyclone made *landfall* in Myanmar on Friday, 2 May 2008, sending a *storm surge* 40 kilometres up the densely populated *Irrawaddy delta, causing catastrophic destruction and at least 1,38,373 fatalities. The Labutta Township* alone was reported to have 80,000 dead, with about 10,000 more deaths in *Bogale. Nargis is the costliest tropical cyclone on record in the North Indian Ocean* at the time, before that record was broken by *Cyclone Amphan* in *2020.*

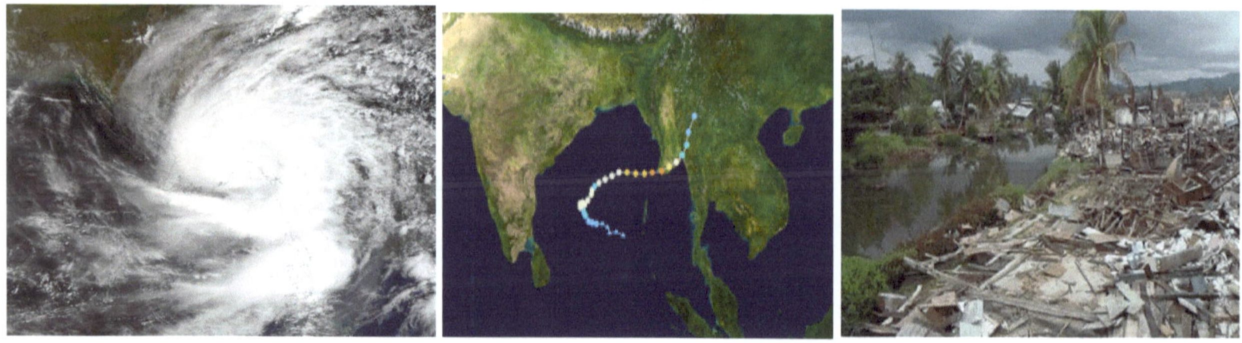

Year	D	DD	SC	SCS	VSCS	ESCS	SuSC	Strongest Cyclone
2010-2019	99	68	43	27	21	10	1	Kyarr

Super Cyclonic Storm Kyarr was an extremely powerful *tropical cyclone* that became the first *super cyclonic storm* in the North Indian Ocean since *Gonu* in *2007*. *It was also the second strongest tropical cyclone in the Arabian Sea. Kyarr developed the Equator. The system organized itself and intensified to a tropical storm on October 24 as it moved eastwards. It became Super Cyclonic Storm on October 27, as it turned westward. On that same day, Kyarr peaked as a Super Cyclonic Storm, with maximum winds of 240 km/h. Afterward, Kyarr gradually began to weaken, while curving westward, and then turning to the southwest. On October 31, Kyarr weakened into a Deep Depression, before turning southward on November 2, passing just to the west of Socotra. Kyarr degenerated into a remnant low later that day, before dissipating on November 3, just off the coast of Somalia. Despite the immense strength of the storm, and many countries being affected by high tides and storm surges, there were no reported fatalities.*

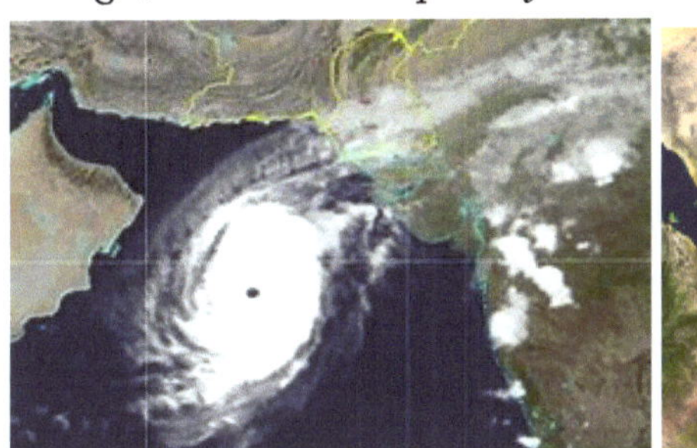 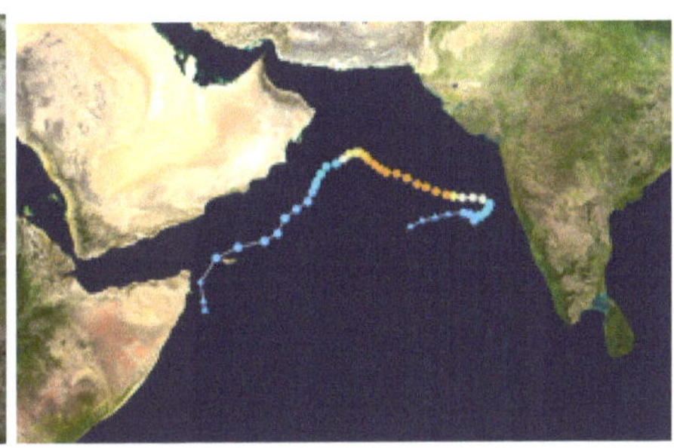

Year	D	DD	SC	SCS	VSCS	ESCS	SuSC	Strongest Cyclone	No. of Death	Remarks
2020	9	6	5	4	3	1	1	SuSC Amphan	269	First SuSC in the Bay of Bengal since <u>1999</u>. Featured the costliest cyclone ever recorded in North Indian Ocean
2021	10	6	5	3	2	1	0	ESCS Tauktae	230	
2022	15	7	3	2	0	0	0	SCS Asni	79	
2023 (Till May 2023)	2	1	1	1	1	1	0	ESCS Mocha		
Total	**36**	**20**	**14**	**10**	**6**	**3**	**1**	**Amphan**		

The Super Cyclonic Storm Amphan *was the first SuCS over the BoB, after the Odisha SuCS of 1999. south Andaman Sea and adjoining southeast Bay of Bengal (BoB) on 13th May. It concentrated into a cyclonic storm 16th May. Moving nearly northwards, it further intensified into a Severe Cyclonic Storm over southeast BoB on 17th May. Further, it intensified into a Very Severe Cyclonic Storm on the same day,, Extremely Severe Cyclonic Storm on 18th and into a Super Cyclonic Storm around noon of 18th May. It maintained the intensity of Super Cyclonic Storm over west-central BoB for nearly 24 hours, before weakening into an Extremely Severe Cyclonic Storm over west-central BoB around noon of 19th May. Thereafter, it weakened slightly and crossed West Bengal – Bangladesh coasts as a Very Severe Cyclonic Storm, across Sundarbans on 20th May with maximum sustained wind speed of 155 – 165 kmph gusting to 185 kmph. Then it gradually moved north-northeast ward. It moved very close to Kolkata during this period. Moving further north-northeas tward, it weakened into an SCS over Bangladesh & adjoining West Bengal. Later it rapidly weakened over Bangladesh on 21st May.*

34. **Forecasting**. With development in various infrastructure like weather satellite, weather radar and advancement in Numerical Weather Prediction (NWP) models, forecasting on the track and intensity of tropical cyclones has been quite accurate these days. Most of the countries located in the tropical regions have established separate Met Station for monitoring the tropical systems. However, the conventional method of analysing various charts like surface chart, upper air chart, prognostic charts, study of prevailing thermodynamic situation, etc are still regarded as important tools for forecasting tropical cyclones.

35. In India, the responsibilities of issue of forecast and warning in respect of tropical storm is carried out by India Met Department (IMD) which is an organization under Ministry of Earth Science. Land-based surface and upper-air stations, Doppler Weather Radars (DWRs), satellites, ships and buoys (Buoys are the anchored floating observation stations installed over sea) are the observational networks that functions under IMD. The cyclone warning organisation of IMD has a three-tier system to cater to the needs of states adjacent to sea at national, regional and local levels. The agencies carry out international responsibility also. The three Area Cyclone Warning Centres (ACWCs) are located at Chennai, Mumbai and Kolkata and four Cyclone Warning Centres (CWCs) functioning under ACWCs are located at Visakhapatnam, Ahmedabad, Bhubaneswar and Thiruvananthapuram. The ultimate responsibility for operational storm warning work for the respective areas rests with the ACWCs and CWCs.

36. Area of Responsibility of ACWC/CWC is shown below: -

ACWC/CWC	Sea Area	Coastal Area (75 Km from Coast Line)	Maritime State
Kolkata ACWC	Bay of Bengal	West Bengal, Andaman & Nicobar Islands	West Bengal, Andaman & Nicobar Islands
Chennai ACWC	--	Tamil Nadu and Pondicherry	Tamil Nadu, Pondicherry and Karaikal
Thiruvananthapuram CWC	--	Kerala and Karnataka	Kerala, Mahe, Karnataka and Lakshadweep
Mumbai ACWC	Arabian Sea	Maharashtra and Goa	Maharashtra and Goa
Bhubaneshwar CWC	--	Odisha	Odisha
Visakhapatnam CWC	--	Andhra Pradesh	Andhra Pradesh and Yanam
Ahmedabad CWC	--	Gujarat, Diu, Daman, Dadra & Nagar Haveli	Gujarat, Diu, Daman, Dadra & Nagar Haveli

37. The World Meteorological Organization (WMO) serves as a platform for the information sources for tropical cyclone forecasters to obtain data and tools which are useful for monitoring and forecasting of tropical cyclones. Regional specialized Meteorological Centres (RSMC) have been established in different parts of the tropical regions from which WMO collects input and utilizes for bringing improvements in the forecasting techniques. The RSMCs are shown in the following picture.

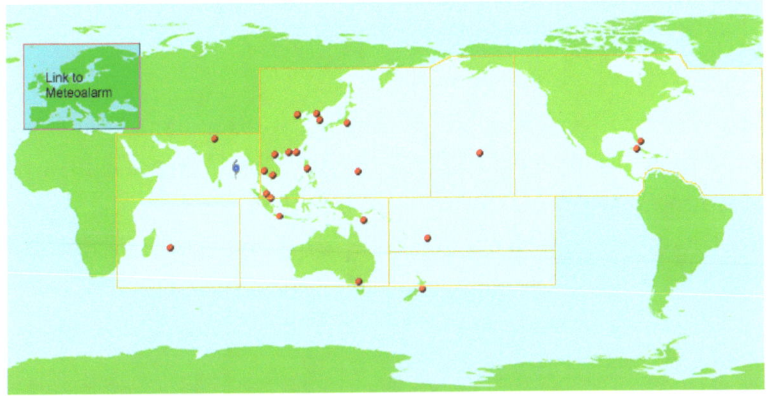

38. The above is a picture available in the WMO (RSMC) site where a link is given to the location of any prevailing tropical cyclone. The image is of 12 May 23 when a tropical cyclone was located over Bay of Bengal. On clicking on the symbol of tropical cyclone all information in respect of intensity, movement, landfall, etc, are shown.

39. Various RSMCs are listed below: -

Area of Responsibility	RSMC
Caribbean Sea, Gulf of Mexico, North Atlantic and eastern North Pacific Oceans	Miami, South Florida
Western North Pacific Ocean and South China Sea	Tokyo, Japan
Bay of Bengal and the Arabian Sea	New Delhi, India
South-West Indian Ocean	La Reunion, Mascarene
South-West Pacific Ocean	Nadi, Fiji
Central North Pacific Ocean	Honolulu

40. The area of responsibility of RSMC- New Delhi covers Sea areas of north Indian Ocean north of equator between 450 E and 1000 E and includes the member countries of WMO/ESCAP Panel on Tropical Cyclones viz, Bangladesh, Maldives, Myanmar, Pakistan, Sri Lanka, Oman, Yemen, Thailand, Iran, Saudi Arabia, Qatar and UAE.

Fig 2.2 Area of responsibility of RSMC- Tropical Cyclones, New Delhi

The area of responsibility of RSMC- New Delhi covers Sea areas of north Indian Ocean north of equator between 45° E and 100° E and includes the member countries of WMO/ESCAP Panel on Tropical Cyclones viz, **Bangladesh, Maldives, Myanmar, Pakistan, Sri Lanka, Oman, Yemen, Thailand, Iran, Saudi Arabia, Qatar and UAE** as shown in Fig.2.2

41. **Nomenclature**. Tropical cyclones can last for a week or more. Therefore, there can be more than one cyclone at a time. Weather forecasters give each tropical cyclone a name to avoid confusion. In general, tropical cyclones are named according to the rules at regional level. In the Atlantic and in the Southern hemisphere (Indian ocean and South Pacific), tropical cyclones receive names in alphabetical order, and women and men's names are alternated. Nations in the Northern Indian ocean began using a new system for naming tropical cyclones in 2000. The names are listed alphabetically country wise, and are neutral gender wise.

The common rule is that the name list is proposed by the National Meteorological and Hydrological Services (NMHSs) of WMO Members of a specific region, and approved by the respective tropical cyclone regional bodies at their annual/biannual sessions.

42. **Nomenclature over North Indian Ocean Region**. Within the North Indian Ocean between 45°E – 100°E, tropical cyclones are named by the India Meteorological Department (RSMC New Delhi) when they are judged to have intensified into cyclonic storms. If a cyclonic storm moves into the basin from the Western Pacific, then it will keep its original name. However, if the system weakens into a deep depression and subsequently re-intensifies after moving into the region, then it will be assigned a new name. In May 2020, the naming of cyclone Amphan exhausted the original list of names established in 2004. A new list of names has been prepared and is being used in alphabetical order for storms after Amphan.

List	Contributing Nation						
	Bangladesh	**India**	**Iran**	**Maldives**	**Myanmar**	**Oman**	**Pakistan**
1	Nisarga	Gati	Nivar	Burevi	Tauktae	Yaas	Gulab
2	Biparjoy	Tej	Hamoon	Midhili	Michaung	Remal	Asna
3	Arnab	Murasu	Akvan	Kaani	Ngamann	Sail	Sahab
4	Upakul	Aag	Sepand	Odi	Kyarthit	Naseem	Afshan
5	Barshon	Vyom	Booran	Kenau	Sapakyee	Muzn	Manahil
6	Rajani	Jhar	Anahita	Endheri	Wetwun	Sadeem	Shujana
7	Nishith	Probaho	Azar	Riyau	Mwaihout	Dima	Parwaz
8	Urmi	Neer	Pooyan	Guruva	Kywe	Manjour	Zannata
9	Meghala	Prabhanjan	Arsham	Kurangi	Pinku	Rukam	Sarsar
10	Samiron	Ghurni	Hengame	Kuredhi	Yinkaung	Watad	Badban
11	Pratikul	Ambud	Savas	Horangu	Linyone	Al-jarz	Sarrab
12	Sarobor	Jaladhi	Tahamtan	Thundi	Kyeekan	Rabab	Gulnar
13	Mahanisha	Vega	Toofan	Faana	Bautphat	Raad	Waseq

List	Contributing Nation					
	Qatar	**Saudi Arabia**	**Sri Lanka**	**Thailand**	**U.A.E.**	**Yemen**
1	Shaheen	Jawad	Asani	Sitrang	Mandous	Mocha
2	Dana	Fengal	Shakhti	Montha	Senyar	Ditwah
3	Lulu	Ghazeer	Gigum	Thianyot	Afoor	Diksam
4	Mouj	Asif	Gagana	Bulan	Nahhaam	Sira
5	Suhail	Sidrah	Verambha	Phutala	Quffal	Bakhur
6	Sadaf	Hareed	Garjana	Aiyara	Daaman	Ghwyzi
7	Reem	Faid	Neeba	Saming	Deem	Hawf
8	Rayhan	Kaseer	Ninnada	Kraison	Khubb	Balhaf
9	Anbar	Nakheel	Viduli	Matcha	Khubb	Brom
10	Oud	Haboob	Ogha	Mahingsa	Degl	Shuqra
11	Bahar	Bareq	Salitha	Phraewa	Athmad	Fartak
12	Seef	Alreem	Rivi	Asuri	Boom	Darsah
13	Fanar	Wabil	Rudu	Thara	Saffar	Samhah

CHAPTER 23

OPTICAL PHENOMENA IN ATMOSPHERE

1. Atmospheric optical phenomena are visual events that take place in Earth's atmosphere as a consequence of light reflection, refraction, and diffraction by solid particles, liquids droplets, and other materials present in the atmosphere. Such phenomena include a wide variety of events like rainbows, halos, coronas, etc. These phenomena have no significance weather wise, nor they have any impact on aviation. However, they occasionally provide useful information on the constitution of cloud particles and enable an aviator in the air to identify cloud types. Brief descriptions of some of the optical phenomena are given in the following paragraphs.

2. **Halos.** A halo is an optical phenomenon produced by the interaction of light from the sun or moon with ice crystals in the atmosphere, resulting in coloured or white arcs, rings or spots in the sky. They are produced by the refraction and reflection of sunlight or moonlight by prismatic ice crystals and hence in most cases by some form of cirrus cloud. The commonest case is a circle of light of angular radius 22° with the Sun or Moon in the centre. When well-developed, the halo round the Sun shows a pure clear red on the inside but other colours are usually difficult to recognise. The presence of a halo signifies the predominance of ice crystals in the cloud. The clouds that give rise to this phenomenon are cirrostratus or thick cirrus which are composed of ice crystals.

3. **Coronas.** In addition to reflection and refraction, the path of a light rays can be altered by yet a third mechanism, diffraction. Diffraction occurs when a light ray passes close to some object. Diffraction may also result in the separation of white light into its coloured components. When light rays from the Sun or the Moon pass through a thin cloud, they may be diffracted. Interference of the various components of white light that generates the colours makes up the corona. The pattern formed by the diffraction rays is a ring around the Sun/Moon. The colours are usually dull. Red appears on the outside and blue on the inside. Coronas are caused by altostratus, nimbostratus and dense altocumulus clouds which predominantly contain water droplets.

4. **Glory.** A glory is similar to a corona but is most commonly observed during an airplane ride. As sunlight passes over the airplane, it may fall on water droplets in a cloud below. The light that is diffracted then forms a series of coloured rings around the airplane's shadow.

5. **Sundogs**. A sundog consists of glowing spots around the sun. They are created by sunlight refracting off plate-shaped ice crystals in the cirrus clouds. Sundogs are some of the most frequently observed optical phenomena and can be observed throughout the year and anywhere in the world. Sundogs tend to be most visible when the Sun is close to the horizon.

6. **Rainbows**. The most common optical phenomenon in the atmosphere is a rainbow. A rainbow is a circle or an arc of circle, of coloured light. The radius of the circle makes an angle of 42°. The bow results from refraction of the Sun's rays both on entering and on leaving a raindrop, together with one total internal reflection. The colours of the rainbow are due to the different angles of refraction for the different colours which constitute white sunlight. The colours are those of the visible spectrum of white light from red on the outside to violet on the inside.

7. A secondary rainbow, concentric with the primary is occasionally seen but with a radius of about 52° and the colours in the reversed order. The picture below shows both primary and the secondary rainbows.

8. **The Blue Sky (Scattering of Light).** White Sun light that bounces off small objects is not reflected uniformly, but is scattered in all directions. When light from the Sun collides with molecules of oxygen and nitrogen, it is scattered. Light with shorter wavelengths (blue, green, indigo, and violet) is scattered more strongly than is light with longer wavelengths (red, orange, and yellow). No matter where a person stands on Earth's surface, he is more likely to see the bluish light scattered by air molecules than the light of other hues. The process of scattering is responsible for the fact that humans observe the sky as blue.

9. **Aurora**. Near the poles, beautiful light shows in the sky can be seen frequently. These lights are called **auroras**. Aurorae are highly colourful and spectacular displays in the upper atmosphere in the height band 65 to 1000 Km in the polar and sub-polar regions. They are, in fact, disturbances in the upper atmosphere caused when the ionised constituents are excited by incoming solar energy particles.

10. Such display of lights takes place in Ionosphere which exists in the lower thermosphere. In Ionosphere, the atmosphere is in a highly ionized state. The concentration of ions in this layer causes reflection of radio waves resulting in the beautiful display of lights. The highest concentration of ions is found at about 110 Km, 160 Km and 250 Km above the Earth's surface.

Concentration of ions (charged particles) causes reflection of radio waves. This enables longwave radio communication possible.

11. **Parhelic Circle**. A parhelic circle is a type of halo, appearing as a horizontal white line on the same altitude as the sun, or occasionally the Moon. If complete, it stretches all around the sky, but more commonly it only appears in sections.

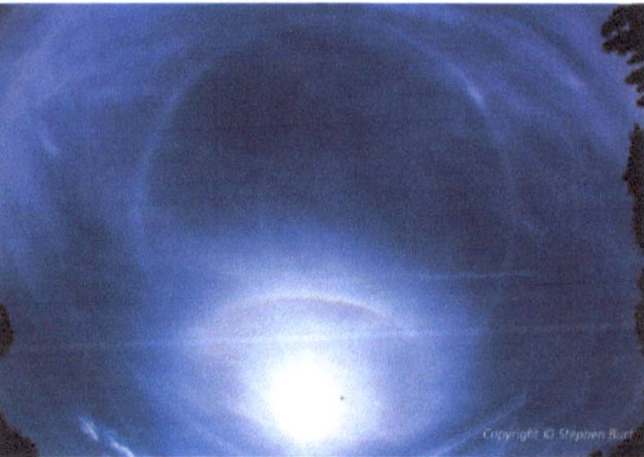

12. **Light or Sun Pillars**. In light pillar a vertical beam of light appears to extend above and/or below a light source. The effect is created by the reflection of light from tiny ice crystals that are suspended in the atmosphere or that comprise high-altitude clouds. If the light comes from the Sun (usually when it is near the horizon), the phenomenon is called a sun pillar. Light pillars can also be caused by the Moon or earthly sources, such as streetlights and erupting volcanoes.

13. **Mirages**. When a layer of air very close to the ground is heated more strongly than the air immediately above it, the light rays pass through two transparent media, the hot & less dense air and the cooler & more dense air and are

refracted. As a result of the refraction, the blue sky appears to be present on Earth's surface; it may look like a body of water and objects such as trees appear to be reflected in that water. This optical phenomenon is commonly seen in the desert areas or other hot areas during day time.

14. **Twinkling of Stars.** Stars are so far away that their light reaches Earth's atmosphere as a single point of light. As that very narrow beam of light passes through Earth's atmosphere, it is refracted and scattered by molecules and larger particles of matter. Sometimes the light travels straight toward an observer and sometimes its path is deflected. To the observer, the star's light appears to alternate many times per second, which produces twinkling. Planets usually do not twinkle because they are closer to Earth and the light that reaches Earth from them consists of wider beams rather than narrow rays.

CLIMATOLOGY

CHAPTER 24

CLIMATE OF INDIA

1. There are a number of climatic regions in India which varies from tropical climate in the south to temperate and alpine climate in the Himalayan north. The elevated regions in the Himalayan north receive sustained winter snowfall. On the other hand, there are varieties of topographical features in the country which influence the climate. The Himalaya and the Thar desert plays the most important role in the climate. The Himalayas protect the country from the cold winds blowing from the Central Asian region. This keep north India warmer than the places of same latitude in other countries. The Thar desert in Rajasthan being warmer than the other places pulls the moisture laden winds from the Arabian Sea during the period from June to September/October due to which most places in the country receives monsoon rain during the period.

2. **Important Topographical and Geographical Features.** Located entirely in the northern hemisphere, the mainland extends to 3214 km from north to south and 2933 km from east to west. The land mass has diversified physiographic condition. Mountains occupy about 10.6%, hills occupy 18.5%, plateaus occupy 27.7% and the plains occupy 33.2%.

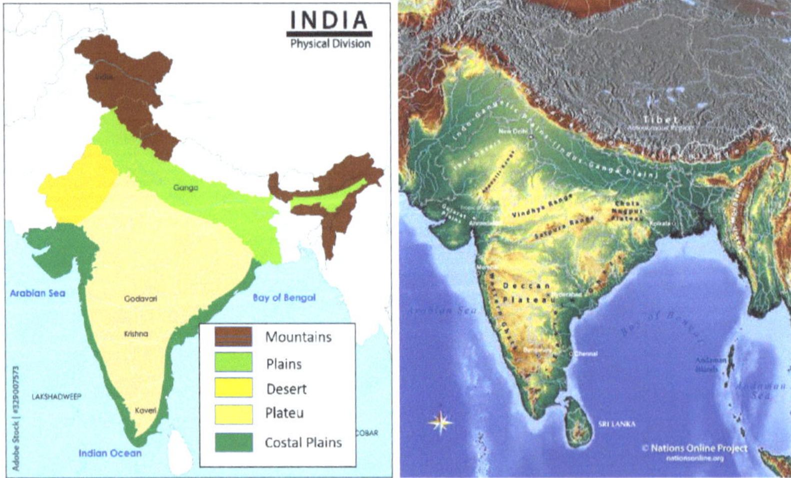

3. The topographical features which plays important role in the climate of India are briefly discussed in the following paragraphs.

4. **The Mountains.**

   a) **The Himalayas.** The Himalaya mountain range in India is split into the Great Himalaya, Middle Himalaya and Outer Himalaya, The Trans Himalaya Karakoram range, Trans-Himalaya Ladakh range, Trans-Himalaya Zanskar range, and the the Purvanchal range.

      i. The Great Himalaya stretches Jammu and Kashmir to Arunachal Pradesh. The highest peaks are Kanchenjunga in Sikkim (Shared with Nepal) and Nanda Devi in Uttarakhand with altitudes of 28,169 and 25,643 Ft respectively.

ii.   The Middle Himalaya mountain range runs parallel to the Great Himalaya on its southern side with the altitudes of the peaks ranging from 5,000 to 20,000 Ft.

iii.   The Outer Himalayas, also known as the Shivalik Range, is regarded as the Himalayan foothills. It separates the mountains from the plains, and consists of valleys and hills that rise about 5,000 feet above sea level. A large part of the range is located in Himachal Pradesh, up to the Beas River. It also encompasses Jammu, Uttarakhand, and sub-Himalayan West Bengal.

iv.   The Trans-Himalaya, to the north of the Great Himalaya in the Union Territory of Ladakh, is India's most isolated and remote mountain range. It's made up of the Karakoram, Zanskar and Ladakh ranges. The Karakoram Range extends north into the Gilgit-Baltistan. It has eight peaks over 24,600 Ft altitude. The tallest peak is K2 which is At 28,251 feet above sea level and is the second highest mountain in the world.

v.   The Ladakh Range lies to the south of the Karakoram Range, between the Nubra Valley and Leh. It runs parallel to the Indus River and extends to India's border with Tibet. The peaks in this range are about 16,400 to 19,700 feet above sea level.

vi.   The Purvanchal Range lies south of the Brahmaputra river in Arunachal Pradesh and forms the boundary between India and Myanmar. It extends along the Northeast Indian states and has a relatively low elevation that decreases towards the south. The average height of peaks in this range is about 9,845 feet above sea level. The highest is Dapha Bum, in the Mishmi hills with an altitude of 15,020 Ft located at the northeastern tip of Arunachal Pradesh. In Nagaland, the highest peak is Saramati in the Naga Hills at 12,550 feet above sea level. In the hills of Manipur, the elevation is generally less than 8,200 feet above sea level. The highest peak in Mizoram is Phawngpui, also known as Blue Mountain, at 7,080 feet above sea level in the Mizo Hills. However, the elevation of the Mizo Hills is generally less than 4,920 feet.

b)   **The Aravalli Range.** The Aravalli Range is about 500-mile-long which runs from Champaner and Palanpur in eastern Gujarat to the outskirts of Delhi. It borders the Thar desert and provides protection from the extreme desert climate. The highest peak is Guru Shikhar at Mount Abu, near the Gujarat border, with an elevation of 5,650 feet above sea level. Most of the hills are concentrated in the area around Udaipur.

c)   **The Vindhya Range.** The Vindhya Range runs across central India on the northern side of the Narmada River in Madhya Pradesh. It extends from Jobat in Gujarat to Sasaram in Bihar which is nearly 700 miles. It is not a single mountain range but chains of hills, ridges and plateaus. The general elevation of the Vindhya Range is around 980-2,100 feet above sea level, with the tallest Kalumar Peak, at 2,467 feet above sea level in the Damoh district of Madhya Pradesh.

d)   **The Satpura Range.** On the southern side of the Narmada River in Madhya Pradesh, the Satpura Range runs parallel to the Vindhya Range in between the Namarda and Tapti rivers. It extends for about 560 miles from the Rajpipla Hills in Gujarat to the Maikala Hills in Chhattisgarh where it meets the Vindhya Range at Amarkantak. The Satpura Range is higher than the Vindhya Range, with peaks reaching over 4,000 feet in the heavily forested Mahadeo Hills at Pachmarhi. The highest one is Dhupgarh, at 4,400 feet above the sea level. This is the tallest peak in central India.

e)   **The Western Ghats.** The lengthy Western Ghats runs for approximately 5,250 miles along the western side of India, separating the coast from the Deccan plains. It extends from near the Satpura Range in Gujarat down through Maharashtra, Goa, Karnataka, Kerala and Tamil Nadu to end at the southernmost tip of India near Kanyakumari. The Western Ghats is made up of multiple mountain ranges, with more than 70 peaks varying in height from 1,713 feet to 8,842 feet above sea level. Almost a third of them are above 6,561 feet, with most of these being in Kerala. The highest is Anamudi, in the Anaimalai Hills on the Kerala-Tamil Nadu border. Other major ranges in the Western Ghats are the Sahyadri mountains in Maharashtra, Cardamom Hills in Kerala, and Nilgiri mountains in Tamil Nadu. These mountains influence India's weather by acting as a barrier against the southwest monsoon clouds and drawing much of the rainfall.

f) **The Eastern Ghats.** The Eastern Ghats separates the coast from the plains on the eastern side of India. It runs through Odisha, Andhra Pradesh and Tamil Nadu where it meets the Western Ghats at the Nilgiri mountains. The Eastern Ghats is flatter than the Western Ghats, and its hills have been divided into several parts by the major rivers like the Godavari, Mahanadi, Krishna, and Kaveri. It has a few peaks over 3,280 feet above sea level and are mainly in the Maliya Range in Odisha and Madugula Konda Range in Andhra Pradesh. The highest is Jindhagada Peak in Andhra Pradesh, with an elevation of 5,545 feet.

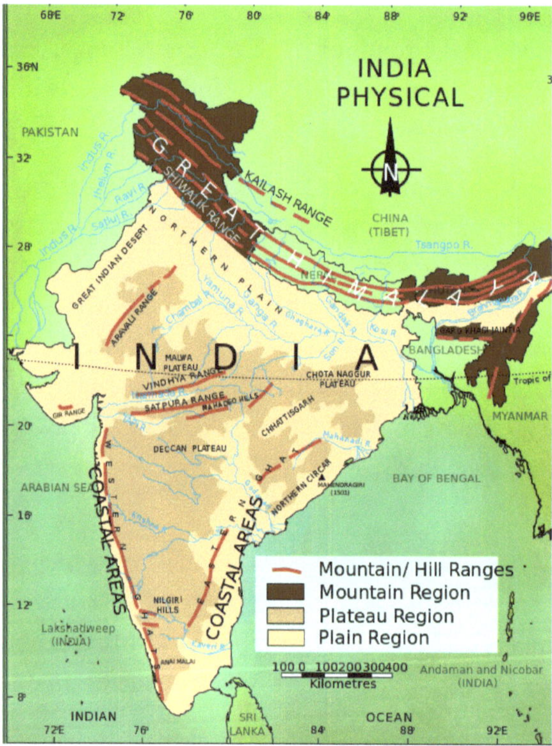

5. **The Indo-Gangetic Plains.** The Indo-Gangetic Plain, also known as the North Indian River Plain, is a 7,00,000 km² fertile plain encompassing northern regions of the Indian subcontinent, including most of northern and eastern India, most of eastern-Pakistan, virtually all of Bangladesh and southern plains of Nepal. Also known as the Indus–Ganga Plain, the region is named after the Indus and the Ganges rivers and encompasses a number of large urban areas. The plain is bound on the north by the Himalayas, which feed its numerous rivers and are the source of the fertile alluvium deposited across the region by the two river systems. The southern edge of the plain is marked by the Deccan Plateau. On the west rises the Iranian Plateau.

6. The Indo-Gangetic Plain is divided into two drainage basins by the Delhi Ridge; the western part drains to the Indus, and the eastern part consists of the Ganga–Brahmaputra drainage systems. This divide is only 350 m above sea level, causing the perception that the Indo-Gangetic Plain appears to be continuous from Sindh in the west to Bengal and Assam in the east.

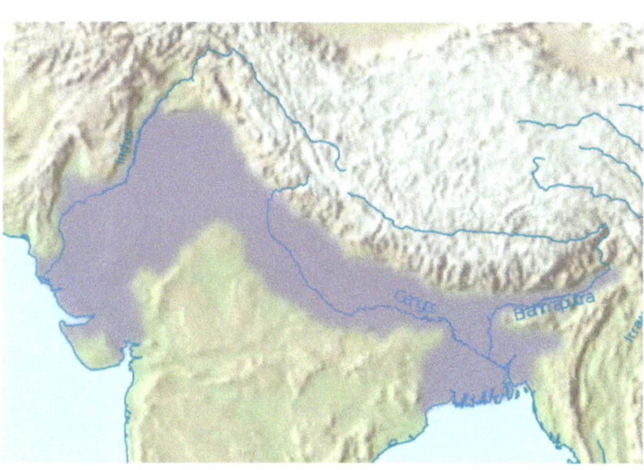

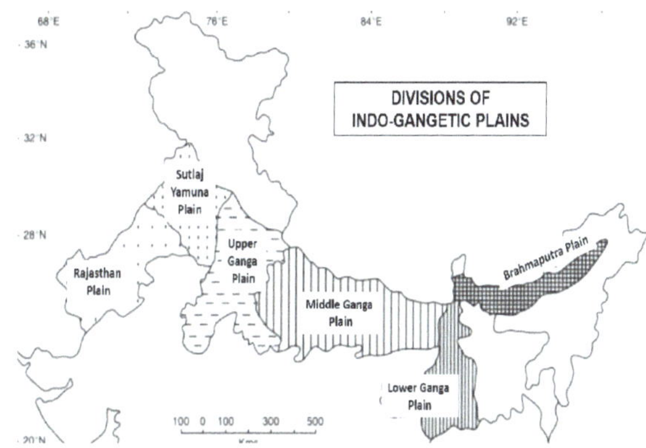

7. **The Thar Desert**. The Thar Desert, also known as the Great Indian Desert, is located in the north-western part of the Indian subcontinent that covers an area of 2,00,000 km² in India and Pakistan. About 85% of the Thar Desert is in India, and about 15% is in Pakistan. More than 60% of the desert lies in Rajasthan. The portion in India also extends into Gujarat, Punjab, and Haryana. The Indo-Gangetic Plain lies to the north, west and northeast of the Thar desert, the Rann of Kutch lies to its south, and the Aravali Range borders the desert to the east.

8. **The Deccan Plateau**. The large Deccan Plateau or the Peninsular Plateau of the Indian Subcontinent is located between the Western Ghats and the Eastern Ghats. It extends to the north up to the Satpura and Vindhya Ranges. The plateau is drier than the coastal region of southern India and is arid in places.

9. **The Coastal belt**. Coastal belt spans from the south west Indian coastline along the Arabian sea from the coastline of the Gulf of Kutch in its westernmost corner and stretches across the Gulf of Khambhat, the Konkan and southwards up to Cape Comorin in the southernmost region of South India. The coastline on the South Eastern part of the Indian Subcontinent along the Bay of Bengal extends from the easternmost Corner of shoreline near the Sunderbans in Coastal East India.

Coastal Belt of India

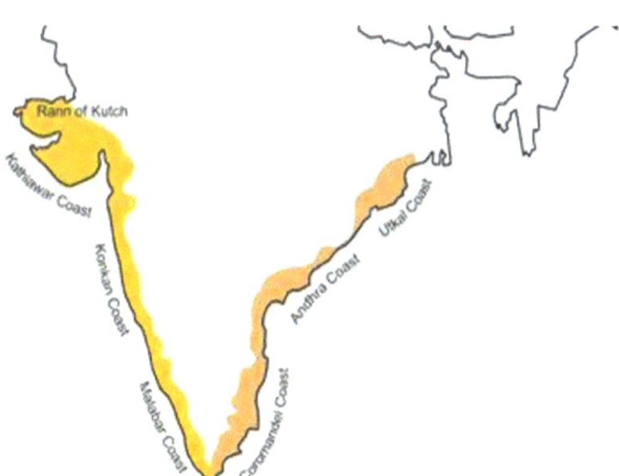

10. **The Rivers**. The rivers of India are classified into four groups which are Himalayan rivers, Deccan rivers, Coastal rivers, and Rivers of the inland drainage basin. The Himalayan Rivers are formed by melting snow and glaciers and therefore, continuously flow throughout the year. The main Himalayan river systems are those of the Indus and the Ganga-Brahmaputra-Meghna system The Deccan Rivers on the other hand are rain fed and therefore fluctuate in volume. Many of these are non-perennial. The Coastal streams, especially on the west coast are short in length and have limited catchment areas. Most of them are non-perennial. The streams of inland drainage basin of western Rajasthan are few. Most of them are short lived. A few rivers in Rajasthan do not drain into the sea. They drain into salt lakes and get lost in sand with no outlet to sea. Besides these, there are the Desert Rivers which flow for

some distance and are lost in the desert. In the Deccan region, most of the major river systems flowing generally in east direction fall into Bay of Bengal. The major east flowing rivers are Godavari, Krishna, Cauvery, Mahanadi, etc. Narmada and Tapti are major West flowing rivers.

11. **The Islands.** There are 234 islands under India of which 204 are in the Bay of Bengal and 33 are in the Arabian Sea. The Andaman and Nicobar Islands and the Lakshadweep Islands are the major island groups.

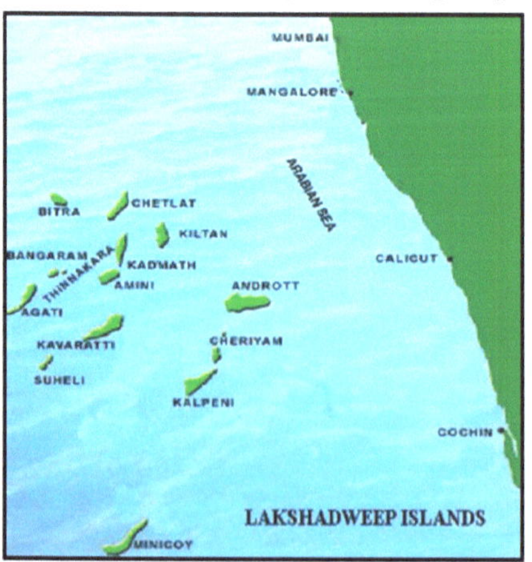

12. **Climate of India.** The climate of India comprises a wide range of climatic conditions across a large geographic area and different topographical features. In general, there are six major climatic subtypes in India, which range from desert in the west, to alpine tundra and glaciers in the north, to humid tropical regions supporting rainforests in the southwest and the island territories. The Himalayas act as a barrier to the cold and dry winds flowing down from central Asia. Thus, North India stays warm or only mildly cold during winter. In summer, the same phenomenon makes India relatively hot. Climate of India is broadly classified as below: -

a) **Tropical Wet Climate.** Also known as tropical monsoon climate, it is the most humid climate that covers a strip of southwestern lowlands adjoining the Malabar Coast, the Western Ghats; southern Assam, Lakshadweep Islands and Andaman and the Nicobar Islands. It is characterised by moderate to high year-round temperature, even in the foothills. The rainfall is seasonal but heavy, typically above 2,000 mm per year. Most rainfall occurs between May and November. December to March are the driest months.

b) **Tropical Wet and Dry Climate.** Most of inland peninsular India except for a semi-arid rain shadow east of the Western Ghats commonly experiences a tropical wet and dry climate. It is significantly drier than tropical wet zones. Long Winter and early summers typically bring dry periods with temperatures averaging above 18 °C (64 °F). The summer season is extremely hot in the low lying areas where maximum temperature may cross 450 C during May leading to heat waves condition.[8] The rainy season lasts from June to September; annual rainfall averages between 750–1500 millimetres across the region. Once the dry northeast monsoon begins in September, most precipitation in India falls on Tamil Nadu, leaving other states comparatively dry.

c) **Tropical Dry.** A tropical arid and semi-arid climate dominates regions where the rate of moisture loss through evapotranspiration exceeds gain of moisture from precipitation. it is subdivided into three climatic subtypes.

    i. A tropical semi-arid steppe climate, predominates over a long stretch of land south of Tropic of Cancer and east of the Western Ghats and the Cardamom Hills. This region, which includes Karnataka, inland Tamil Nadu, western Andhra Pradesh & Telangana, and central Maharashtra, gets between 400–750 millimetres of rainfall annually. The regions are drought-prone due to delay in monsoon or weak monsoon conditions. North of the Krishna river, the summer monsoon brings most the rainfall. To the south, significant post-monsoon rainfall also occurs in October and November. In December, the coldest month, temperatures are still around 20–24°C. March to May experience hot and dry weather with mean monthly temperatures around 32°C, with 320 mm of precipitation. Without artificial irrigation, that region proves unsuitable for agriculture.

    ii. Most of western Rajasthan experiences an arid climatic regime. The annual rainfall over the region is mainly from the Induced Western Disturbances, occasional westward shifting of monsoon activities and on rare occasion, movement of tropical systems from the Arabian Sea. Mean annual rainfall is less than 300 mm. The summer months of May and June prove exceptionally hot with mean monthly temperatures in the region being around 35 °C. The maximum temperature during this period occasionally exceeds 50 °C. During winters, temperatures in some areas can drop below freezing due to waves of cold air from Central Asia. In both summer and winter, the diurnal variation of temperature is very high. It is of the order of about 140 C in summer and in winter, the variation is even higher.

    iii. East of the Thar Desert, the region running from Punjab and Haryana to Kathiawar experiences a tropical and sub-tropical steppe climate. The zone, a transitional climatic region separating tropical desert from humid sub-tropical savanna and forests, experiences temperatures less extreme than those of the desert. Average annual rainfall measures 300–650 mm. However, there is wide variation in the different years. Daily summer temperature maxima rise to around 40 °C.

e) **Sub-Tropical Humid.** Most of Northeast India and much of North India are subject to this climate. Though they experience hot summers, temperature during the coldest months may fall to as low as 0°C. Annual rainfall ranges from less than 1,000 mm in the west to over 2,500 mm in parts of the northeast. As most of this region is far from ocean, there is wide temperature swings characterising continental climate. The winter season is generally dry owing to powerful anticyclonic and katabatic (downward-flowing) winds from Central Asia. Winter rainfall is associated with Western Disturbances. Powerful thunderstorm associated with rain occurs during the pre-monsoon season and most of the rainfall collection takes place during the monsoon months.

f) **Montane or Alpine Climate**. India's northernmost areas are subject to a montane or alpine climate. In the Himalayas, sharp temperature contrasts exist between sunny and shady slopes, high diurnal temperature variability, temperature inversions and altitude-dependent variability in rainfall are also common. The northern side of western Himalayas have a cold desert type of climate. It is a region of barren, arid, frigid and wind-blown wastelands.

Areas south of Himalayas are largely protected from cold winter winds coming in from the Asian interior. The leeward side (northern face) of the mountains receives less rain. The southern slopes of western Himalayas, well-exposed to monsoon, receive heavy rainfall. Areas situated at elevations of 1070 m to 2290 m receive the heaviest rainfall, which decreases rapidly at elevations above 2290 metres. Most precipitation occurs as snowfall during late winter and spring months. Himalayas experience their heaviest snowfall between December and February and at elevations above 1500 m. Snowfall increases with elevation by up to several dozen millimetres per 100 metres increase. Elevations above 6000 metres never experience rain since all precipitation falls as snow.

13. **Factors Affecting Climate of India**.

    a) **Location**. The Indian subcontinent stretches from 8°N to 37°N and is located to the north of the equator. Tropic of Cancer passes across the center of the country. Hence, the southern areas are closer to the equator and experience higher temperatures. While the northern parts of the country experience lower temperatures comparatively. However, during the hot season (Pre-monsoon) the dry and hot continental westerly or north-westerly continental winds raises the temperature over north India often leading to heat wave or severe heat wave conditions. The temperatures are quite low during the winters. The presence of the Arabian Sea and the Bay of Bengal, cause the east and west coasts of the country to be humid and mild.

    b) **Effect of Sea** The closer the regions to the sea, the more humid is the climate. They experience moderate summers and mild winters. However, areas situated far away from the coastline, have little influence of the sea and hence, experience extreme climatic conditions. The climate becomes more humid as one gets closer to the sea.

    c) **The Himalayan Mountains**. These mountains are a climatic divider between Central Asia and the Indian subcontinent. They do not allow the cold Central Asian winds to enter the continent and keep it warmer than other regions. The mountain ranges also keep the south-west monsoon winds confined to south of them.

    d) **Atmospheric Pressure and Wind pattern**. Mainly two major pressure and wind pattern prevail over the country in a year. During winter season the high-pressure belt prevails over the middle latitudes and the low pressure area prevail to the south over the Indian Ocean. The country experiences north-easterly winds flowing from the subtropical high-pressure belt of North towards the equatorial low-pressure areas during the winter season. These winds carry very little moisture since they flow only over land. Hence, they do not bring any rain to the county. However, these continental winds, while travelling across the Bay of Bengal, get saturated with moisture and cause rainfall over south east coast, that is, southern portion of the east coast including the southern peninsular region during the winter season. On the other hand, these dry continental winds are also cold as they flow from the colder mid-latitude regions. Due to this the north Indian regions remain comparatively colder during the winter season, at times leading to cold or severe cold wave conditions.

14. In summers this reverses and low pressure is created in interior Asia. Hence the southwest monsoon winds are originated and because these winds flow over the warm oceans they collect moisture and bring the majority of rainfall in the country.

CHAPTER 25

SEASONS OF INDIA

1. **The Monsoon**. The term monsoon refers to the wind pattern which undergoes a periodic reversal. Monsoon circulations affect weather over vast areas. In the northern summer (July), monsoons are prevalent over a large part of the Indian, Pacific and Atlantic Oceans, with southern hemispheric air penetrating well into the northern hemisphere in India, Southeast Asia and parts of Africa. In the northern winter (January), northern hemispheric air penetrates into South America (Brazil), Eastern Africa and Northwest Australia. The monsoon regions are the areas where the air of the colder hemisphere penetrates into the warmer hemisphere. The main areas thus influenced are Indian subcontinent, Southeast Asia, Japan, Southeast China, West Africa, Eastern Africa, Northwest Australia, Southeast USA.

2. **The Asian Monsoons**. The Asian monsoon region stretches from latitudes 10°S to 50 °N. On close examination, it is seen that three distinct circulations hold sway over this region. These can be termed as Indian, Malayan and Japanese monsoons. The seasonal reversal of winter continental winds and summer maritime winds gives rise to changes in the temperature and precipitation patterns. The countries affected are India, Pakistan, Bangladesh, Myanmar, Thailand, Cambodia, Laos, Vietnam, Malaya, Singapore, China and Japan.

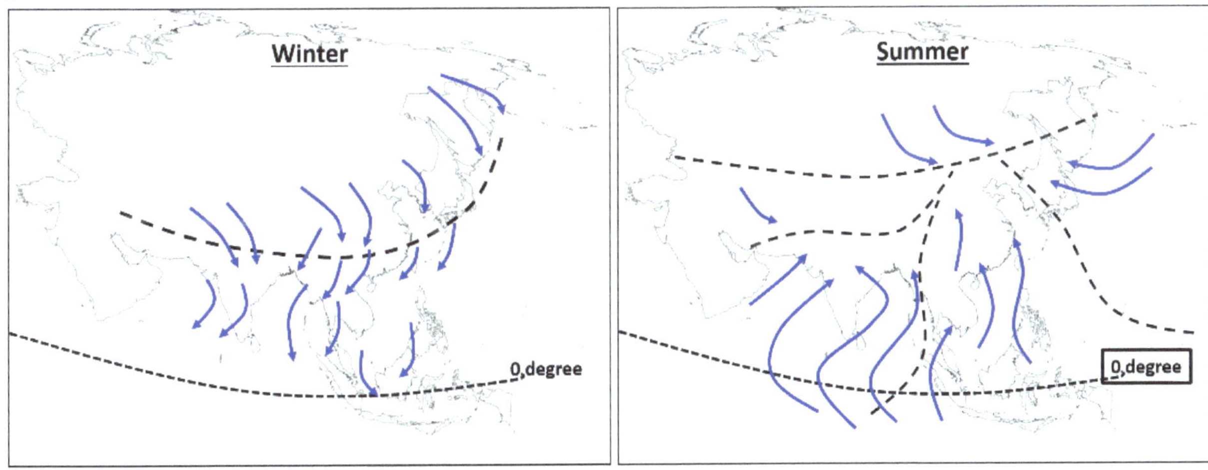

3. **The Indian Monsoon**. To the general public in India, monsoon means the rainy season. In the Indian subcontinent there are two monsoons, the southwest monsoon which affects practically the whole subcontinent and the northeast monsoon which causes rainfall only in the south-east Peninsular region, though occasionally, entire south peninsula receives rainfall. The Indian monsoon is a part of the Asian monsoon and it needs to be viewed in the context of the entire Indian Ocean. The Indian Ocean area can be considered to be an enclosed region with the Asian massif on the north and extending to the icy Antarctica in the south. The African land mass encloses it on the west while it is partly enclosed by Australia and the Malayan archipelago to the east. The unique build-up of the monsoon is caused by this geography with open seas from the tropic of cancer to the Antarctic and the Himalayan and Burmese mountain ranges to the north and the east.

4. **The** most prominent of the world's monsoon systems is the Indian Monsoon, which primarily affects India and its surrounding regions. It blows from the northeast during cooler months and reverses direction to blow from the southwest during the warm months of the year. This process brings large amounts of rainfall to the region during June to September.

5. The Indian Ocean area can be considered to be an enclosed region with the Asian landmass on the north and extending to the Antarctica in the south. To the west is the African land mass and it is partly enclosed by Australia and the Malayan group of islands to the east. The unique buildup of the monsoon is caused by this geography with open seas from the tropic of cancer to the Antarctic and the Himalayan and Burmese mountain ranges to the north and the east. These monsoon systems are shown in the figures below.

SW Monsoon   NE Monsoon

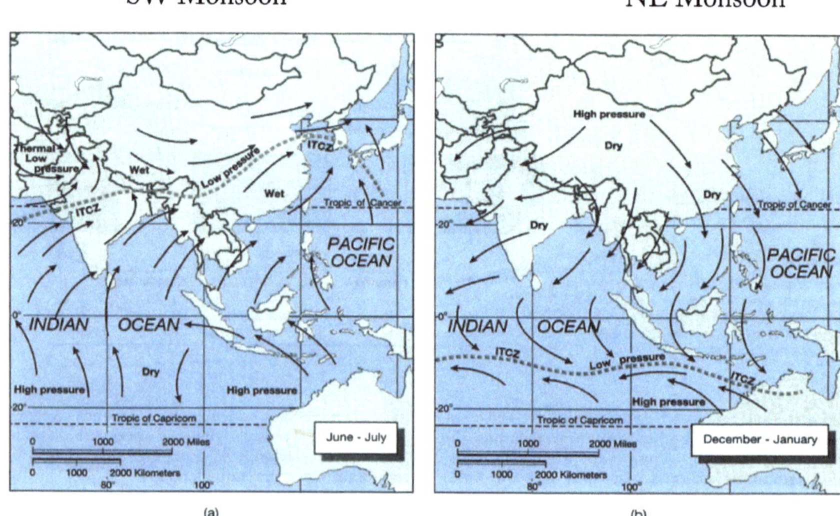

(a)   (b)

CHAPTER 26

SOUTHWEST MONSOON

1. The name 'Southwest Monsoon' is used both for the phenomena of rains and southwesterly surface winds and the period during which they occur. It is believed that the root word of monsoon is 'Mausam' in Arabic which simply means 'Season. However, in Meteorology, monsoon or monsoonal effect has come to mean the seasonal reversal of wind. Over most of India and southeast Asia, vast differences are noticed from preceding or following months of the winter (also called 'Northeast Monsoon'). The mean rainfall over the Indian plains in the southwest monsoon period of June to September is above 900 mm and for the rest of the year it is less than 150 mm. Winds are southwesterly over the Bay and the Arabian Sea in July compared to northeasterly in January. The monsoon air mass is maritime and moist in great depth as against the winter dry continental air. Temperatures in the troposphere decrease to north in winter, while monsoon reverses this gradient. The westerly jet stream of the other seasons is replaced by the easterly jet stream. Most of the annual rainfall of the densely populated areas of the world, viz. India, southeast Asia, Japan and parts of China is due to the phenomenon of the southwest monsoon.

2. **Surface Pressure Distribution.** Development of a low due to increased heating over India starts in March itself. By April, low pressure areas begin to establish roughly along 10° N in north Africa and about the Tropic of Cancer in the Indian region and Myanmar. The Peninsular India south of 15^0 N remains under the influence of maritime winds and hence, the formation of the low pressure is prominently seen north of 20^0 N. The sub–tropical high in the south Indian Ocean is along 30° S. By May, the summer continental low pressure areas completely dominate north Africa and Asia. Its main centre over India is near 30° N, 75° E with an extension as a trough up to Odisha and further to northeast Bay of Bengal. The heat low is more marked in June with the main centre over Pakistan. Monsoon activity is maximum in July when the low-pressure area over central Pakistan and adjoining regions is most intense. A trough lies over north India with axis from Sriganganagar to the Head Bay, which is referred to as the 'monsoon trough'. Pressure gradient is strong south of this trough. The Indian Ocean 'High' has strengthened and is centred at about 30° S, 60° E. Pressure continuously decreases over the Indian Ocean northwards of this high-pressure belt. In August, the intensity of the Afro–Asian low decreases. By September, pressures rise north of 40° N over Asia. The pressure gradient south of monsoon trough also decreases significantly in September. By October, with further increase in pressure, the pressure field is flat over the country. The Asian 'High' is establishes along 50° N and is centred at about 90° E.

3. It can be summarized by stating that there are three chief features during the southwest monsoon season. They are, the low-pressure area over Pakistan, the associated trough extending southeast ward from the low pressure and the strong pressure gradient

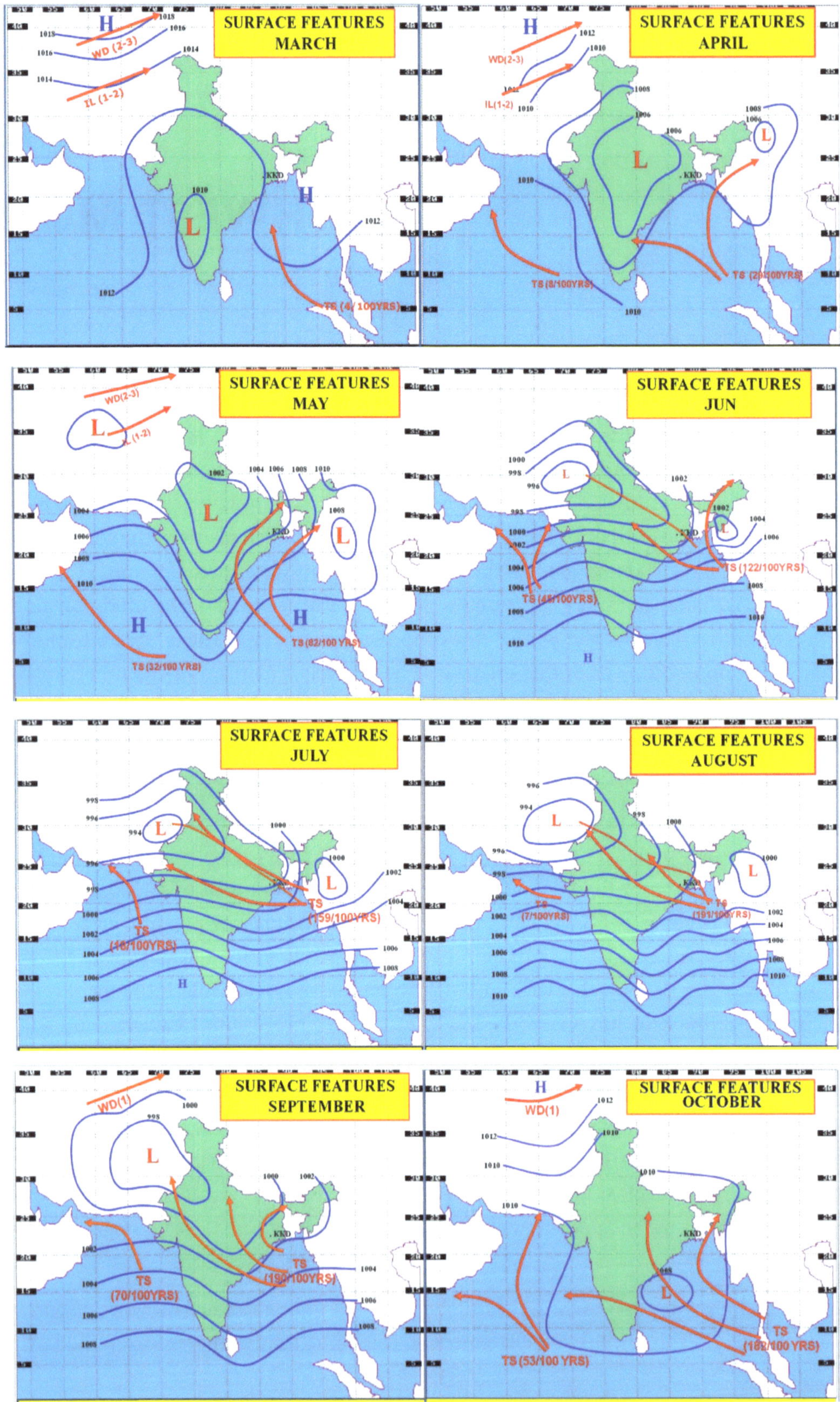
SURFACE FEATURES
MARCH
SURFACE FEATURES
APRIL
SURFACE FEATURES
MAY
SURFACE FEATURES
JUN
SURFACE FEATURES
JULY
SURFACE FEATURES
AUGUST
SURFACE FEATURES
SEPTEMBER
SURFACE FEATURES
OCTOBER
Isobars
Normal Track of Depression/Cyclonic Storm
Monsoon Trough

4. **Wind Pattern**. In June, winds at surface are from east, to the north of the Axis of Monsoon Trough (AMT) extending normally from the Low-Pressure area over Pakistan to Head Bay/Southeast Bay. Over Gangetic West Bengal winds are mainly from south. Over the rest of India, winds blow from west or southwest being more westerly off the west coast of the Peninsula and more southwesterly in the Bay. Between 5° N and 15° N and west of 65° E, the speed is about 20 kt in the Arabian Sea. Elsewhere the range is 10 to 15 kt. In July, easterlies prevail over the country to the north of the AMT. To the south of this line, southerlies occur over West Bengal and south-westerlies to westerlies elsewhere. Nearer the west coast of the Peninsula, the direction is west-south-westerly to westerly, except along the Kerala coast where north– westerlies are observed. Mean speeds over land are not more than 10 kt. They are between 10 kt and 20 kt along the west coast. In the Arabian Sea west of 68o E and between 10° N and 20° N, it is over 25 kt. Over most of the Bay and the rest of the Arabian Sea, the speed is about 15 kt. By September, there is a weakening in the wind speed due to weakening of pressure gradient.

5. The surface winds experienced in various parts of the country during monsoon season are shown in the following figure.

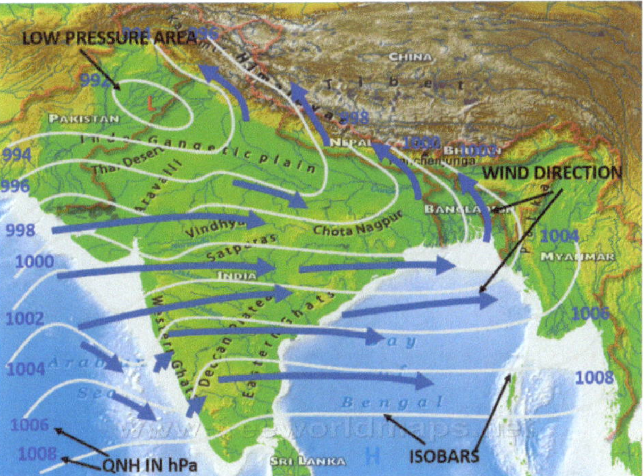

6. Normal wind patterns for the months of June to September at lower troposphere (5,000' ASL), which have similarity with the surface pattern are shown below.

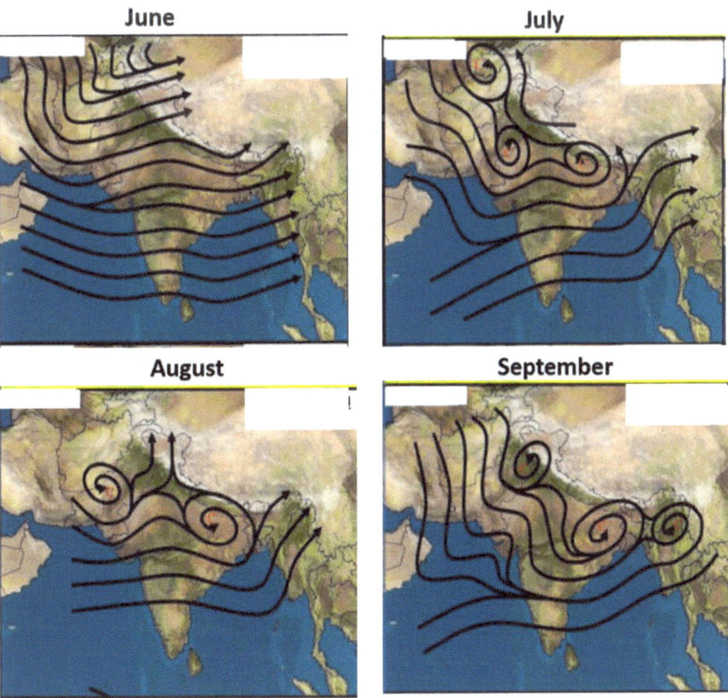

7. **Air Temperature at Surface**. Land gets progressively heated after January and by April, temperatures are of the order of 33 °C to 35 °C over the Peninsula and adjoining regions up to about 24° N. Temperatures along the coasts are between 28° C and 30° C. The highest temperatures over India and Myanmar occur at about 20° N. In July, the southwest monsoon causes extensive cloudiness. Clouding is heavy between 17° N and 24° N in the central regions, west of 77° E and south of 17° N in the Peninsula, and to the east of approximately 85° E in northeast India. Temperatures are even in these regions, being generally between 28° C and 29° C; but on the west coast, temperatures of the order of 26° C to 27° C are experienced. The region of highest temperatures shifts from near 20° N in April to 28° N by July.

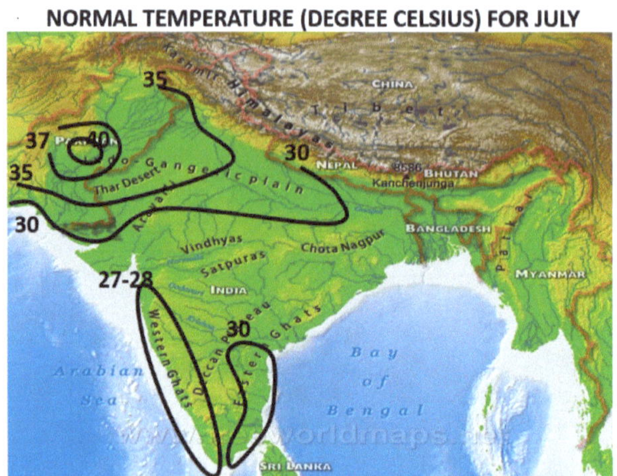

8. **Rainfall Distribution**. The rainfall distribution in the principal rainy season of India, the southwest monsoon period (June to September), is shown in Figure below. Except in Kashmir and neighbourhood, the southeast Peninsula and the east coast areas, the annual rain is mainly accounted for by the falls in this season. Orographic influence is dominant in the distribution of rainfall in this season, as the prevailing winds blow almost at right angles against the western Ghats and the Khasi–Jaintia hills. There is rapid increase of rainfall to the north of a line running from Ahmednagar to Masulipatnam up to the southern slopes of the Vindhyas.

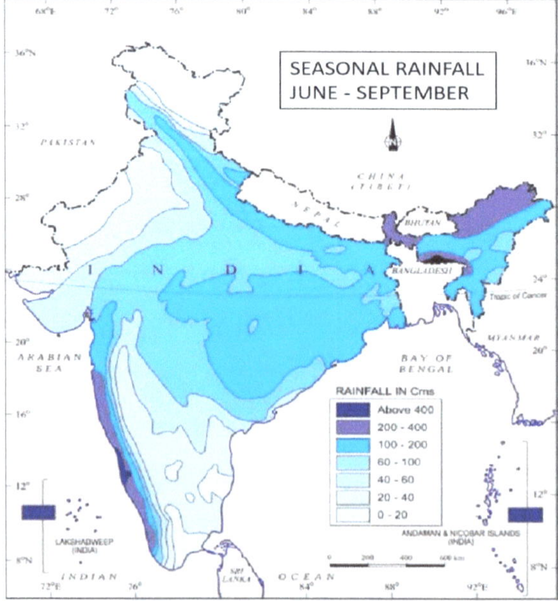

9. In the north Indian plains, a minimum rainfall belt runs from northwest Rajasthan to the central parts of West Bengal, practically along the axis of the monsoon trough. Rainfall decreases very rapidly southwards along west coast from 9.5° N to Kanyakumari. The rainfall at Kanyakumari in this season is about the same as in the Great

Indian Desert. To the east of the western Ghats between 8° N and 10° N, rainfall decreases considerably with a very steep gradient across the eastern slopes. Rainfall is only 2 cm in some places in the coastal strip in extreme south Tamil Nadu.

10. Hills and mountain ranges cause significant variations in rainfall distribution. On the southern slopes of the Khasi–Jaintia hills rainfall is over 800 cm while to the north, in the Brahmaputra valley, it drops to about 120 cm. Cherrapunji's annual rainfall of 1142 cm is due to orographic lifting. From the west coast, rainfall increases along the slopes of the western Ghats and rapidly decreases on the eastern lee side. From the coast of West Bengal and the hills of Orissa, rainfall decreases inland. Further westwards, the Chota Nagpur hills, the Maikala Range and the Mahadeo hills cause an increase of rainfall. The Gir hills in Kathiawar have more rainfall than the neighbourhood. Mount Abu in Aravalli has a rainfall of 169 cm while the surrounding plains have only 60 to 80 cm.

11. Across northern India, a line of rainfall minimum runs roughly along the. Area to south of this rainfall minimum falls in the track of monsoon depressions which are responsible for much of the rainfall. In tracts further to north, there is the influence of the Himalayas in increasing the rainfall. Apart from this, there is also a decrease of rainfall from east to west, from about 120 cm in West Bengal to less than 20 cm in the Great Indian Desert in west Rajasthan. In the eastern Himalayas, rainfall is more than in the western portions. In the east, annual rainfall of 400 cm has been recorded but less than 200 cm in the west.

12. Rainfall in the Andaman and Nicobar Islands in the southwest monsoon season is about 140 to 190 cm, while in Lakshadweep, it is only about 100 cm though both the groups are in the same latitude belt. Calicut on the mainland in the west coast, however, gets 235 cm, more than the Bay Islands.

13. Mean monthly rainfall amounts are not uniform during this period. Broadly, maximum fall is received in July. But, Arabian Sea Islands get more rain in June than in July, while Kerala has about the same rainfall in the two months. In Assam and sub–Himalayan West Bengal, June and July are equally rainy, decreasing thereafter. Some parts of Assam get slightly less rain in July. In Bihar, Uttar Pradesh, Gangetic West Bengal, east Madhya Pradesh and parts of Orissa, July and August have the same amounts. In the Peninsula between 19° N and 16° N and east of 76° E, an increase in September is noticed, apparently due to the effect of depressions and lows forming at lower latitudes.

14. The day to day variation in rainfall is still higher owing to the day to day variation of some of the important synoptic features which will be discussed separately. Active monsoon conditions decrease sharply after July in Kerala, Coastal Karnataka and Konkan. In west Uttar Pradesh, such periods are more in August. In Madhya Pradesh, July to September are equally active. North Assam and Vidarbha show an increase again in September. In Kerala and Arabian Sea Islands, monsoon is only 'normal' on about half the days.

15. Heavy rain of the order of 25–30 cm is not infrequent in this period, but they are more probable north of 15° N along west coast and north of 20° N in the rest of the country, apart from the western Ghats. In areas of poor rainfall like Rajasthan, Saurashtra and Kutch, whole season's normal rain may occur on one day. Khasi–Jayantia hills are noted for heavy falls of over 75 cm in 24 hours on the windward side. In Assam, heavy rains (>25 cm) are most frequent in June, the frequency dropping to one–third in August, in spite of the 'breaks'. In Uttar Pradesh and Bihar, heavy rains are more frequent in August and September. The high frequency of heavy rains in these areas and in Punjab and Haryana in September, is promoted by interaction with extra– tropical systems. In Madhya Pradesh, all months are equally liable to heavy falls but along the west coast, the frequency drops to a quarter after July.

16. **Onset and Withdrawal of Monsoon.** The onset of monsoon is associated with the following noticeable changes.

    a)   Winds from about southwest or west to the south of the ITCZ (AMT) and easterly to the north.

    b)   Decrease in temperature from the heat of April and May.

    c)   Increase in rainfall.

17. Dates of onset of monsoon can be fixed by the changes in any of these features for any year or from climatological means of these elements for a sufficiently long period. However, onset of monsoon over Kerala and its further north and east ward progress is declared based on the guidelines issued by India Met Department (IMD). The broad guidelines are as follows:

    a) If 60% of the fourteen Met observatories located in Lakshadweep and Kerala-Karnataka coast (Minicoy, Amini, Thiruvananthapuram, Punalur, Kollam, Allapuzha, Kottayam, Kochi, Trissur, Kozhikode, Talassery, Cannur, Kasargode and Mangalore) report at least 2.5 mm of rainfall for two consecutive days after 10th May, onset of monsoon over Kerala is declared on the second day. However, occurrences of the following criteria are also necessary.

    b) The monsoon westerlies/south-westerlies should extend up to about 14,000' ASL (600 hPa) in the area from the equator to Lat 10°N and Long 55 to 80°E. The zonal wind speed over the area bounded by Lat 5 to10°N, Long 70 to 80°E should be of the order of 15-20 kt at 3000' (925 hPa).

    c) Outgoing Longwave Radiation (OLR) value derived from INSAT should be less than 200 W/m2 in the area from Lat 5 to 10°N and Long 70 to 75°E.

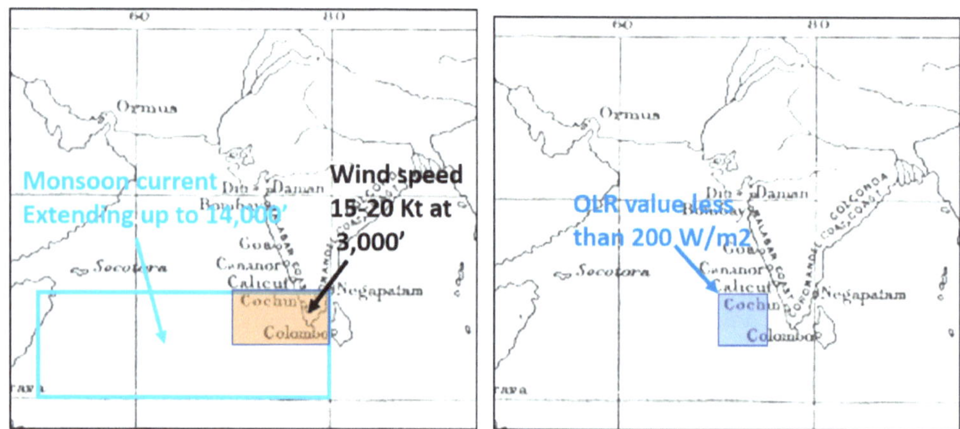

Outgoing Longwave Radiation (OLR) measures the amount of terrestrial radiation released into space, as well as the amount of cloud cover and water vapor that intercepts that radiation in the atmosphere.

18. After onset of monsoon over Kerala, further advancement is declared based on the following criteria.

    a) Along the west coast, position of maximum cloud zone, as inferred from the satellite imageries may be considered.

    b) The satellite water vapour (WV) imageries may be monitored to assess the extent of moisture incursion.

19. **Northern Limit of Monsoon (NLM).** Southwest monsoon normally sets in over Kerala around 1st June. It advances northwards, usually in surges, and covers the entire country around 8th July. The NLM is the northern most limit of monsoon up to which it has advanced on any given day.

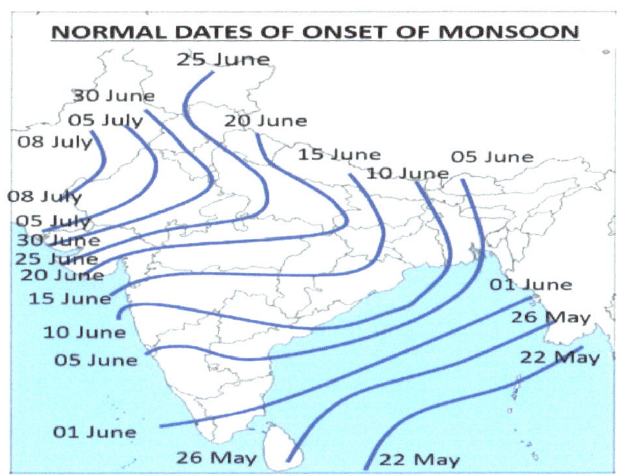

20. **Progress of Monsoon**. As shown in the above figure, by the end of June SW monsoon covers Andaman and Nicobar Islands and almost half of the Bay of Bengal. By first June it hits Kerala. Thereafter, it gradually advances northward. By 8th July the entire country is under the sway of monsoon.

21. **The Branches of Monsoon**. The monsoon current reaches Sri Lanka, Andaman Sea and Tenasserim coast towards the third week of May. It advances into south Kerala, mid Bay of Bengal, eastern Bangladesh and Tripura by the beginning of June. From here on the monsoon current progresses in two distinct branches. The Arabian Sea branch moves up and reaches Maharashtra by the 10th June and up to Saurashtra and Madhya Pradesh by the 20th June. Meanwhile, the Bay branch spreads over the Bay of Bengal, Bangladesh, Meghalaya, Assam and central and north Bengal. From there on the Bay branch is channeled by the Himalayas and the deflected current extends westwards over the Indo-Gangetic plains. Later both the branches merge in the Gangetic basin roughly along the Axis of Monsoon Trough (AMT) and gradually extend to Uttar Pradesh, Haryana, Punjab and Rajasthan. By fourth week of June the current extends to Jammu and Kashmir.

Arabian Sea and Bay of Bengal Branches

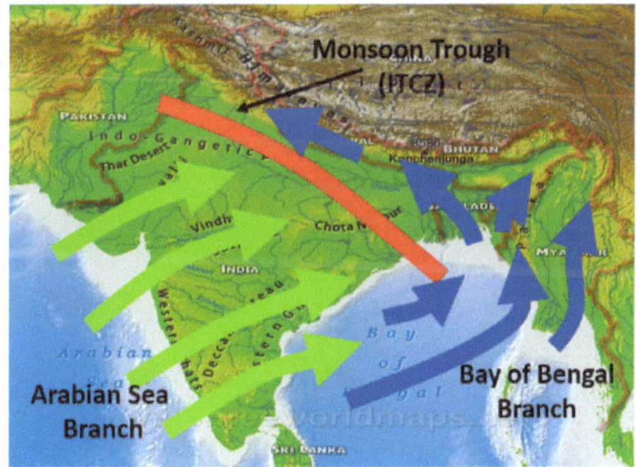

22. **The Monsoon Front**. Monsoon front is a discontinuity which is established along the AMT over the Gangetic basin. This is the meeting place of the Arabian Sea branch (Em- west/south-westerlies) and the deflected monsoon current (Bay of Bengal branch - Em or EmTm) which are east/south-easterlies. There is some difference between the two air masses, the fresh monsoon air (Arabian branch) being slightly cooler and more moist. In the northwest, the surface heated continental air (Tc) also flows into the trough. The axis of the monsoon trough or the monsoon front more or less coincides with the position of the ITCZ. In the northeastern portions of the subcontinent, in the Brahmaputra valley, a subsidiary or secondary front between the Bay Branch of the monsoon air (nEm) and the Himalayan katabatic easterlies (nTm) also exists. This can often be seen separated from the main monsoon front, which may not extend to this area. On some occasions, these two coincide. The orientation, position and intensity of the monsoon front varies on a day-to-day basis they determine the distribution and intensity of the rainfall in

northern portions of the country.

23. This monsoon trough is, practically a part of the ITCZ which shift northward during norther hemisphere summer season.

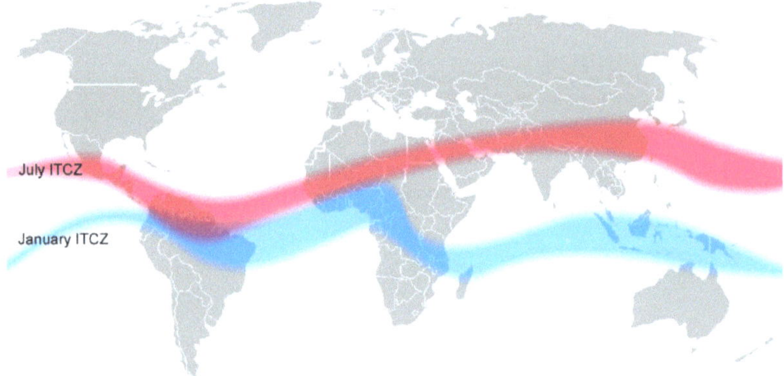

24. **Withdrawal of Monsoon.** During the course of October there is once again a change in the circulation pattern of the southwest monsoon over the Indian area but the rains increase over south India. This change in circulation should be rightly regarded as the end of the southwest monsoon and it is not appropriate to identify the much later decrease of rainfall in south India as the withdrawal of the southwest monsoon. The displacement of the monsoon air by continental air mass and development of anti–cyclonic flow would determine the dates of withdrawal over north and central India. However, there is a requirement of fixing the normal dates of withdrawal of monsoon as the dominance of the continental air takes place from the extreme northwest and advances southward in a gradual manner.

25. However, as per guidelines issued by IMD, withdrawal from extreme North-western parts of the country should not be declared before 1st September. After 1st September, cessation of rainfall over the area for 5 continuous days, establishment of anticyclone in the lower troposphere (5,000' ASL or below), and considerable reduction in moisture content are to be considered for declaration of withdrawal of monsoon. Further, monsoon should be reported as withdrawn from the entire country only after 1st Oct when there is an indication of a change over from the south-westerly wind regime.

26. Based on above, the normal dates of withdrawal of southwest monsoon has been set by IMD as follows: -

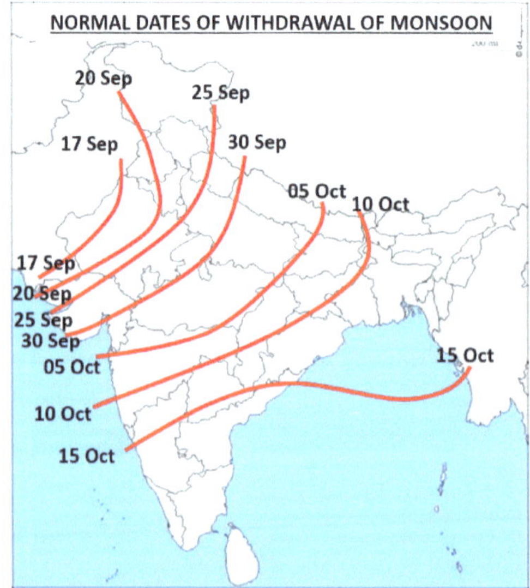

Semi-Permanent Systems of Southwest Monsoon.

27. **Mascarene high**. This is high-pressure region located near the Mascarene Islands in the southern Indian Ocean with its centre near 30°S, 50°E. In one way, it is a source of Southwest monsoon in India. Normally, the high-pressure region starts forming by mid-April and its strength is an important factor which determines the intensity of monsoon in India. A stronger high pressure will produce stronger winds or monsoon current. The intensification of Mascarene high strengthens the cross-equatorial flow in the form of East African Low-Level jet and the corresponding monsoon current over Arabian Sea. The intensity of Mascarene High is also found to be associated with the onset of monsoon over India as well as the subsequent fluctuations in its activity.

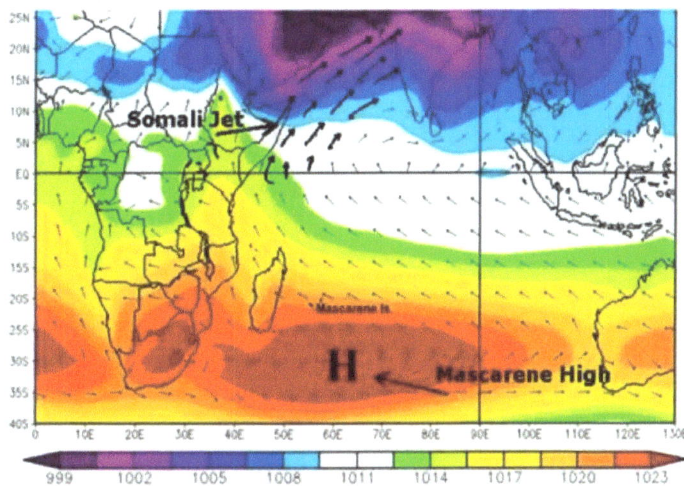

28. The broad belt of high pressure around the Mascarene Islands generates a cross-equatorial flow known as the Somali Jet which travels to Arabian Sea as the Low Level Jet of monsoon flow. These jets bring good monsoon shower to the west coast of India. A strong, low level jet usually means a strong monsoon over peninsular India. Winds from Mascarene High head in a north-westerly direction towards the east coast of Africa (Somalia). Here, the topography of Somalia deflects the winds towards the east. Also, after crossing the equator, these winds experience the Coriolis Effect. Hence, these monsoon winds get deflected eastwards and now they blow from south-west to the north-east.

29. **Heat Low.** As mentioned earlier, as the solar equator shifts northward after the winter season, the entire land mass of Indian sub-continent gets heated up and a low-pressure area forms due to heat. This low-pressure area, which can be marked from the end of February over southern Peninsula, gradually moves northward and become deeper. The progressive development of this heat low over the sub-continent and its location over central parts of Pakistan in July is perhaps the most important causative factor of the monsoon.

30. The heat low is a part of the low-pressure belt extending from Sahara to central Asia across Arabia, Iran, Afghanistan, Pakistan and northwest India with off-shoots of troughs in various directions. The normal location of the centres of low from April to June are as follows: -

April – East Madhya Pradesh

May - Punjab

June to July - Central Pakistan

31. There is a little retreat to east and south in August and September. The centres of the heat lows over land areas are located near regions of maximum heating, out of reach of maritime air mass. The development of heat low causes the circulation which brings cooler air mass in favourable area offsetting the effect of solar heating. The tapering shape of the Peninsula and the Himalayan barrier to north and Assam and Myanmar hills to east which make the

maritime air mass pervade over most of India displace the centre of the low to the extreme northwest of the sub-continent.

32. Blocking of cold air incursions from the north by the Himalayas in the lower troposphere must have made the heat low over the Indian sub-continent more intense and is the deepest low in the northern hemisphere during summer. However, this area, that is central Pakistan and adjoining areas is not the place of the highest temperature. Sahara records much higher temperatures. The configuration of mountain ranges in the northwest corner of the sub-continent and the adjustment of wind field and consequently the pressure pattern throws up the low pressure centre over the central parts of Pakistan. As far as the Indian monsoon is concerned, what is important is the existence of low pressure centre in the northwest of the sub-continent and not the intensity or deepness of the low.

33. The heat low is only in the first 1.5 km and is over-lain by a well-marked ridge extending up to upper troposphere, which is a part of the subtropical high-pressure belt. Hence, there is subsidence of air over the region above 1.5 Km. This is why, the area of the low pressure remains dry during the season.

34. The intensity of the heat low can be correlated. Below normal pressures in the heat low region and above normal pressures in the Peninsula are regarded as favourable for monsoon activity. Pressure gradient would then be strong over the Peninsula, which is conducive to monsoon rains. The heat low may also strengthen when the ridge aloft weakens under the influence of westerly troughs moving further north. Some westerly troughs can cause formation of weak lows over northern India and lead to increase of rainfall.

35. **The Monsoon Trough**. From the seasonal low over Pakistan and neighbourhood, a trough extends southeastwards to Gangetic West Bengal and further to Head Bay, North Bay or Northeast Bay. The trough line normally runs at surface from Ganganagar to Calcutta through Allahabad, with west to southwesterly winds to south and easterlies to the north of the trough line. Mean surface wind at Calcutta is mainly from south in spite of the depressions which form in the north Bay of Bengal. This trough is seen in the upper air also up to about 6 km, sloping southwards with height. At 4 km it runs from Bombay to Sambalpur. The air mass to the south of the trough line is the Arabian Sea monsoon while the air to the north may have had some travel over the Bay. Temperature is about 2° C higher to the north.

36. The monsoon trough is regarded as the equatorial trough of the northern summer in the Indian region. This is also the ITCZ which shifted northward during the northern hemisphere summer. The position of the trough line varies from day to day and has a vital bearing on the monsoon rains. No other semi-permanent feature has such a control on monsoon activity. Position of the trough line close to the foot-hills has come to be referred as 'Break in Monsoon' on account of drastic decrease in rains over the country, though the Himalayan mountain belt experiences heavy falls which can cause floods in the rivers originating there. Swing of the trough to the central parts is with monsoon depressions from the north Bay moving west to west-northwest ward across the country.

37. The position and orientation of the trough changes on a day-day basis. When the trough rapidly shifts north or south, which can be even 5 latitude in a day, monsoon activity is enhanced in that area. The eastern portion of the monsoon trough shifts southwards into north Bay before a depression forms. In the rear, the trough swings back northwards over northeast India. The monsoon trough having shifted to a northerly position, can cause increased rains over Assam; a fresh trough may form near the normal position. In September, the monsoon trough takes North-northwest/south-southeast orientation, when a depression is forming at a lower latitude

38. **Tibetan High**. In July, at about 700 mb and aloft, a ridge lies over Pakistan and northwest India to the west of about 75° E, with its axis along 30° N. Another high appears to the east of 80° E at 500 mb with axis near about 28° N. This high has its centre at 28° N, 98° E, and is distinct from the Pacific High at 140° E. At this level, the high covers the Tibetan Plateau while the centre of the high is at its eastern periphery. There is only a broad high-pressure belt at 100 mb from even 30° E to 150° E along 35° N over Indian region. *This high over Tibet from 500 mb upward,*

centred near Tibet at 500 mb and over Tibet at 300 mb and 200 mb, has come to be known as the 'Tibetan High'. In June the axis of the anticyclonic belt is at about 25° N, at 300 and 200 mb, and near 30° N only at 100 mb, at the southern periphery of Tibet. From June to September, there is some shift in the position of the centre. In August the feature is more pronounced.

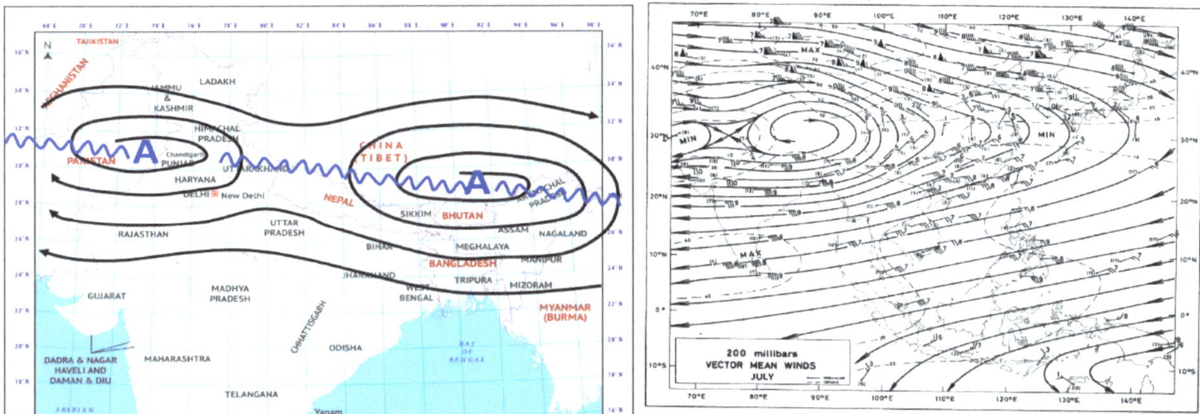

39. Well-distributed rainfall over India is associated with well-pronounced and east–west oriented anticyclone over Tibet at 500 and 300 mb levels, and a pronounced high index circulation over Siberia, Mongolia and north China. The Tibetan 'High' may sometimes shift much to the west of its usual position. In such a situation, the monsoon may extend further westward into Pakistan and in extreme cases into north Iran, though such a westward position of the Tibetan 'High' would be against its having origin in the Tibetan Plateau.

40. When the centre of the Tibetan High shift east ward to north Myanmar, the extratropical east ward moving systems penetrates into the monsoon regime. Such situation generally weakens the monsoon flow and results in Weak Monsoon condition.

41. **Tropical Easterly Jet (TEJ).**South of the sub-tropical ridge over Asia, the easterly flow concentrates into jet stream centred near about the latitude of Chennai at 100 hPa in July. The jet stream runs from the east coast of Vietnam to the west coast of Africa. Over Africa, the location is at 10° N. Normally, the jet is at an accelerating stage from the south China sea to south India and decelerates thereafter. Position and speed of the Jet fluctuate from day to day. Over most of the Atlantic Ocean, continental America and Pacific Ocean, the easterly jet is not generally found.

42. TEJ is an important upper tropospheric feature of summer monsoon circulation over south Asia. The north-south movement as well as the structure of Tropical Easterly Jet is observed to be closely linked to monsoon activity. The strength of the Jet gradually increases from south China Sea to south India. Thereafter, the speed starts decreasing towards west. TEJ can be marked over the Peninsular region from June to August roughly along $12 - 15\ {}^0$N with it's axis at 200 to 100 hPa. The core speed of the Jet is generally between 60 and 85 Kt. However, maximum speed of 150 Kt has been observed over Indian Peninsular region.

43. **Low Level Jet (LLJ).** The monsoon low-level jet (MLLJ) originates at Mascarene high and after traveling thousands of kilometers enters India from the western boundary causing deep clouding and rainfall. Although its core lies at 850 hPa, it may have more vertical extent. This low-level jet, also known as "Somali Jet", is a relatively narrow wind stream along the East African Coast off the coast of Somalia over the Indian Ocean and peninsular India during the summer monsoon season. It is associated with the strongest cross equatorial flow from the southern to northern hemisphere. The Low-level jet is generally located in the lowest 1-2 km of the troposphere and is strongly influenced by local factors such as orography, friction and diurnal cycle of heating. It is normally strongest in July and August. The average peak speed is 35 to 45 Kt. However, mush stronger core has also been observed. LLJ brings about active monsoon condition along the west coast. With the Bay branch of monsoon, some of the jets also travel up to Bay of Bengal as far east as the Andaman Sea.

44. LLJ exhibits systematic diurnal variation. Maximum winds of the synoptically induced large-scale monsoon jet occur during the daytime, and the orographic channeled winds through the mountains of Western Ghats in the night-time. These diurnal changes in monsoon winds modulate the moisture convergence process and the associated evolution of rainfall over India. The LLJ exhibits an increasing trend in wind speed on both seasonal and monthly scales, except for August which shows a decreasing trend. The weakening of the LLJ in August has a profound influence on the number of monsoon depressions forming over the Bay of Bengal (which are decreasing), and on the number of break days (which are increasing) and associated precipitation reduction over the central Indian region.

Other Systems.

45. **Monsoon Depression**. Monsoon depressions are the synoptic features that cause most of the monsoon rains. These depressions which form in the Bay of Bengal north of $18°$ N, move west-northwest ward at least up to the central parts of the country before weakening or filling up, and give widespread rains in the southwest quadrant with many heavy falls. A second rain belt often develops outside the depression field to the west and fairly distinct from the southwest sector, due to convergence between northwest flow, northeast/easterly flow around the depression and westerlies to south, in the lower troposphere

46. These low-pressure systems are referred to as depressions when surface winds are up to 33 kt (while over the sea) and cyclonic storms when higher speeds prevail. Weaker systems with only one closed isobar and wind speeds less than 17 kt, are called lows.

47. In June, July and August, depressions and storms generally form in the Bay north of $18°$ N and west of $92°$ E, and a few up to $15°$ N. Most of the formation is to the north of $20°$ N. In September, however, the formation extends up to $14°$ N. In the Arabian Sea, the systems originate in June within five degrees of the coast north of $12°$ N and in rare cases very close to Gujarat coast in later months. Land depressions develop mostly over northeast India.

48. Depressions and storms of July move mainly west to west-northwest ward over the Bay and across the country up to $25°$ N, and west-northwards in August. Sometimes, when the systems interact with the WDs moving across higher latitudes, they re-curve towards northwest, north or even northeast ward. Depressions early in June, with advancing monsoon, may take a north-northeast track and usher the monsoon into West Bengal and Assam and adjacent states. Even in July and August, some initial northerly track may occasionally be seen before systems change to the usual west-northwest ward movement. During June and September, the systems are more spread out, while in July and August, they are within a narrow belt.

49. Depressions mostly develop out of three types of situations. In half the number of cases, they form out of a diffuse pressure field that develops over north Bay. In 15 percent cases, the signs appear as an upper air cyclonic circulation at any level up to mid–troposphere. One third of the depressions are born out of diffuse lows that travel across Myanmar into north and adjoining central Bay, some of them having been remnants of typhoons that moved from south China Sea.

50. Monsoon depressions have a profound influence on the rainfall over most of the country, much more than their numbers would suggest. Monsoon activity in one part or the other is influenced by these systems, starting from the precursors to even during the weakening stage of low. Their travel across a large tract distributes rainfall over a wide area. Such effect of monsoon depressions prevails on the average for about one third of days in the period of the monsoon.

51. An interesting effect of monsoon depression is that while forming in northwest and adjoining parts of west central Bay, it strengthens rainfall along Konkan Coast. Once the depression has formed, rainfall decreases, unless some other favourable synoptic pattern has developed. A second spell of rains may occur in this area as the depression

moves to the west of 77° E but remaining south of 23° N. However, during the period of formation of the depression results in dry condition over Assam and adjoining states.

52. **Low Pressure Areas**. Low pressure areas, less intense than monsoon depressions, also form quite frequently during the monsoon and are responsible for causing substantial rainfall. Excluding the residuary lows out of weakening monsoon depressions, the lows fall into two distinct categories: one which form over the Bay just like depressions but do not develop up to depression intensity and the other that develops in situ over northern India.

53. The lows form with just one closed isobar over the Bay of Bengal, hardly give prior indication of their formation. Their movement is generally not regular. Howevere, their vertical extension is comparable to that of depressions.

54. The other type of lows forms over land north of Lat. 22° N, mostly over Bihar, north Madhya Pradesh and south Uttar Pradesh. They generally form under the influence of trough in middle latitude westerlies, the mode of formation being similar to the induced lows of the winter season. Such lows dissipate when the westerly system moves away eastwards.

55. These land lows are generally smaller in extent than the lows arriving from the sea. Formation is mainly to the west of Long. 85° E. Life period is about two to three days. They give concentrated rainfall when the system underlies the forward portion of the mid–tropospheric trough. Preferred sectors of rainfall have not been noticed. The system shows slow movement eastward.

56. **Trough in Monsoon Westerlies**. During weak monsoon conditions, in the westerly sweep across northern India, weak troughs are noticed to develop in the lower troposphere over the plains, mainly to the east of the longitude 80° E. They cause increased rains for about 2° longitude on either side and farther to the east. Rainfall is not widespread and heavy as in the field of depressions and lows. But the increase of rainfall from a 'break' situation is quite marked, scattered heavy falls also occurring. These troughs are more numerous, in July than in other monsoon months.

57. **Break in Monsoon**. There are periods when the monsoon trough is located close to the foot of the Himalayas which leads to striking decrease of rainfall over most of the country but increase along the Himalayas and parts of northeast India and southern Peninsula. Such a synoptic situation is referred to as 'Break in Monsoon'. This situation is more common during mid-August. The beginning of July and end of August have lesser number of break days.

58. In a break situation, the monsoon trough is at the foot of the Himalayas or not noticed at all. The surface winds become westerlies. Similar conditions may prevail in upper air. This is in contrast to the mean position of the trough from Ganganagar to Calcutta at sea level and further south aloft.

59. During 'breaks' rainfall almost ceases over most parts of the country. Heavy falls occur in and near Himalayas but not simultaneously along the whole length. Influence of eastward moving troughs in westerlies further north seems to cause these patches of heavy rains. Himalayas to the east of 85° E are susceptible to much heavier falls than to the west. In northeast India, this above normal rainfall during 'breaks' extends to plains, well into Bihar, West Bengal and Assam. In the south, Tamil Nadu and Rayalaseema get more rains, as thundershowers. Areas not affected by high rainfall deficiency (>50 %) during 'breaks' are Jammu and Kashmir, Rajasthan, Gujarat State, Vidarbha, Madhya Maharashtra, Madhya Pradesh, Telangana and Odisha. In prolonged 'breaks', the air mass even over western Himalayas also becomes dry and rains decrease.

60. The monsoon trough shifts towards the foot–hills of the Himalayas when a depression moves to these mountains, which is associated with more southerly travel of extra-tropical systems. Such a movement of depression always takes the monsoon trough to the foot-hills. Monsoon revives only when the monsoon trough swings back to the normal position. This happens when a fresh low or depression forms over north Bay.

61. **Effect of Middle Latitude Westerly Systems on Monsoon**. When the southwest monsoon is fully established over India, middle latitude westerlies prevail to the north of Lat. 30° N only. Still systems in these westerlies seem to

exercise considerable influence on the monsoon weather over northern India. The extra–tropical systems themselves consist of troughs with ridges to the rear. It may be presumed that the ridges in rear can sometimes develop into a warm high. Following are their effects on monsoon: -

a) Intensifying or developing lower tropospheric lows or troughs,

b) Enhancing rainfall in pre-existing systems,

c) Causing recurving of depressions and lows

d) Leading to onset of break conditions.

62. The effect of westerly troughs on rainfall seems to be more over north west India. This rainfall is quite often moderate to heavy and forms a distinct zone separate from the rainfall belts further south. The rainfall peaks are found to be associated with passage of low-pressure systems to the north of this country in the middle latitude westerlies. This type of influence of westerly troughs on rainfall can occur without the monsoon trough shifting towards Himalayas. Monsoon can be even active over the central parts of the country, sometimes with a depression moving along the usual track.

63. Under their influence of an induced WD developed over northern parts of Pakistan and also over southeast Baluchistan, Sind & neighbourhood monsoon gets activated over northwestern parts of the country as far south as Gujarat. As the parent WD (moving across higher latitude 45-55 ^{0}N) moves further to east, it triggers off depression over central or north India in the pre-existing monsoon trough.

64. Presence of a a well–marked low around 42° N north of India with an associated upper trough reached up to 25° N often leads to re-curving of an existing monsoon depression or low towards north-west, north or even north-east ward. Besides, weather produced by monsoon depressions could be accentuated by a WD moving eastward at a higher latitude.

65. When the monsoon trough is along the Himalayas and favourable for monsoon activity in the neighbourhood, rainfall over north Assam and sub-Himalayan West Bengal is more when a middle latitude westerly trough is extending up to Tibet.

66. As discussed earlier, occasionally, when no monsoon depression exists and a WD moves across northern latitudes, the AMT is pulled to north and it passes along the foot-hills. In such cases, the monsoon currents over north India are replaced by dry and warm westerlies or north-westerlies. During this period, rain fall ceases over the country outside the areas adjoining Himalayas and south Peninsula. This condition is called 'Break Monsoon' condition. The AMT will swing back only when a fresh depression or low forms over north Bay of Bengal.

67. **Trough off the West Coast**. A trough to the west of west coast quite frequently develops, anywhere from north Kerala to south Gujarat during the period of southwest monsoon which is responsible for the strengthening of the monsoon in terms of rainfall, in the adjacent parts of the coastal belt. A small closed circulation may be embedded in some troughs. The troughs form more often near Coastal Karnataka and slowly shift 2° latitude per day northward, though they may also appear and disappear in situ over any area. Upper winds are affected only in the lower troposphere and that too, often below 1 km. However, as the troughs move to Maharashtra coast, upper winds are drawn into the circulation to a greater depth.

68. Nearly half the number of active to vigorous monsoon situations in Konkan and three quarters of such occasions in Coastal Karnataka are with troughs off the west coast.

69. **Mid-Tropospheric Cyclonic Circulation (MTCC)**. These are middle tropospheric cyclonic vortices which form over NE Arabian Sea, adjoining Gujarat and north Maharashtra coast. These are responsible for heavy rainfall over northern sectors of west coast during southwest monsoon season. Heavy to very heavy precipitation is concentrated

in the southwest sector of an MTCC. Existence of such systems accounts for most of the monsoon rains of western India and northeast Arabian Sea. In these systems, colder and drier north-westerlies merge with the cooler and moist easterlies & south-easterlies at middle levels. Such convergence leads to great instability over the area and causes heavy rainfall, mostly convective in nature. When the dry north-westerlies overweighs the easterly component, the feature weakens and disappears.

70. **Upper Easterly Waves.** In the upper troposphere, easterlies prevail during south-west monsoon period, increasing in speed with height, reaching jet speeds near 150/100 hPa over some parts of Peninsular region. These winds are dynamically unstable and are subject to perturbations like troughs or wind maxima travelling from east to west. Wind speeds are greater than the rate of movement of these trough lines and there is convergence to the rear of the trough line and divergence in advance. Weather is caused by the upper divergence in advance. These easterly troughs are most marked between 400 and 200 hPa and have a wave length. of 20 longitude and a speed of 10 to 15 kt. Coldest air is near the trough–line with relatively warmer air ahead as well as in the rear. It is seen that monsoon activity over a region increases one day prior or during passage this feature.

71. **Orographic Effect on Monsoon Activity.** The hilly features spread all around the country play a vital role in the monsoon activity and rainfall distribution. The northern part of the Indian sub–continent is bounded by the Himalayas in the north, the hill ranges from Arakan Yoma to the Patkai hills in the and the ranges in the west from the Hindukush mountains to the Kirthar range. Further, the Tibetan Plateau lies to the north of the Himalayas at an elevation between 4000 and 5000 m. Actually, a chain of ranges higher than 2000 m runs between Latitudes 30° and 40° N from about 40° E to 100° E. What would have been the circulation over the Indian sub-continent if the east-west mountain ranges had not been in existence, is difficult to visualize.

72. The effect of orography on rainfall is strikingly brought out in the monsoon season. Hills and mountain ranges get more rainfall than the neighbouring plains.

73. In Khasi hills of Meghalaya, is located the rainiest place Cherrapunji. The highest rainfall in a day at this station is of the range 0f more than 1000 mm. The mean air flow is from south up to 3 km, becoming easterly aloft. The Baral range to the east prevents air flow around. Rainfall in the monsoon season in plains, at Mymensingh 50 km to south of the land-rise (but to the west of Cherrapunji longitude) is about 1600 mm. Shillong (1500 m) just to the north of the crest of the hill range, 35 km north of Cherrapunji, has a rainfall of only 1590 mm, Tura (370 m) on the western slope of the Garo hills (of peak of 1400 m) has 2470 mm rain, more than Shillong. Mawphlang (25°27' N, 91°46' E, height > 1500 m) near the crest but still on the windward side in relation to southerly flow, gets 2490 mm. Jowai (1390 m) just below the crest on the leeward side has 2560 mm of rain. The phenomenal rains at Cherrapunji prevail over a short length of the range, up a small portion of the slope. The funnel shaped catchment opening to south on either side of which Cherrapunji and Mawsynram are located, seems to increase the convergence in air rushing from south. It is also seen that it is on the central slope of the hill ranges that the forced uplift and rainfall are maximum. Towards the crest, this is much less effective. Here the decrease of rainfall on the leeward side is not so drastic as in some other cases. However, Lanka (25°55' N, 92°58' E) in a bowl to the northeast, records only 790 mm. This bowl of elevation less than 150 m is surrounded by hill ranges on all sides except for an opening 20 km wide to northwest.

74. Rainfall enhancement on account of orography may depend on slope of ground, height, configuration of land around and synoptic systems causing rainfall. Height alone is not a decisive factor. Agumbe (659 m) in the Western Ghats records about 7180 mm of rain in June to September as compared with 8370 mm at Cherrapunji (1313 m). In the Ghats themselves, Mahabaleshwar at nearly twice the height of Agumbe records 16 % less rain. The slope near Agumbe is 1/6, about the same as at Cherrapunji, while it is 1/15 at Mahabaleshwar. The rainfall at Matheran is quite high for its location on the plateau of a minor range almost at the beginning of its lee side. Colaba and Santacruz

differ in rainfall amounts by 25 per cent though within 30 km. A hillock of 300 m within 5 km of Santacruz apparently accounts for this increase.

75. Within 80 to 100 km leeward from the crest of the Western Ghats, rainfall decreases to 300 mm. Guwahati on the lee-side of the Khasi hills has 1210 mm of rain and Lanka 790 mm. The lee-side decrease is much more in the case of the Western Ghats. Certain synoptic patterns are quite favourable for causing rain in the Brahmaputra valley without flow pattern being affected by lee-side effects, as in break monsoon situations. Similarly, rainfall on the lee-side of the Western Ghats to the north of 19° N is a little more than to the south, as this area can get rains from the travel of monsoon depressions, particularly in September, un-influenced by lee-side effects on the flow by the Western Ghats. South of 11° N, the rainfall on the lee-side is still less but this is also a continuation of the extreme dry belt from Tuticorin, which on the average, drier in this period than even the western Indian desert.

76. **El Nino and Southern Oscillation (ENSO),**   El Nino is the warm phase of the El Nino Southern Oscillation (ENSO), a climate event that happens when warm water in the Pacific Ocean interacts with the atmosphere, causing various weather events around the world, from droughts to floods. An El Nino arrives on a cycle of about every three to seven years. El Nino meaning, The Little Boy, or Christ Child was originally recognized by fishermen of the coast of South America in the 1600s, with the appearance of unusually warm water in the Pacific Ocean. The name was chosen based on the

time of year (around December) during which these warm water events tended to occur.

77. To understand how these phenomena affect the Monsoon system, we must first know

about the Normal and Abnormal conditions of Oceanic Currents.

78. In normal condition, The Peruvian coast has relatively high pressure than the areas near north Australia and South-East Asia. During normal year two things are strong - Cold Peru Current and Trade Winds. As a result, cold water is dragged from Peru towards Australia.

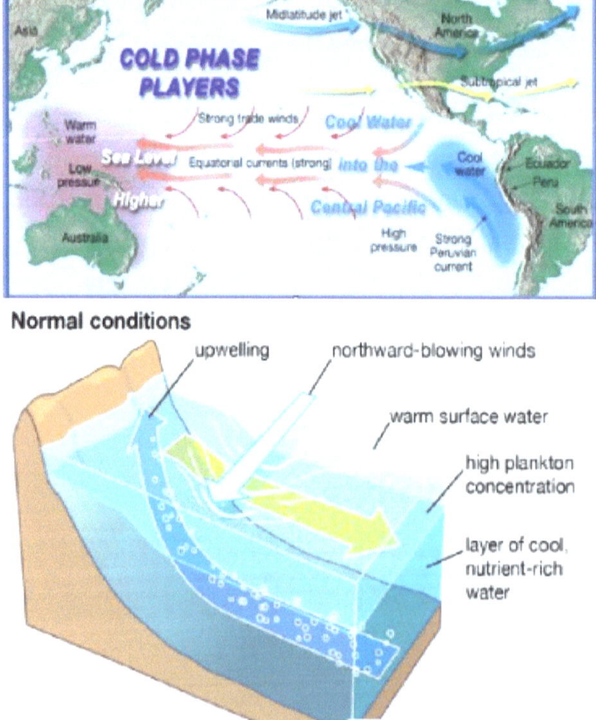

79. In above image, the red (warm) water region around Australia is called Western Pacific Pool (WPP). WPP is a zone of low pressure so, warm air ascends and results in cloud formation and rain over North Australia. This air also

joins walker cell and begins descending near Peru. Descending colder air results in formation of anti-cyclonic (High pressure region) condition over the region which subsequently give rise to dry weather over that region including Atkama desert.

80. During an El - Nino Condition two things become weak - Cold Peru Current and Trade Winds. As result, cold water is not dragged from Peru to Australia but reverse of the normal condition happens. Warm water is dragged from Australia towards Peru and consequently, warm water & low pressure condition develops in the Eastern Pacific (Peru) and Cold condition & high pressure in Western Pacific (Australia). In such situation, there will be rain in Peru, Atkama and even Southern USA. Northern Australia, Indonesia, etc. will experience less rainfall or draught like condition. Storms and Hurricanes in East Pacific will increase.

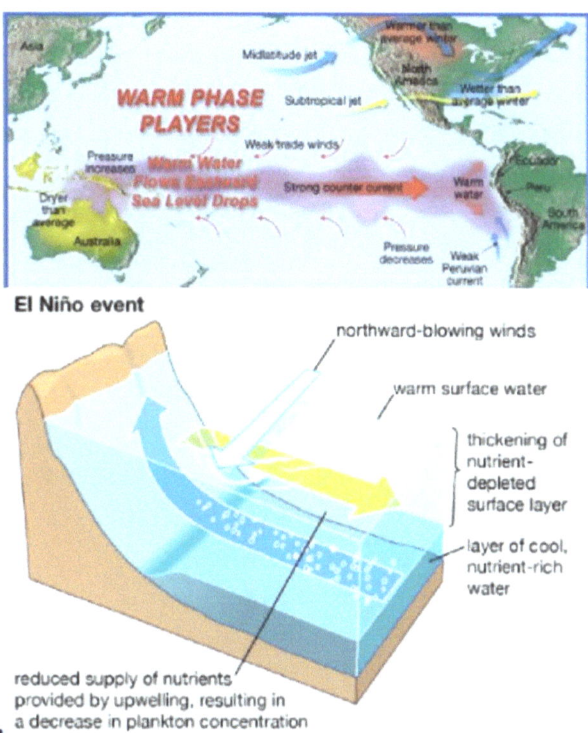

81. **La-Nina.**   La Nina, "anti-El Nino" or simply "a cold event" is the cooling of water in the Eastern Pacific Ocean. The water in Eastern Pacific, which is otherwise cool gets colder than normal. There is no reversal of the trade winds but it causes strong high pressure over the eastern equatorial Pacific. On the other hand, low pressure is caused over Western Pacific and Asia.

82. In the above situation, there will be high pressure zone over Ecuador and Peru region and these regions will experience lesser rainfall or drought like condition. On the other hand, low-pressure area will form over the eastern Pacific region and due to ascend of the warm air there will be rainfall activity over eastern Pacific and Asia.

83. For India, an El Nino is often a cause of concern because of its adverse impact on the south-west monsoon. A La Nina, on the other hand, is often beneficial for the monsoon. The La Nina that appeared in the Pacific in 2010 probably helped 2010's south- west monsoon end on a favorable note. However, it also contributed to the deluge in Australia, which resulted in one of that country's worst natural disasters with large parts of Queensland either under water from floods of unusual proportions or being battered by tropical cyclones.

84. **How does El Nino and La-Nina affect Indian Monsoon**. El Nino-Southern Oscillation (ENSO) water circulation happens between Australia and Peru but the wind movement is part of larger atmospheric circulation. Hence,

affects the monsoon flow and rainfall distribution over India. During normal condition, the sea surface over south America coast is cool which, eventually cools the overlying atmosphere creating a high-pressure zone. This high-pressure zone also includes the area of Mascarene High. Low pressure is create over western Pacific, north of Australia. Winds over western Pacific rises and descends over the eastern Pacific high-pressure zone. This situation is favourable for the Indian monsoon as there is normal flow of monsoon current from Mascarene High towards Arabian Sea.

85. Problem for the monsoon, or in the global circulation as a whole arises when there is rise in temperature over eastern Pacific. Such rise in the sea surface temperature in the eastern Pacific is called the El-Nino situation, due to which the normal high-pressure zone turns into a low-pressure zone. Now, the air ascends over eastern Pacific and descends over western Pasific creating a reverse of the normal circulation. In this condition, the monsoon current over the Mascarene region weakens and the Asian monsoon weakens.

86. The situation other way round happens when there is below normal temperature over eastern Pacific. That is, the temperature over the region is colder than the normal. This situation is called the La-Nina condition due to which, the high-pressure zone over eastern pacific region including the Mascarene region stronger and eventually results in better flow of monsoon current from the Mascarene high region.

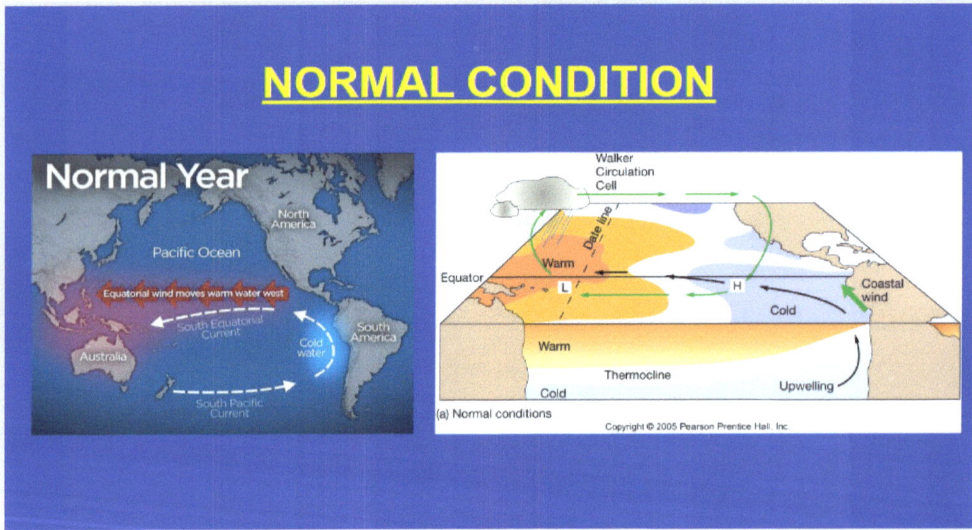

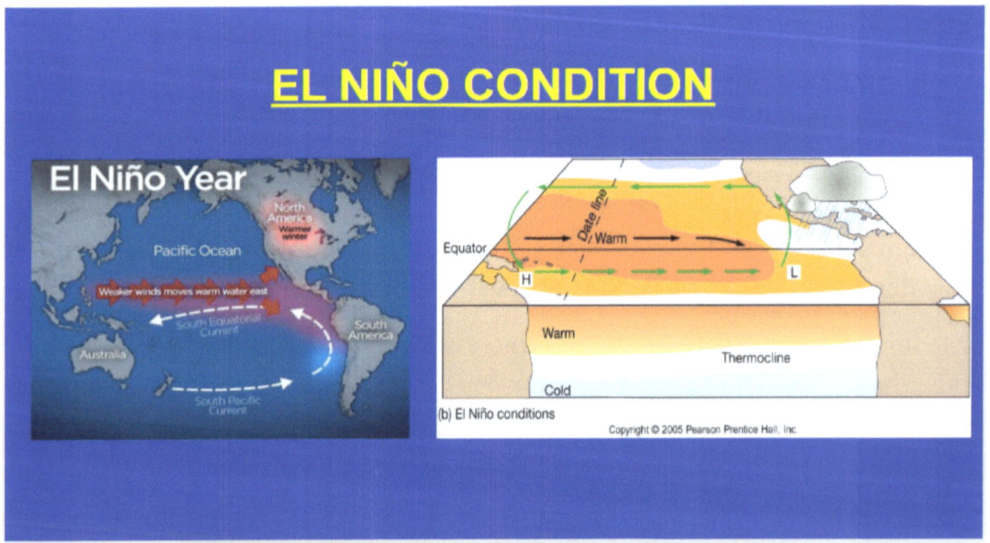

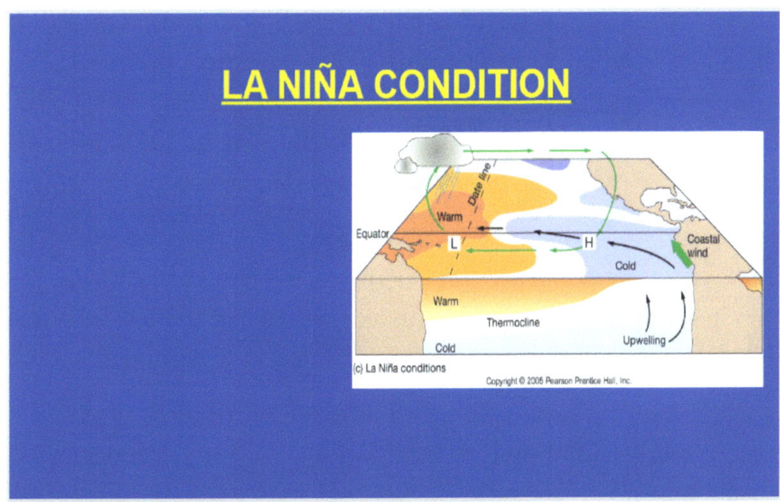

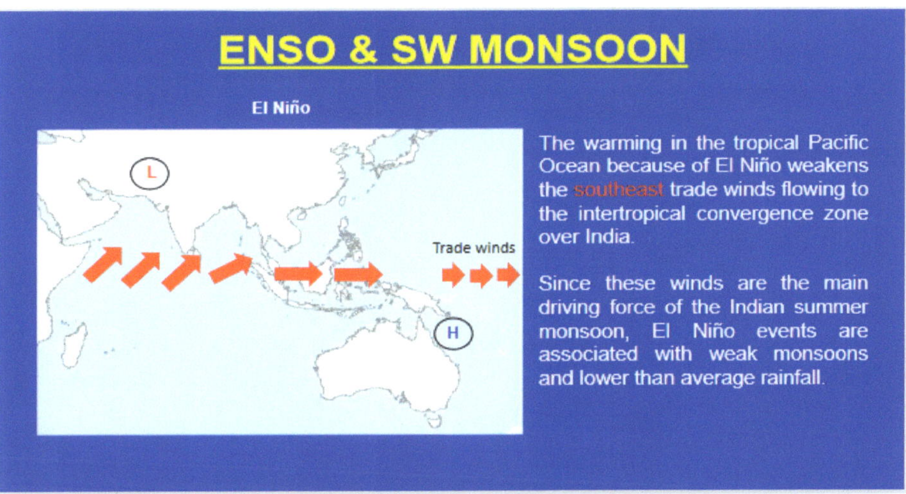

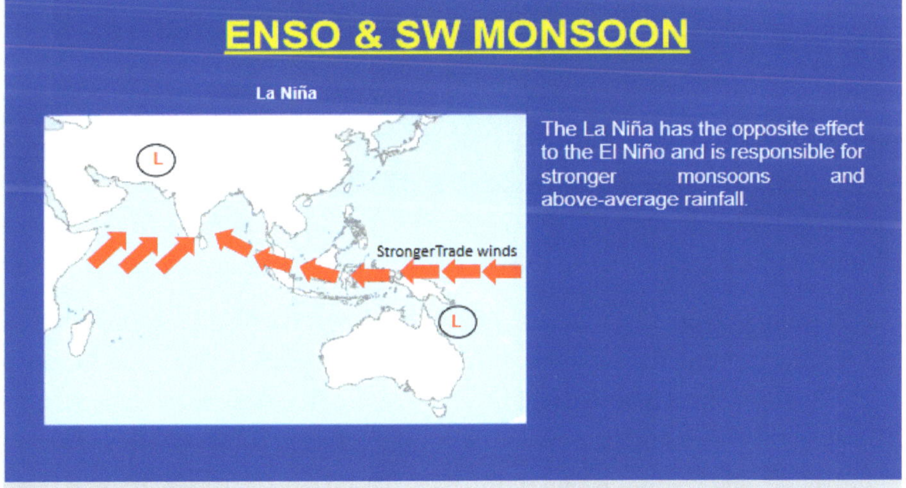

87. **The Indian Ocean Dipole (IOD)**. The difference in sea surface temperature between two areas (or poles, hence a dipole), a western pole in the Arabian Sea (western Indian Ocean) and an eastern pole in the eastern Indian Ocean south of Indonesia has been termed as the Indian Ocean Dipole (IOD). The IOD affects the climate of Australia and other countries that surround the Indian Ocean Basin, and is a significant contributor to rainfall variability in this region.

88. Like ENSO, the change in temperature gradients across the Indian Ocean results in changes in the preferred regions of rising and descending moisture and air. IOD is a coupled ocean and atmosphere phenomenon, similar to

ENSO but in the equatorial Indian Ocean. IOD has a link with ENSO events through an extension of the Walker Circulation to the west and associated Indonesian throughflow (the flow of warm tropical ocean water from the Pacific into the Indian Ocean). Hence, positive IOD events are often associated with El Niño and negative events with La Niña. When the IOD and ENSO are in phase the impacts of El Niño and La Niña events are often most extreme over Australia, while when they are out of phase the impacts of El Niño and La Niña events can be diminished.

89. During positive event, the sea surface temperature (SST) in the western Indian Ocean is relatively warmer than the east. There are anomalies in easterly wind across the Indian Ocean and there is less cloudiness over north-west of Australia. Less rainfall is also experienced over southern and top end of Australia.

90. During negative event, the reverse is the situation. SST over western Indian Ocean is cooler in compared to the east. Winds become more westerly and bring increased clouding to Australia's north-west region. More rainfall is experienced in southern and top end of Australia.

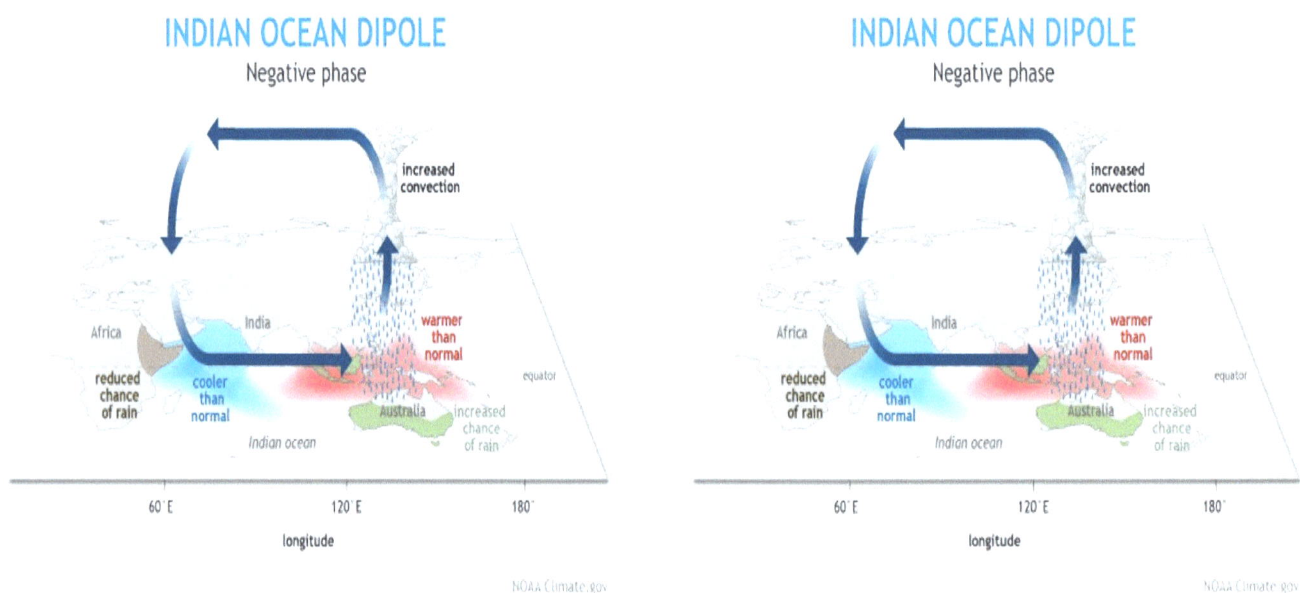

91. IOD involves a periodic oscillation of SST, between "positive", "neutral" and "negative" phases. Positive IOD results in more cyclones than usual in Arabian Sea. Negative IOD results in stronger than usual formation of Tropical Cyclones in Bay of Bengal. The strength of IOD is given in terms of its Dipole Mode Index (DMI) which is the difference of SST between west Indian Ocean (50-70°E; 10°S-10°N) and east Indian Ocean (90-110°E; 10°S-0°).

92. **Madden-Julian Oscillation (MJO).** The Madden-Julian Oscillation is a phenomenon in which there is an eastward spread of large regions of enhanced and suppressed tropical rainfall observed over the Indian and Pacific Ocean. It was discovered in 1971 by Dr Roland Madden and Dr Paul Julian of the American National Centre for Atmospheric Research (NCAR), when they were studying tropical wind and pressure patterns. They observed regular oscillations in winds between Singapore and Canton Island in the west central equatorial Pacific.

93. First, an area of enhanced tropical rainfall is seen over the western Indian Oceans, which spreads eastwards into the warm waters of the tropical Pacific. This pattern of tropical rainfall tends to lose its identity as it moves over the cooler waters of the eastern Pacific, before reappearing at some point over the Indian Ocean again. A wet phase of enhanced rainfall and convection is followed by a dry phase, where thunderstorm activity decreases. Each cycle lasts approximately 1-2 months and there are 8 phases.

Phase 1 – Enhanced convection (rainfall) develops over the western Indian Ocean.

Phase 2 and 3 – Enhanced convection (rainfall) moves slowly eastwards over Africa, the Indian Ocean and parts of the Indian subcontinent.

Phase 4 and 5 – Enhanced convection (rainfall) has reached the Maritime Continent (Indonesia and West Pacific)

Phase 6, 7 and 8 – Enhanced rainfall moves further eastward over the western Pacific, eventually dying out in the central Pacific.

94. The next cycle (with Phase 1) begins after this. Following a region of enhanced convection (rainfall) is a region of suppressed convection (no rainfall). During the MJO cycle there is a 'dipole' (a stark contrast) in rainfall anomalies. For example, In phase 6 there is enhanced convection over the western Pacific and suppressed convection over the Indian Ocean. In phase 2, it is the opposite way around.

95. **Effect of MJO on World Weather**. The MJO creates favourable conditions for tropical cyclone activity, which makes the MJO. The enhanced rainfall phase of the MJO can also bring the onset of the Monsoon seasons around the globe. Conversely, the suppressed convection phase can delay the onset of the Monsoon season. There is evidence that the MJO influences the El Nino Southern Oscillation (ENSO) cycle. It does not cause El Nino or La Nina, but it can contribute to the speed of development and intensity of El Nino and La Nina episodes. The MJO appears to be more active during neutral and weak ENSO years.

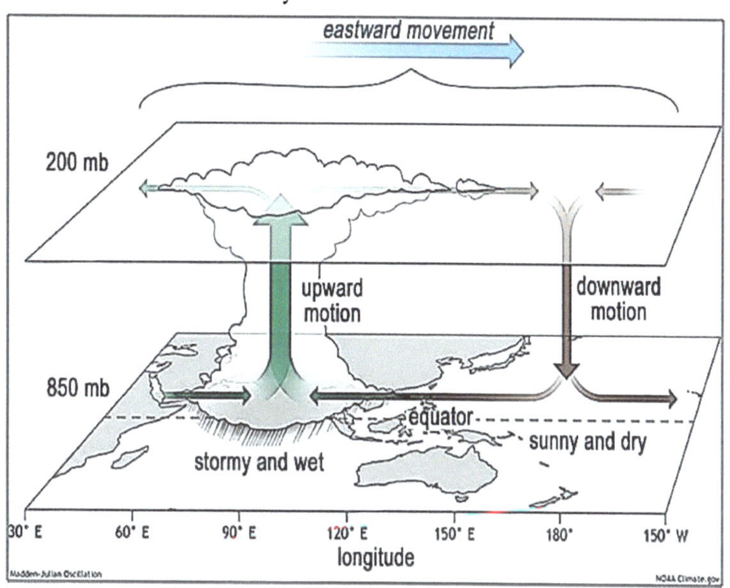

CHAPTER 27

NORTH EAST MONSOON

1. As discussed earlier, there is a complete reversal of the wind pattern in the lower levels in NE monsoon season from the SW monsoon season. The reversal process begins in October. During the southwest monsoon season, the low-pressure area is to the north and the high-pressure zone is located over the Indian Ocean. In NE monsoon season the low-pressure area over north is replaced by the sub-tropical high-pressure zone and the low-pressure area prevails over the Indian Ocean resulting in flow of north-easterly wind flow over Indian sub-continent. This change in surface pressure gradient and lower tropospheric winds is associated with the southward movement of the continental tropical convergence zone (CTCZ) and the subtropical anticyclone in the upper troposphere. The NE monsoon is often described as the retreating phase of the southwest monsoon.

2. During NE monsoon the northern hemispheric winter circulation is established which is dominated by a strong surface high pressure region over Siberia, a low over eastern equatorial Pacific region and secondary shallow lows over the north Indian ocean. The low-level northeasterly flow over the Indian subcontinent and southeast Asia is a salient feature of winter circulation. There are similarities in circulation and rainfall patterns of southeastern peninsular India and other South / Southeast Asian regions and also the established roles of Siberian High, sub-tropical westerly jet stream and the teleconnection influences of global parameters such as ENSO on the NE Monsoon rainfall. There is an asynhronicity in the establishment of north-easterlies over the Bay of Bengal and the South China Sea; the north-easterlies appear over the Bay of Bengal in late October, a month later than over the South China Sea.

3. NE monsoon winds have less vertical extension and are comparatively dry and stable than the SW monsoon current. The India Meteorological Department (IMD) refers to the October to December period as the northeast monsoon, which is a part of the northeast trades.

4. There is diurnal, synoptic, intra-seasonal and annual variation in NE monsoon rainfall. The rainfall is also highly variable both spatially and Temporally. The coefficient of variation (inter-annual) of the north-east monsoon rain 25%, which is more than that of the southwest monsoon rainfall which is 10%.

5. South peninsula is well known for large year to year variability. In 2015, the city of Chennai experienced heavy rainfall causing flood, which resulted significant loss of life and property. Four years later, in 2019, Chennai city experienced severe water shortage and crisis.

6. South peninsula is predominantly influenced by NE monsoon and the area consists of five meteorological sub-divisions, which are Coastal Andhra Pradesh, Rayalaseema, Tamil Nadu, South Interior Karnataka and Kerala. NE monsoon also brings good rainfall over Sri Lanka and Maldives. In India, the NE monsoon season contributes about 11% of its annual rainfall over as a whole. South peninsula receives much more rainfall compared to northern parts of the country. Many districts over the south peninsula receive 30–60% of the annual rainfall during this

season. This season is also termed the retreating monsoon season or the post-monsoon season in which the zone of maximum rainfall migrates to southern parts of India, Sri Lanka and the neighboring sea.

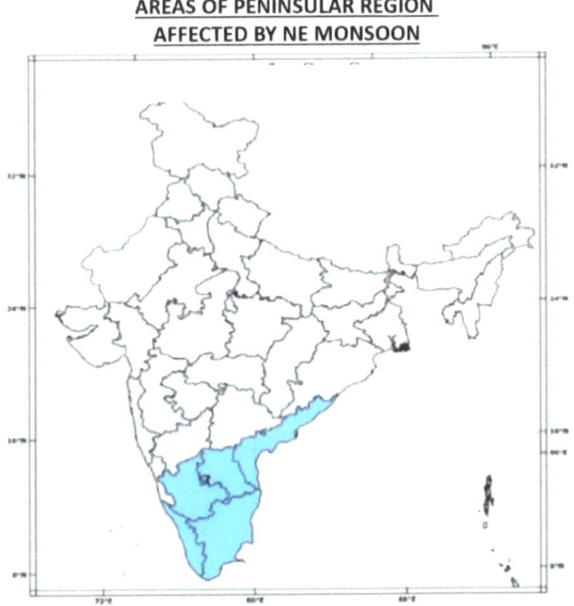

7. With the withdrawal of the Southwest monsoon from the northern parts of India, the mean sea level pressure and upper tropospheric wind circulation patterns over India change rapidly from the summer monsoon type to winter type. By October, the Inter-tropical Convergence Zone (ITCZ) or the monsoon convergence zone which is positioned over northern parts of India starts shifting southwards.

8. NE monsoon season coincides with this retreating phase of the Inter Tropical Convergence Zone (ITCZ). There are also associated rapid changes in upper air circulation and sea level pressure patterns. During the October–December season, there is a clear evidence of rainfall maximum in the Southern Hemisphere.

9. **Sea Level Pressure**. By October, a low-pressure area gets established over the Central and South Bay of Bengal and adjoining east coast, which shifts to south Bay in November and further southwards, close to the equator in December. This low-pressure area is more marked over the southwest Bay in October and November. In the Arabian Sea, the low-pressure area is not well marked during these months. However, an east-west oriented trough of low pressure is observed in nearly the corresponding latitudes as in the Bay of Bengal during October and November. In October, the pressure gradient is generally weak, which gets strengthened by November. While the low-pressure area gets shifted southwards, the high-pressure area associated with the Siberian High also gets strengthened over the northern parts of the country. The surface pressure gradient (high in the north and low in the south) over the Bay of Bengal also strengthens. During November and December, the isobars are nearly parallel to the equator, suggesting stronger surface easterlies/north-easterlies over the Bay of Bengal.

10. **Upper Air Features**. During October, an east-west trough is seen extending from the lower levels to 700 hPa from South Bay to the South Arabian Sea. This east-west trough is seen shifting southwards from October to December, consistent with the equatorward shifting of the ITCZ. By December, the east-west trough is seen close to 15 ^{0}N. This east-west trough is the region with positive vorticity and convergence, thus causing abundant rainfall over this region. This shear zone in the lower troposphere contributes to the genesis of low-pressure systems over the Bay of Bengal and the Arabian Sea.

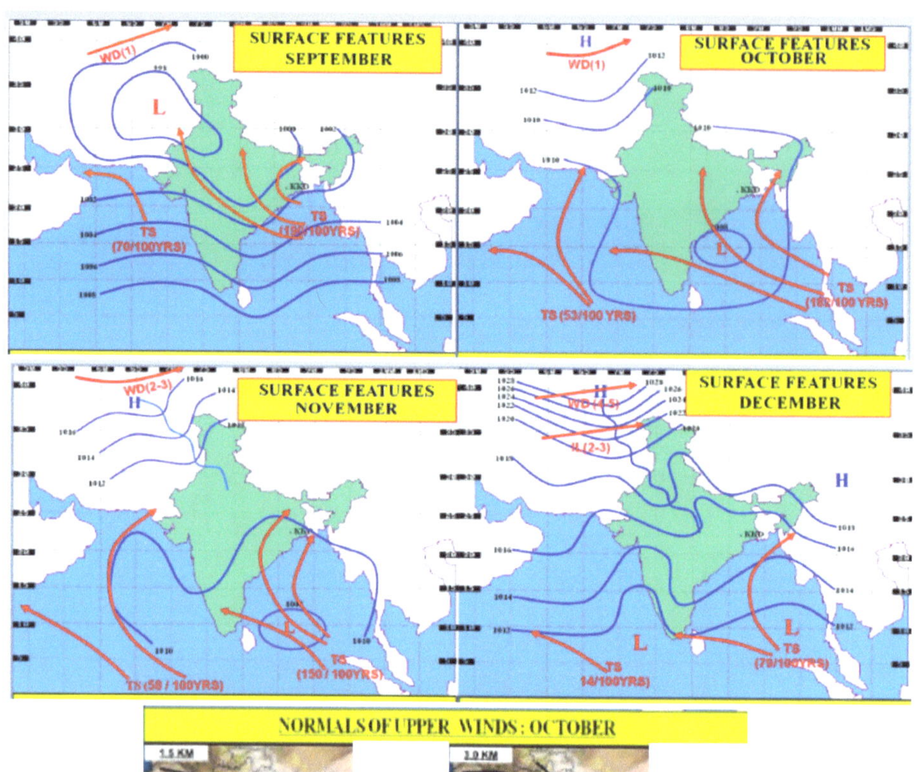
SURFACE FEATURES
SEPTEMBER
SURFACE FEATURES
OCTOBER
SURFACE FEATURES
NOVEMBER
SURFACE FEATURES
DECEMBER
WD(1)
H
WD(1)
WD(2-3)
H
H
WD(4-5)
H (2-3)
L
L
L
L
L
L
H
TS (70/100YRS)
TS (162/100YRS)
TS (53/100 YRS)
TS (162/100 YRS)
TS (58 / 100YRS)
TS (150/100YRS)
TS 14/100YRS)
TS (79/100YRS)

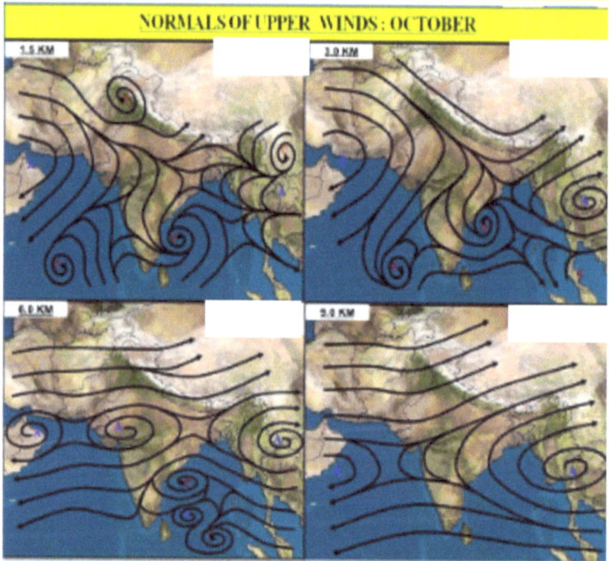
NORMALS OF UPPER WINDS : OCTOBER
1.5 KM
3.0 KM
6.0 KM
9.0 KM

NORMALS OF UPPER WINDS: NOVEMBER
NORMALS OF UPPER WINDS: DECEMBER
1.5 KM
3.0 KM
1.5 KM
3.0 KM
6.0 KM
9.0 KM
6.0 KM
9.0 KM
STJ 110-140 KMH

11. In the middle and upper troposphere (6 Km & 12 Km), easterlies prevail over the south Peninsula and adjoining seas. The sub-tropical ridge at these levels also shifts southwards from October to December. The sub-tropical ridge in the upper levels is an area of divergence and contributes to the intensification of low-pressure systems over the region. The Tropical Easterly Jet (TEJ), which is one of the semi-permanent systems during the southwest monsoon totally disintegrates once the southwest monsoon is withdrawn from the country. From October to December, the sub-tropical westerlies over the northern parts of the country also get strengthened and start moving to lower latitudes. An upper tropospheric ridge gets established over the southern parts of India by October and November at the 12 Km level.

12. **Rainfall**. October is the rainiest month for the South Peninsula. By December, the rainy season is practically confined over extreme south Peninsula including Tamil Nadu.

13. Coastal Andhra Pradesh gets rainfall during both the southwest and northeast monsoon seasons. During the NE Monsoon season, Coastal Andhra Pradesh receives about 320 mm of seasonal rainfall, with October contributing 56% of seasonal rainfall. About 34% of seasonal rainfall occurs during November. The NE monsoon season contributes about 31% of annual rainfall. Over Rayalaseema, the NE monsoon season contributes about 240 mm, which is 33% of the annual total.

14. Tamil Nadu receives more rainfall during the NE monsoon season compared to that during the Southwest monsoon season. There is a considerable increase in rainfall activity from September to October and November. The seasonal rainfall during the NE monsoon season is around 440 mm which is about to 48% of its annual rainfall. Over South-interior Karnataka (SIK) seasonal rainfall is around 200 mm, which is about 19% of the annual total. Over SIK, October contributes maximum rainfall during the season, while December hardly contributes to the seasonal total.

15. Over Kerala, the NE monsoon season contributes about 490 mm, which is about 17% of the annual total. During the NE monsoon season, Kerala receives maximum rainfall, even slightly more than Tamil Nadu. October contributes maximum rainfall over Kerala, which reduces in November and December.

16. **Cyclonic Storms (Depression and Above)**. In October, north Tamil Nadu, Coastal Andhra Pradesh, Odisha and West Bengal experience the landfall of depressions or cyclonic storms. Some storms after moving northwestwards recurve towards northeast. North Tamil Nadu and coastal Andhra Pradesh experience Tropical cyclonic storms. In November, most of the storms forming over the Bay of Bengal move northwestwards and make landfall over Tamil Nadu and the coastal Andhra Pradesh. In November, a few storms recurve and make landfall over West Bengal and Bangladesh. In December, cyclonic storm activity generally decreases. In December also, when storms form, they move northwestwards and make landfall over Tamil Nadu. A few storms recurve and make landfall over coastal Andhra Pradesh and Bangladesh. There are a few storms, which form over the Bay of Bengal, cross the south peninsula and emerge in the Arabian Sea, thus making longer lifetime. The systems forming over the Arabian Sea, either move westwards towards the Arabian sub-continent or towards Gujarat/Pakistan. The storms forming over the Arabian sea generally do not make landfall over the south peninsula. They tend move northwestwards. A few storms However, move north and affect the Gujarat coasts.

17. **Easterly Waves/Troughs in Easterlies**. Easterly wave is a wave within the broad easterly current and moves from east to west, generally more slowly than the current in which it is embedded. Although best described in terms of its wave like characteristics in the wind field, it also consists of a weak trough of low pressure. To the west of the trough line in an easterly wave, there is generally found divergence, a shallow moist layer, and exceptionally fine weather. The moist layer rises rapidly near the trough line; in and to the east of the trough line intense convergence, cloudiness and heavy rain showers prevail. Easterly waves occasionally intensify into tropical cyclones over the Bay of Bengal.

18. So far, there has not been adequate study on easterly wave over Indian Ocean region. However, some of the few studies carried out suggests passage of westward propagating disturbances as predecessors to formation of monsoon lows and depressions. It is also found that the period of the wave is about 5 days, speed of about 6 ms^{-1} and wavelength of 2300 km. One of the studies by Balachandran (1998) finds that the easterly waves are associated with inverted V-shaped cloud mass. Using satellite imageries, he also attempted to find the speed of a wave (8.2 kt).

19. In general, northerly meridional winds, subsidence, divergence and fair weather are the general atmospheric characteristics ahead of an approaching easterly wave trough and southerly meridional winds, rising motion, convergence and active weather are the characteristics behind the wave trough.

20. It is seen that many rainfall episodes over Tamil Nadu during NE monsoon season have been associated with easterly waves and during episodes of large deficiency in rainfall over Tamil Nadu, the easterly waves have been found to be confined to east Bay of Bengal.

21. Broadly, Easterly Waves can be defined as a perturbation in the easterlies close to the ITCZ propagating from east to west. These waves are normally seen in lower and middle troposphere. However, over Indian region it is seen in the middle and upper troposphere moving from Bay of Bengal to the Arabian Sea through southern Peninsula.

22. The formation of a trough in the geostrophic wind, concave towards the lower pressure, is called a trough. As the wind crosses a reference latitude towards the south, a curve is observed towards high pressure and it is called a ridge in the wave. Ahead of a trough where the wind in the wave is slowing down and converging, some air gets pushed upward away from the surface, producing low pressure area near the surface. Conversely, ahead of an upper air ridge, where the air is speeding up and diverging, air gets 'sucked down' into the long wave producing subsidence and higher pressure near the surface. Lower level divergence, subsidence and fair weather condition are found ahead of the trough axis. Convergence, ascending motion and rainfall activity are found to the rear.

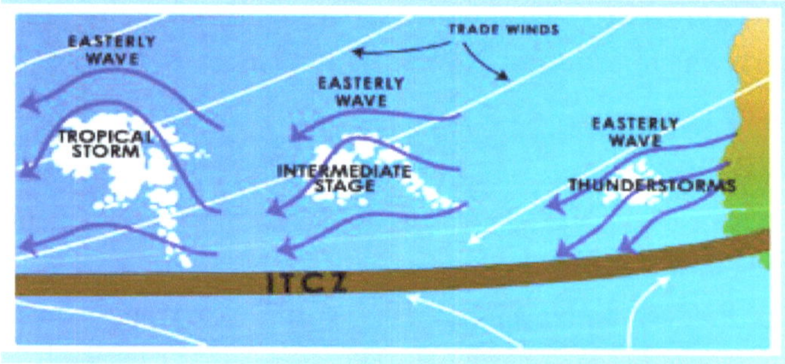

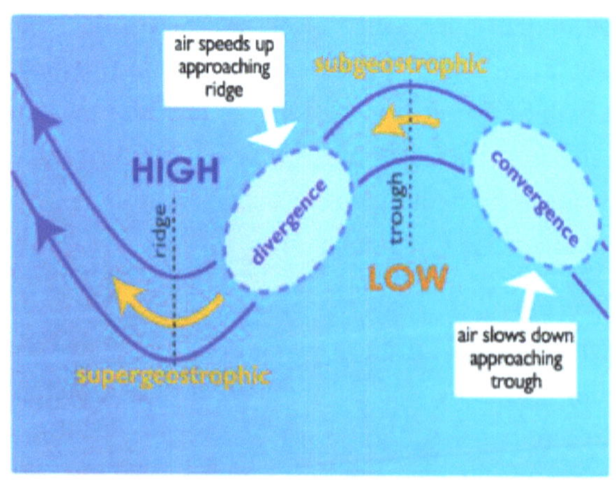

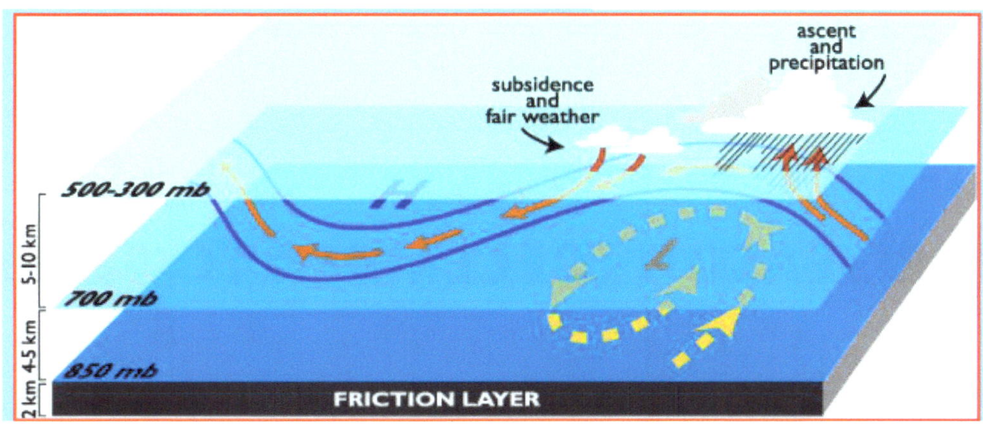

Locations of ascent and subsidence in an easterly wave

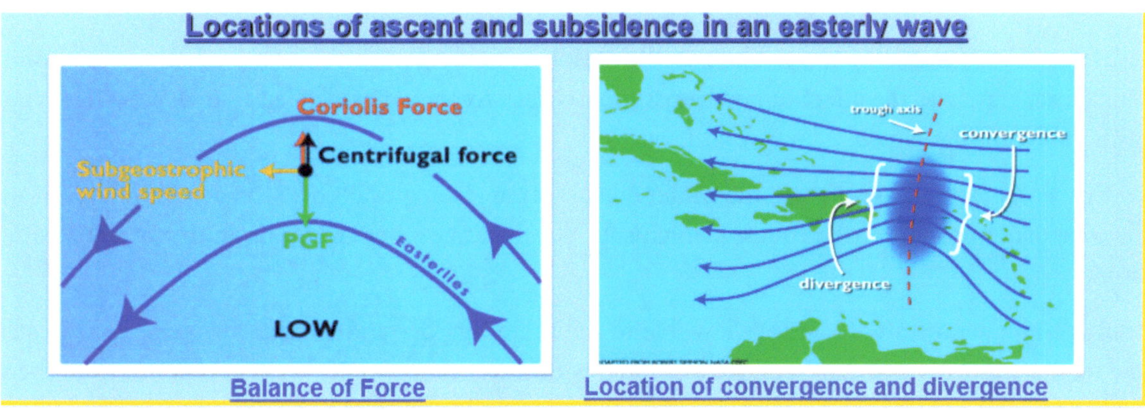

Balance of Force

Location of convergence and divergence

WINTER SEASON (DECEMBER TO FEBRUARY)

1. Winter is the coldest season of India. Though, the season coincides with the north-east monsoon season, it needs to discussed separately as the effect of north-east monsoon in respect of associated rainfall activities is confined to the southern Peninsular region.

2. During this season, as discussed before, the winter northeast monsoon prevails over the subcontinent. An anticyclone prevails over major portion of the country which subsides the atmosphere resulting in dry weather with clear sky condition.

3. Over north India, temperatures are very low due to the prevailing dry and cold north-westerly to westerly winds over the regions. The dry and cold spell are disrupted with travel about 4 to 5 WDs per month from the west. The WDs cause increase in clouding, rainfall over plains and snowfall over the hills.

4. **Surface Pressure Pattern**. There is large high-pressure belt extending from Sahara to Siberia. The high-pressure belt is, however, not continuous but breaks up into smaller cells due to the effect of topography like land, sea and hills. Hence, there are regions of low-pressure zone within the high-pressure belt, mainly over the areas like the Arabia, Black Sea, Caspian Sea, Nile region, Mediterranean Sea, Red Sea and Persian Gulf. These low-pressure areas deepen within the high-pressure belt and move towards east-north-east or north-east in form of low or depression, sometimes with their secondary systems to their south. The primary disturbances move across much northern latitudes.

5. Over Indian region, the secondary systems that form in the coastal areas of Nile, Upper Egypt, Sudan, Red Sea, Gulf of Oman and Northeast Arabian Sea are of importance. These move north-east ward and affect north Indian regions. The pressure distribution for January is shown in the map below: -

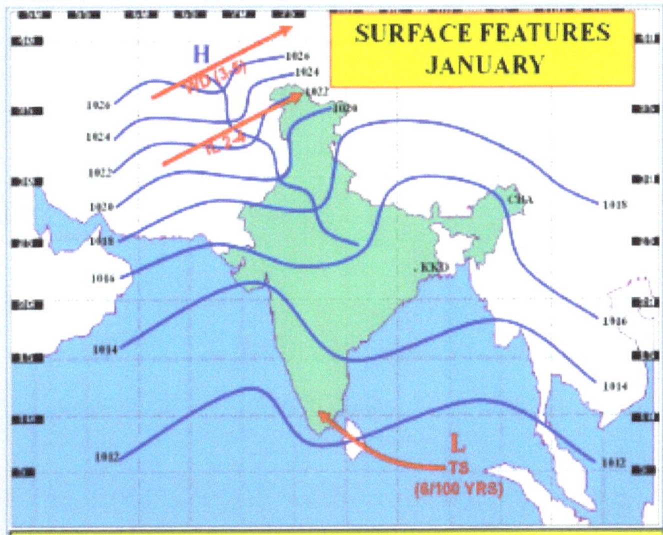

6. **Upper Air Circulation**. At lower levels, a CYCIR can be marked on most of the days over north-west part of the country or/and adjoining north Pakistan which are associated with WDs. Cyclonic circulations can also be marked over Indian ocean adjacent to south Arabian Sea and south Bay of Bengal. Another CYCIR can sometimes be marked at lower levels over Assam, mostly associated with secondary WDs.

7. What is more significant to the cold season is the lower level anticyclone that predominantly prevails over central parts of the country. This anticyclone is mainly responsible for setting up the north-east monsoon mechanism over the Peninsular region. At middle levels, this anticyclone shifts south ward and joins the sub-tropical ridge, which also shift south ward with height.

8. The STJ shifts well to the south of Himalayas. In February, it can be marked roughly along 22^0 N latitude. The Tropical easterly jet stream (TEJ) disappears during the previous season.

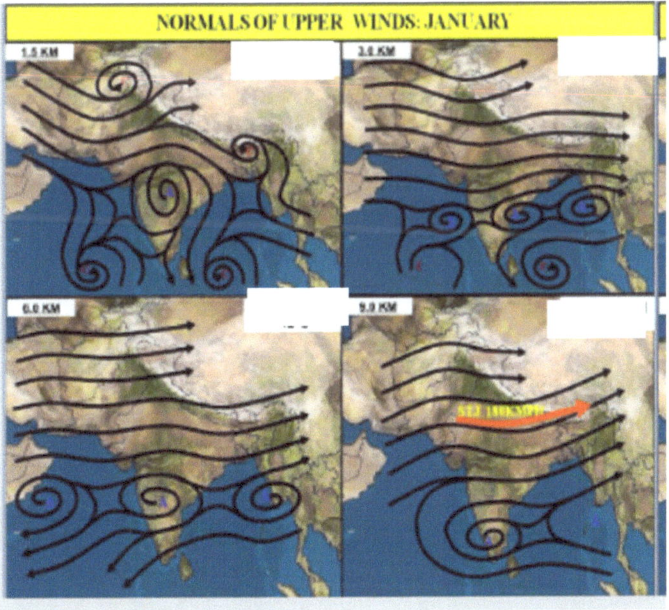

9. **Weather**. Outside the southern Peninsular region mainly cold dry north-easterly winds prevail at surface during the season. Low temperature, clear sky and lower humidity are the main characteristics of the season. There is large diurnal variation in temperature in this season.

10. The dry spell over north Indi are broken with movement of westerly systems which leads to increase in moisture, increase in clouding and rainfall activities. Frequency of WDs is about 4 to 5 per month. Their southernmost tack can be seen in January and February.

11. Most of the WDs that affect Indian region are in occluded stage. Appearance of high cloud, fall in pressure, rise in temperature indicates approach of the WDs of warm front nature. On approach, the clouds lower to medium and low clouds, and then drizzle and rain commence. Passage of the WDs is indicated by change in wind direction, formation of convective clouds, convective precipitation, fall in temperature and rise in pressure. After the passage of the system, cold and dry sets in and cause cold wave condition. Prolonged widespread fog also occurs due to rapid clearance of clouds and availability of moisture due to recent rain.

12. In general, fog remains up to late morning or mid-day. However, very dense fog sometimes may prolong for days together.

13. Thermal inversion in the lower tropospheric levels continue to persist. However, there is decrease in the pollution level over northwest part of the country. The main reason for this decrease in pollution is decrease in massive burning of agricultural waste. The other important factor is the rainfall episodes over the region due to movement of WDs.

14. Under the influence of the stronger WDs, induced systems form at lower latitudes (around central Pakistan or Rajasthan. Such features move east-north-east ward and affect weather over north MP, UP, Bihar, GWB, and even up to NE India.

15. Sometimes, a trough extends from Kerala to Gujarat along the west coast of the Peninsula. Under favourable conditions, this trough extends right up to Rajasthan. Often, a cut-off low forms and moves in eastward causing rainfall in the central parts of the country and south Uttar Pradesh.

16. South of latitude 18°N, the north-easterlies, after picking up moisture from the Bay of Bengal and the far easterly marine winds joins the circulation of north-east monsoon, which in turn brings precipitation over the southern Peninsular region. Also over the Peninsula, westward moving easterly waves cause cloudiness and precipitation. Sometimes a cyclonic storm also may cause heavy rain.

17. The figures below show the rainfall distribution and normal minimum temperature over the country during winter season. (Source- IMD.

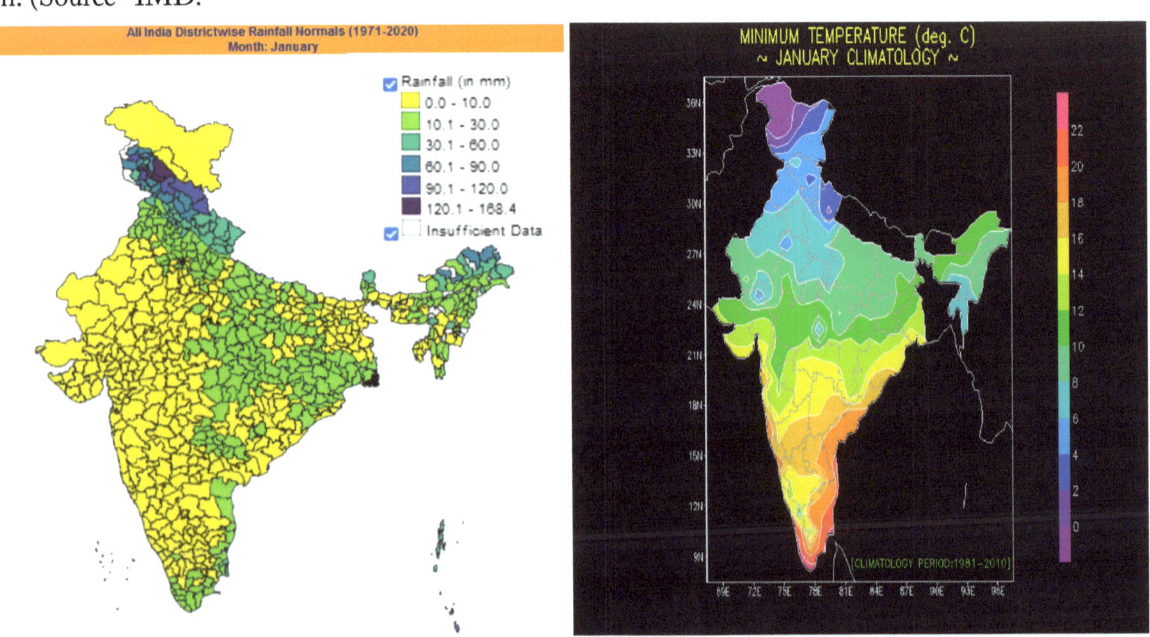

CHAPTER 29

PRE-MONSOON SEASON (MARCH TO MAY)

1. This is the season when the pressure and wind patterns of the winter season tend to change into the south west monsoon patterns and can be called as the transitional season or the intermittent season. Prior to the establishment of the monsoon pattern, the pressure and wind patterns get disrupted. Widespread dust haze over north India and extreme high temperatures over most parts of the country are the main characterises of this season. North-west and north India is affected by Andhis and the eastern part is affected by severe thunderstorms which are called the norwesters or 'Kalbaisakhis'. Most parts of the peninsular region also experience thunderstorm accompanied by precipitation. NE India experiences severe thunderstorm with heavy downpour which, sometimes cause flood in Brahmaputra or Barak valley much before the arrival of the monsoon. The thunderstorm activity is actually triggered by incursion of moisture from the seas into the dry and hot regions due to movement of WDs across middle latitudes and location of anticyclone over or in the vicinity of sea which pushes cool moist air into the lands. Anabatic and Katabatick winds become prominent over the hilly regions, particularly over north-eastern states resulting in thunderstorm during evening or late-night periods. The path of the westerly systems shifts northward. Land and sea breeze is prominently felt over the coastal regions.

2. **Surface Pressure Pattern.** As mentioned before, this is the season when the pressure pattern of the winter season tends to change over to the pattern of the monsoon season. The pressure gradient (north-south) gets disturbed. As the Sun (Solar Equator) crosses the equator and move towards north, the entire land mass of India gets heated up. Due to continuous and rapid rise in temperature, a low-pressure area forms over the Peninsular region by beginning of March or towards end of February. This low-pressure area shifts northward as the season progresses. By April the feature is seen over central India around Madhya Pradesh and by May the low further shifts northward and then settles over Rajasthan and adjoining Pakistan by June. Sometimes, the low can also be marked over eastern sector around Bihar during April/May.

3. Pressure gradient decreases rapidly as the season progresses. The difference between extreme north and south of the country becomes very less (About 2-3 hPa). Hence, winds are light over most parts of the country. However, towards end of May or beginning of June, steeper pressure gradient build-up around the seasonal Low over north India resulting in prevalence of strong hot and dry winds (Loo).

4. A trough of low covers almost the central region of the country and comparatively high-pressure cells exist over the Arabian Sea and Bay of Bengal.

5. Increase in sea surface temperature over the Bay of Bengal and Arabian Sea results in formation of tropical cyclones. Over the Bay, the cyclones generally form over south-east part between 10^0 to 15^0 N. They initially move north or north-west ward, but often re-curve towards north-east ward and hit coasts of Andhra Pradesh, Odisha, West Bengal, Bangladesh or Myanmar causing extensive damage and devastation. Some of the storms, however, move northwest and cross the Tamil Nadu coast. Some of these, after travelling over the Peninsula as weak depressions or

lows, emerge into the Arabian Sea and revive. Over Arabian Sea, the storms form over south-east of the sea around Lakshadweep region and generally move north-west ward. A few of the storms may re-curve towards north or north-east and make land fall over Gujarat or Maharashtra coasts.

6. The normal pressure pattern for the month of April is shown below.

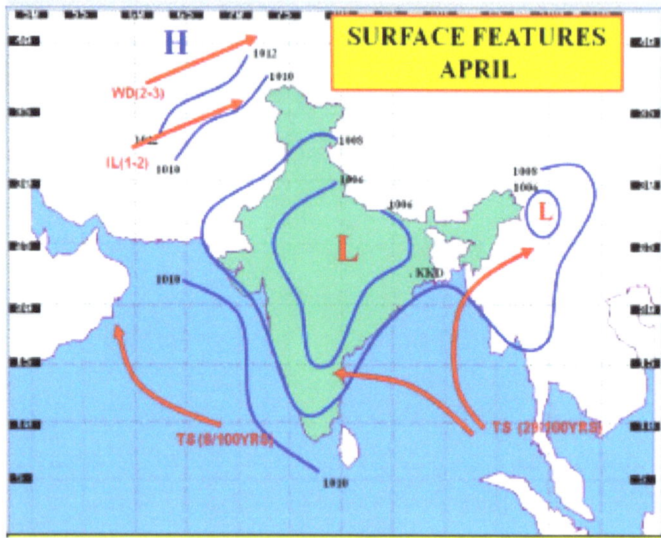

7. **Upper Air Pattern**. The main upper air features are as follows: -

a) At lower levels, mainly north-westerly to westerly winds (dry and hot) prevail over north India and easterly to north-easterly over southern Peninsula. However, during afternoon/evening, local sea breeze prevails over the coastal region. The sea breeze during the season is the strongest and often penetrate much deeper into the land area.

b) The sub-tropical ridge (anti-cyclone over south India shifts southward with height.

c) The sub-tropical Jetstream moves north of Himalaya.

d) On most of the days a north-south line of wind discontinuity can be marked over Peninsular region at lower levels (up to 900 M to 1.5 Km) with its location and orientation changing on a day-day basis. Sometimes this line of discontinuity can be marked much to the north (Bihar/Jharkhand. This feature, where winds from almost opposite directions and with different characteristics meet, is responsible for the thunderstorm activity around it.

e) The low-level anticyclones normally located over Bay of Bengal and Arabian Sea sometimes lead to moisture incursion into east and north-west regions respectively. Such moisture incursion causes severe thunderstorm over eastern parts and dust storm over north-west/northern parts.

f) Low-level winds over north-east India are generally westerly. However, change in the orientation of the anti-cyclone over north Bay, occasionally supply moisture to the region and results in thunderstorm activities.

8. **Weather**. The eastern sector experiences maximum rainfall, which is about 50 to 250 mm per month. Rainfall events are mostly accompanied by thunderstorm. Thunderstorm activity increases and become widespread whenever there is moisture incursion from the sea under any favourable synoptic situation. Bengal, Assam, Orissa and east Bihar are the areas which are most prone to thunderstorm. When influx of moisture is less like in north-western part of the country, dust storm occurs.

9. Occurrence of dust haze is common over north and central India. With dry loose soil and strong winds, on most of the days an extensive area covering Pakistan, Rajasthan, Jammu, Punjab, Haryana, north Madhya Pradesh and

Uttar Pradesh gets covered in dust. At times the suspended dust migrates east ward with the zonal westerlies. Dusty weather is experienced as far east as upper Assam across Bihar and West Bengal.

10. Sometimes, due to strong surface winds of the order of 20 Kt or more occurs over north Indian region under the influence of steeper pressure gradient. Such strong winds raise dust and other loose articles making the weather condition very unpleasant and unhealthy.

11. Over southern Peninsula, Tamil Nadu, Karnataka, Kerala and Telangana prominently experience pre-monsoon thunderstorm.

12. Normal maximum temperature and rainfall for the month of May are shown below (Source-IMD).

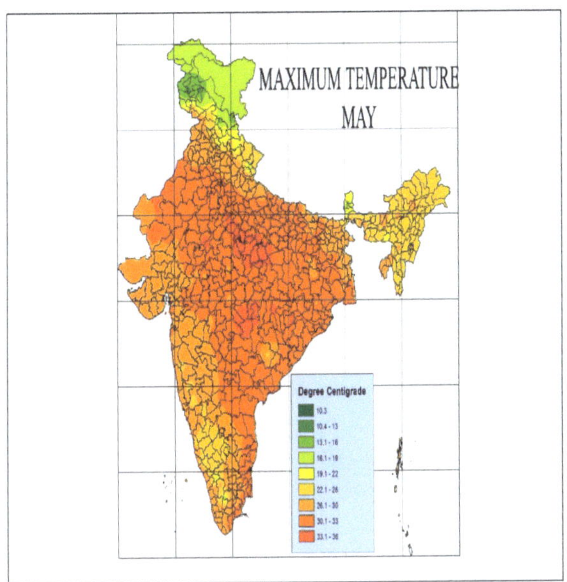

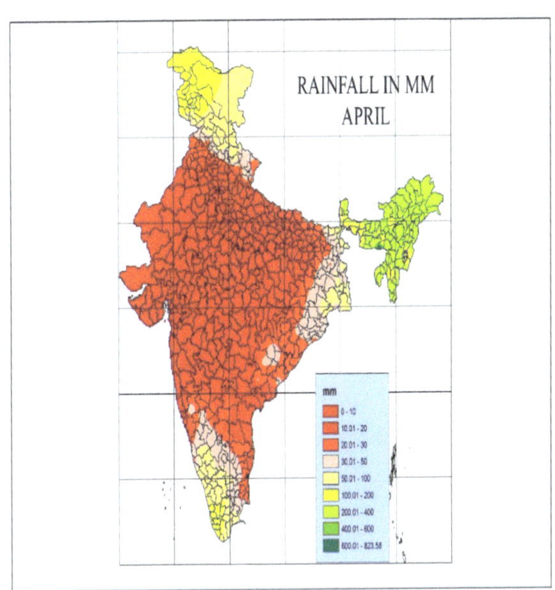

CHAPTER 30

POST-MONSOON SEASON (OCTOBER TO NOVEMBER)

1. Post-monsoon season commences over the country with the withdrawal of south-west monsoon which begins from the extreme north-west part by September. With the withdrawal of monsoon, the monsoon currents are replaced by dry north-westerly to north-easterly winds. Withdrawal takes place in stages towards east and south. By mid of October monsoon withdraws from the entire country and the north-east monsoon sets in over southern Peninsula. This season is the transitional period in which the summer monsoon pressure and wind pattern tend to changeover to the winter pattern which is opposite to the summer pattern. This season is also called the retreating monsoon season. Gradual fall in temperature, rise in pressure, dry weather and clear sky condition over north India and rainfall over south peninsula are the main characteristics of the post-monsoon season.

2. **Pressure Pattern**. It being a transitional season, a diffused and ill-defined pressure pattern prevails over the country during this season. Like the pre-monsoon season, the pressure difference between extreme north and south of the country is just 2 to 3 hPa.

3. The track of the extra tropical systems shift south ward during this season and the extreme norther parts are affected by the systems.

4. During October, there is maximum frequency of tropical cyclone over Bay of Bengal. The storms generally form over or in the vicinity of Andaman Sea region and move towards Andhra, Odisha or Tamil Nadu coast. Severe cyclones form in the region during November also. These storms generally hit Tamil Nadu coast. Some of the storms re-curve towards north or north-east and strike West Bengal or Bangladesh coast. Also, some of these cyclones may enter into the Arabian Sea, revive, and then strike Gujarat or Maharashtra coast. The normal surface chart for October and November and the upper air chart are shown below

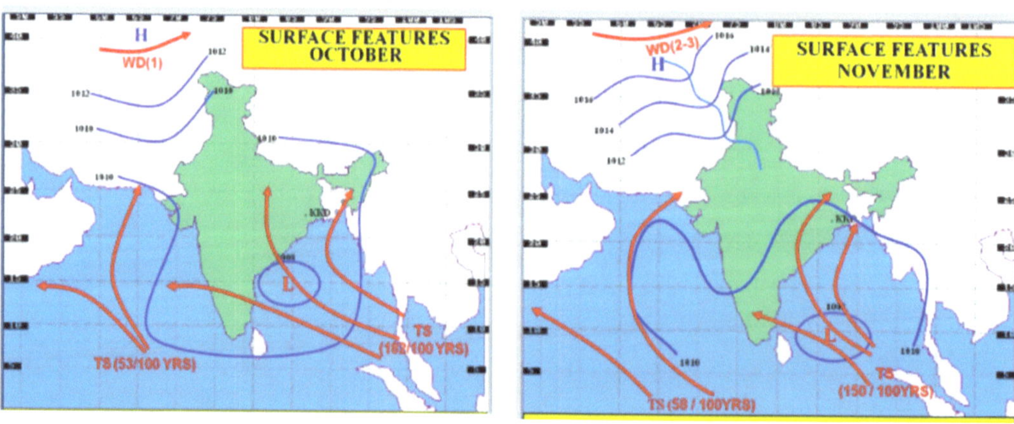

5. **Upper Air Pattern**.

    a) As mentioned before, mainly north-easterly winds prevail over most part of the country at lower tropospheric levels. Over extreme north-west and west, winds of westerly components can be seen. Over southern Peninsula, winds from Arabian Sea and Bay of Bengal converge. A CYCIR can be marked over south-west Bay off Tamil Nadu coast.

    b) In upper levels, the zonal westerlies along with the sub-tropical jet stream shifts south ward and during November the jet can be marked well to the south of Himalayas. As the season progresses, the zonal westerlies become stronger. The easterly jet in the southern latitudes disappears. The sub-tropical ridge starts shifting southward and during October it can be marked roughly along 18^0 N latitude.

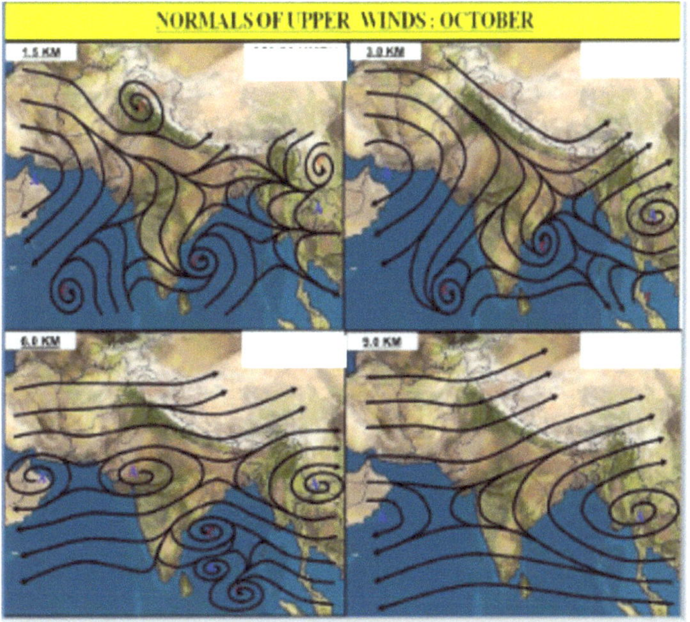

6. **Weather**.

    a) As the southwest monsoon retreats, cold dry continental air sets in over northwest India, resulting in fine clear sky condition. However, as the season progresses and temperature falls, north-western part of the country starts experiencing air pollution due to suspended smoke particles. The dawn and dusks are characterized by mist. The smoke particles in the lower levels mix with the mist and give the worse look of pollution.

    b) About 1 to 2 western disturbances affect the extreme north during the season. Precipitation may occur, particularly in Kashmir and Punjab. The high- altitude regions over J&K and Himachal experience snowfall as early as October.

    c) By mid of October, in northeast India also the weather generally improves. However, when the Bay of Bengal storms cross the Bengal or Bangladesh coast, influx of moisture causes clouding and weather in this region also.

    d) Southern Peninsular region experience NE monsoon rainfall activities. When the Bay storms cross either the Tamil Nadu coast or move northeast skirting the Andhra Pradesh and Odisha coast, weather in the whole coastal belt gets affected. Widespread rainfall with thunderstorm and squall are associated with the storms. Some of these storms causes extensive damage in the Andhra and Tamil Nadu coastal belt.

7. **Temperature Inversion During Post-Monsoon and Winter Seasons**. Temperature inversion or thermal inversion is a case when temperature in certain layers of troposphere increases with height or exhibit a lesser rate of decrease in temperature with height. Normally, temperature decreases by about 2 0C per 1000 Ft in the troposphere. However,

due to certain circumstances, the reverse may happen within a certain layer. Decrease in temperature in the normal rate with height results in keeping the higher layer cooler and denser. This maintains the lighter air in the lower levels to rise in a normal vertical speed. This subject has been discussed elaborately in the 'Stability and Instability of the Atmosphere'.

8. During late post-monsoon season and winter season, the surface of the earth cools rapidly during night due to outgoing radiation. With cooling of the surface, the air layer close to the surface also cools nearly at the same magnitude. However, this cooling process effect only the air layer which is very close to the surface, say up to about 3,000' above. However, the layers above remain unaffected by the radiational cooling and remain warmer than the layer below. This results in arrest of the vertical movement of the air close to the ground. In other words, the colder air near the ground remain settled at the bottom as the layer above it is warmer and lighter. The cold air warms up only when the sunlight penetrates and heat it up. However, during winter season, due to occurrence of smog or fog, there is a delay in penetration of sunlight to the surface, which make the inversion state to remain longer.

9. The smoke particles produced by local pollution or migration of smoke from north-west over Delhi and neighbouring areas, thus, remain locked in the air layer close to the surface, which in normal thermal behaviour of atmosphere, would have ascended up. This situation arises in Delhi and neighbourhood every year from late October to November. The situation improves by December when smoke generation by burning of agricultural waste ceases or decreases and a few spells of rain due to movement of WDs affect the region.

10. There are some other reasons for inversion. However, the reason pertaining to increase in air pollution over Delhi and neighbourhood during post-monsoon season has only been discussed here.

11. Normal rainfall and minimum temperature charts for the month of November are displayed below (Source – IMD).

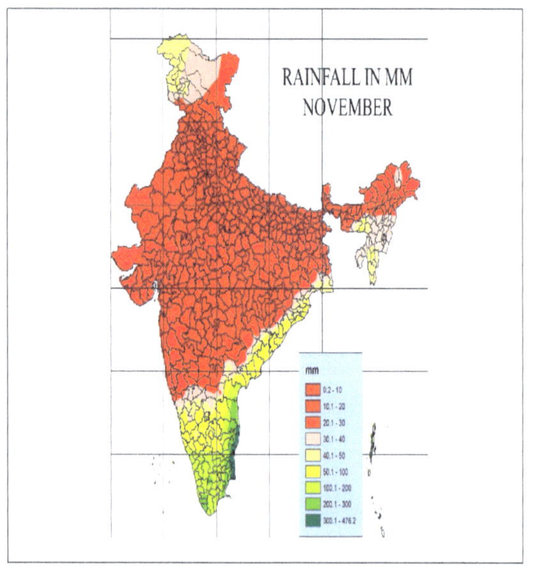

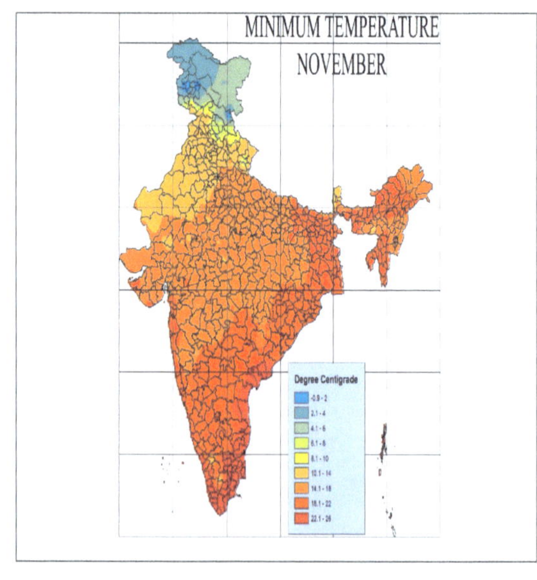

CHAPTER 31

REGIONAL CLIMATE BRIEF

1. Weather conditions over most of the states in India are not homogeneous across the states in a specific period. It is mainly due to the varied geographical features and topography of the different places. For example, a part of a state if adjacent to the sea generally experiences different weather condition than the interior parts. Likewise, a place close to the hills generally experiences more thunderstorm, and hence, more rainfall (if located at windward side) than those parts which are away from the hills or lies in the lee ward side. Therefore, a climatological summary prepared briefly may not represent all the areas appropriately. However, an attempt has been made to describe the climatological conditions of the states aptly. Wherever possible, states have been divided into more than one region.

Ladakh

2. Ladakh is known as the cold desert of India. This region receives rain/snow fall mostly under the influence of western disturbances. However, it being located to the east of the high western Himalayas ranges, is a rain shadow area in respect of the eastward moving western disturbances and hence, receives scarce rain/snow fall. Monsoon current or the pre-monsoon thunderstorm rarely affects the region. Hence, the annual precipitation received in the region is scarce. The altitude being very high (6000 M ASL in average) and mostly being influenced by dry and cold continental winds, the region experiences cold condition in most of the months.

3. **Winter Season (December to March)**. During winter Ladakh region falls under the extensive high-pressure belt which extends from Sahara to Siberia. During this season the temperature is very low. The weather pattern sometimes resembles Arctic weather condition. Generally, Fair weather condition prevails over the region when not affected by western disturbance. On arrival of western disturbance, clouding increases and snowfall occurs. When the western disturbance passes away the region is affected by strong cold and dry continental air (N/NW'ly) leading to further drop in temperature. About 5 to 7 western disturbances per month (December to March) affect the region during the season.

4. **Pre – Monsoon Season (April – June)**. During this season as the sun crosses equator and moves northwards, temperatures start rising and pressure falls gradually. Weather generally remains fair during this season. Frequency of western disturbance decreases to 4 to 5 per month during the season. Winds at surface are light or variable during morning hours and they pick up strength during afternoon to around 25 Kt. Wind speed of 35 to 40 Kt is also common during afternoon/evening hours.

5. **Monsoon Season (July – August)**. SW Monsoon and Monsoon Depression generally do not affect Ladakh. The prevailing mountain ranges which are mainly northwest-southeast oriented, cut off the moisture before it reaches Ladakh. Occasionally, recurving Monsoon Depression breaks up over Pir Panjal range. In such condition prolonged clouding and heavy rainfall occurs over south and southeast sector of Ladakh. Frequency of WD further decreases to 4 to 5 per month. Wind pattern at surface is somewhat similar to that of pre-monsoon month.

6. **Post Monsoon Season (September – Nov).** Temperature starts falling and pressure starts increasing during this season. Average frequency of WD is 3 to 5 per month. There is decrease in the wind speed at surface (10-15 Kt during evening).

7. **Rain/Snowfall**.

Month	Frequency of Rain / Snowfall (Days)	Duration in a day (Hour)	Monthly Total Precipitation (mm)
Jan	7	2-3	18
Feb	6	1-2	6-7
Mar	6	3-4	5
Apr	6	6-7	3-4
May	8	6-7	3
Jun	8	2-3	5
Jul	9	5-6	13
Aug	10	4-5	15
Sep	6	4-5	8
Oct	4	6-7	4
Nov	1-2	4-5	2
Dec	4	2-3	4-5

8. **Temperature(^{0}C)**.

Month	Jan	Feb	Mar	Apr	May	Jun	Jul	Aug	Sep	Oct	Nov	Dec
Mean Temp	-9	-6	0	5	10	14	17	17	13	6	0	-5
Mean Min Temp	-14	-12	-6	-1	3	7	10	10	5	-1	-7	-11
Mean Max Temp	-3	0	6	12	17	21	25	24	20	14	7	1

Kashmir

9. The Kashmir region is bounded to the west by the western end of the Himalaya and by the Pir Panjal range to the east. To the south is Himachal Pradesh and Jammu region. The climate of the region is mostly influenced by western disturbances and cold and dry continental air. However, the Himalayan ranges saves the region from extreme cold polar air mass.

10. **Winter Season (December- March).** Mostly dry and cold continental winds prevail over the region. The dry spells are disrupted due to passage of western disturbances which brings the maximum of the annual rain/snow fall to the region. On approach of the western disturbances, winds become westerly, on arrival the winds veer to south-westerly to easterly and when passes away, again dry continental north-westerly prevails. The winter western disturbance enhances the beauty of the region with massive snowfall. Thunderstorm and hailstorm also occur occasionally. The number of rain/snow days increases as the season progresses from December to March. Maximum precipitation takes place in the month of March.

11. **Pre-Monsoon (April- June).** Weather generally warms up during this season. Snowfall still occurs over the hills under passing western disturbances. The number of rainy days decreases as the season advances. The amount of precipitation reduces as the season progresses from April to June. There is a definite increase in thunderstorm activity in pre-monsoon season.

12. **Monsoon (July- September).** Onset of Southwest monsoon over Jammu and Kashmir is in early part of July and withdraws by the middle of September. In this season, lows and depressions sometimes reach Rajasthan and Punjab. Under the influence of these depressions, South Eastern part of the region experience flood. As the monsoon current is practically blocked by the Pir Panjal range, the region hardly comes under the sway of monsoon. A few western disturbances still affect the region during this season.

13. **Post Monsoon (October - November).** During this transitional period between monsoon and winter, Kashmir region experiences the best weather condition. The sky remains mostly cloud free. There is gradual decrease in temperature making it a pleasant season for the locals (Though the tourists from the other parts of the country may feel it chilling). As the season advances, westerly waves passing to the north of the region start affecting the region and there is increase in clouding with occasional rainfall. Upper reaches of the mountains start getting snowfall. However, as a whole this is the season of the least precipitation.

Annual Time Series of Weather Elements

14. **Rain/Snowfall.** Maximum number of rainfall days is 16 days in the month of May and the minimum of about 04 days in November. Maximum rainfall (Also includes snowfall) of about 112 mm is received in the month of March while minimum of about 25 mm received in November. Average annual rainfall is about 735 mm. About 45% of rainfall is received during winter season, and about 23% is received during monsoon season. Post monsoon season adds only about 7% of annual rainfall.

15. **Temperature over(^{0}C) and Rainfall (mm): Kashmir valley.**

Month	Jan	Feb	Mar	Apr	May	Jun	Jul	Aug	Sep	Oct	Nov	Dec
Mean Temp	0	2	7	12	17	20	21	21	18	13	8	3
Mean Min Temp	-4	-2	2	7	11	15	17	17	13	8	3	-2
Mean Max Temp	5	6	11	17	21	24	25	25	23	19	14	9
Rainfall	241	321	323	279	183	125	440	412	171	92	110	156
Rainy days	11	12	13	12	10	8	14	14	8	7	6	7

Jammu

16. Jammu region lies within the humid sub-tropical zone like the rest of the north-western India. The remarkable topography of this region comprising of the flat terrain to the south and the high mountain ranges to its north is responsible for the extreme contrast in climate. Summer temperatures over this region are above 45 ^{0}C whereas winter season experiences near zero and occasionally sub-zero temperatures. Maximum rainfall is received during monsoon season. However, a good amount of rainfall is also received during winter. Due to the effect of local topography and interaction between westward and eastward moving systems, monsoon rains are mainly convective in nature. The onset of South West Monsoon over Jammu and adjoining region takes place normally during the first day of July.

17. **Rainfall.** There is a gradual increase in rainfall from June to August. It significantly decreases from September. Jammu receives about 65% of rainfall during monsoon season and maximum rainfall is in the month of August. Least rainfall is received in post-monsoon months. In an average July and August experiences about 19 and 18 rainy days respectively, while in November there is only 03 rainy days.

Punjab

18. The state has a balanced amalgamation of high temperature in summer (Pre-monsoon), rain in monsoon and low temperature in winter. The seasons are so distinctly distributed that you can enjoy each season. Punjab experiences both summer and winter to its extreme. It even receives abundant rainfall, which makes the land very fertile. The region lying near the foot hills of Himalayas receive heavy rainfall whereas the regions away from the hills experience less rainfall, the rainfall is scanty and the temperature is high. In other words, rainfall decreases from north to south.

19. **Winter Season (December – March).** Generally dry north-westerly continental winds prevail at surface and lower tropospheric levels over Punjab during winter season. Temperature falls as the season progresses till end of January. Dry weather with chilling cold condition accompanied by widespread prolonged dense fog commonly prevails over the state. The dry spells are broken whenever an induced western disturbance passes across north or central Pakistan. These systems change the winds to south-westerly to westerly, raise the temperature and causes rainfall. After the system passes away, the sky clears up rapidly, leads to rapid fall in temperature and formation of dense fog again.

20. **Pre-Monsoon Season (April – June).** As the solar equator shifts northward after the winter season, temperature over the entire country rises up. Punjab being located at the extreme interior of the land mass, continue to get heated up till the end of June. The atmosphere remains very unstable due to local heating. Incursion of slight moisture due to movement of extra tropical or induced disturbances, Punjab experience severe thunderstorm in form of Andhi on about 09 days in a month. Thunderstorms are often accompanied by strong winds with gale force. Occasionally widespread hailstorm also occurs which damages crops extensively.

21. **Southwest Monsoon Season (Jul – Sep).** Normally Punjab remains south of the monsoon trough and hence, experiences easterly to south-easterly winds at surface and lower levels. In such condition normal monsoon activity with lot of low clouds and moderate rainfall during morning hours. During 'Break Monsoon' condition the monsoon trough shifts northward. In such condition parts of Punjab, specially the places close to the foot hills experience heavy rainfall and thunderstorm. The dry and hot weather condition suddenly becomes cool on arrival of monsoon. However, when not raining, the weather uncomfortably becomes sultry. Whenever a monsoon system recurves north ward from Rajasthan, Punjab may experience heavy fall causing flood.

22. **Post Monsoon Season (Oct – Nov).** SW Monsoon withdraws from Punjab by end of September. Departure of monsoon leads to clear sky condition and gradual drop in temperature. This is the season with the most pleasant weather condition in the state. Second half of November resembles with winter season with chilling temperature and development of fog during morning hours.

23. **Rainfall.** July and August are the wettest months with about 16 rainy days of and October to December are the driest months with 03 days or less rainy days. North Punjab receives about 245 mm of rainfall in July and about 185 mm in August. Total annual rainfall is about 800 mm.

Haryana

24. Haryana is also characterized by very hot summer and very cold winter. Temperatures are more extreme over southern parts, which are adjacent to Rajasthan and are mostly semi-arid. Monsoon season extends from first week of July to third week of September. However, Haryana also receives significant amount of rainfall during winter months due to passage of western disturbance passing across west Rajasthan. The state being away from the hills experiences lesser thunderstorm than Punjab. However, frequency of Andhi (Duststorm) is more than that over Punjab during pre-monsoon season.

25. **Winter Season (December to February).** During this season, generally cold dry N/NW'ly continental air prevails over Haryana. Clear skies, low humidity and low temperatures with large diurnal change are the characteristic features. The dry spells are broken at intervals when migratory extra-tropical disturbances move across northern India. Widespread dense fog also prevails on most of the days.

26. 26. **Pre – Monsoon Season (March to May).** High temperature, widespread Dust Haze, Dust Raising Winds and violent Thunderstorms / Dust storms are the main characteristic weather features of the season. The vertical extent of dust may extend up to mid tropospheric level. Thunderstorm / Dust storm which are characteristic features of pre-monsoon season occur under the influence of Induced western disturbances passing across west Rajasthan.

27. **Southwest Monsoon Season (June to September).** After the onset of SW Monsoon, the wide spread Dust Haze of summer months gets settled down. Frequency of Dust storm / Dust Haze decreases significantly. Intensity of rainfall depends on the proximity of Axis of Monsoon Trough. Rain / Thundershowers are the main characteristics weather features of the season. Prolonged break monsoon condition may again cause Dust Haze. Heavy rainfall is experienced when Depression / Well Marked Low pressure (Monsoon Depression) is located over East Rajasthan and adjoining North West MP and at the same time a western disturbance is observed over North Pakistan and adjoining J&K. Also, at times under the influence of western disturbance when a monsoon systems re-curves from Rajasthan/Northwest MP, Haryana experiences widespread Rain/T' Showers. The south west monsoon withdraws by the second or third week of September.

28. **Post Monsoon Season (October to November).** After the withdrawal of SW Monsoon, by the end of September, dry continental air mass sets in over the state. This leads to clear sky condition. With the advance of the season, atmospheric obscurity phenomena such as Haze / Mist / Fog in the morning and evening hours cause deterioration in visibility. Fog activity generally starts by the end of October. Smoke Haze/smog during the season due to stubble burning in this season has been a great health and aviation hazard and has been a matter of great concern.

29. **Rainfall.** Frequency of rainfall increases from May to July with highest frequency in July and thereafter gradually decreasing. July and August are the wettest months with about 120 mm and 85 mm of rainfall respectively over south Haryana. November and December are the driest month with less than 5 mm of monthly rainfall. Annual rainfall received over south Haryana is around 418 mm which is much less than that over Punjab and north Haryana.

North and East Rajasthan

30. North and east Rajasthan falls under Semi-arid climatic condition as part of the region falls within the Thar desert. The region receives lesser amount of rainfall than the rest of NW Indian regions like west UP, Haryana and Punjab. However, it is seen in the recent years that this region has been receiving above normal rainfall in both monsoon and winter season. Though no concrete reason has yet been drawn in support of the above normal rainfall, it can be attributed to the factors like global climate change, change in the monsoon pattern, variation in the path of western disturbance and positive forestation measures taken by the citizens and the government of Rajasthan.

31. **Winter Season (December- March).** During winter season, generally good weather condition prevails over the region. Visibility condition is also much better than the rest part of the north-west India and hence, it is very favourable for the aviation industry and operation of military aircrafts. Winter days are comfortably warm and the nights are chilled. This is the best season for the tourists to visit the region. Thunderstorm and precipitation in association with western disturbance may occur for 1-2 days in a month. The cold wave conditions prevail for 2-3 days at the rear of WD.

32. **Pre- Monsoon Season (April-June).** The pre-monsoon season is characterized by high temperatures and unstable atmosphere. Dust storm/Thunder storm accompanied with Gale winds occurs in this season. They generally occur in association with movement of Westerly system and moisture incursion from the Arabian Sea. Dust haze and Dust Raising Winds are also common features of this season. Temperatures during day time are extremely high and remain above 40^0 C on most of the days.

33. **South West Monsoon Season (July-September).** The SW Monsoon sets in over the region in the 1st week of July and withdraws in the second week of September. Rain/Thunder showers occur due to north-south oscillation of western end of Axis of Monsoon Trough. The region experiences heavy rains whenever a monsoon depression moves north-west ward from north-west MP. Presence of westerly system over J&K or North Pakistan enhances the rainfall activity over the region.

34. **Post Monsoon Season (October-November).** The post monsoon season again establishes good weather conditions over the region. Weather mostly remains cool and dry and attracts the tourist during this season. Occasionally dust storm may occur at a few places under the influence of western disturbance moving across J & K.

35. **Rainfall.** Maximum rainfall is received in the month of July and it is minimum in the month of November. Average rainfall during July is close to 100 mm and in November it is less than 1 mm. Number of rainy days in July is above 10-12 days. Total annual rainfall received by the region is about 330 - 350 mm.

West Rajasthan

36. West Rajasthan falls under Hot Desert climatic condition as the region falls under the Thar desert. During monsoon season the seasonal low-pressure area (Heat Low) is located over this region and adjoining areas of Pakistan. Over the low-pressure area, there is a cyclonic circulation up to about 5,000 Ft ASL. However, this cyclonic circulation is overlaid by an anti-cyclone which keep the atmosphere stable and resists formation of rain clouds. Due to this reason, west Rajasthan receives the least rainfall in India. Occasionally, under some situation when monsoon current extends west ward, the region including the adjoing areas of central Pakistan receives good monsoon rain.

37. **Winter Season (December – February).** The Winter Season is associated with chilling cold condition during night and early morning hours. Temperature often remains close to zero degree and occasionally becomes sub-zero during January. Generally good weather with good visibility conditions prevail during this season. Western Disturbances moving across J&K, Himachal and Punjab results in partly cloudy skies with medium and high cloud over West Rajasthan. Weather of over the region is affected when the western disturbances move along a more southerly track or with the movement of induced low from South Pakistan across West Rajasthan. Fog occurs occasionally during morning hours after the passage of western disturbance.

38. **Pre – Monsoon Season (March – June).** The Pre-Monsoon Season is characterized by gradual rise of temperatures especially during second half of the season. Temperatures more than 40°C are common during the month of May and June. In April frequency of temperature more than 40°C is 10 days. During first half of this season, 5 - 7 WDs move across North West India which reduces to 2 - 3 during second half of the season. As the season progresses WD shifts northward and maintain more Northerly track and do not affect this region. During first half 2 - 3 induced low / cyclonic circulation form over South Pakistan and adjoining Rajasthan and move North Eastwards/ Eastwards resulting increase in clouding and often associated with dust storms or thunderstorm. The normal date of onset of Monsoon over Jaisalmer is 15 July. Hence, the pre-monsoon condition continues to prevail over the region till mid of July.

39. **South West Monsoon Season (July - September).** The South West Monsoon season over west Rajasthan is between 15 July and 05 September. The variability of South West Monsoon over this region is very high. This season accounts for more than 70 % of the annual rainfall. The temperatures start falling with the advent of the

South West Monsoon and progressively decrease as the season progresses. Monsoon starts withdrawing from the first week of September. Thus, the South West Monsoon lasts for less than three months.

40. **Post- Monsoon Season (Oct – Nov).** This is the most pleasant season in the entire year. It is characterized by warm sunny days and clear cool nights. The weather is generally fine during this season except during the passage of WD across NW India.

41. **Rainfall.** About 70% of rainfall is received during monsoon season. Total annual rainfall is about 240 mm. The wettest month is August with about 75 mm of rainfall and November is the driest month with less than 2 mm of rainfall. In an average west Rajasthan receives less than 10 mm of rainfall from October to April and less than 5 mm of rainfall from November to march.

Himachal Pradesh

42. Himachal Pradesh is located at the extreme north of the Indian sub-continent occupying a region of scenic splendour in the western Himalayas. The state consists of 12 districts. It is bounded by Jammu and Kashmir on north, Nepal and Tibet on east, Uttarakhand on southeast, Haryana on south and Punjab on west and southwest.

43. The state may be broadly divided into 3 geographical regions, viz. outer Himalayas, the lesser Himalayas and the greater Himalayas or the Alpines. The outer Himalayas includes the districts of Bilaspur, Hamirpur, Kangra, Una and the lower parts of Mandi, Sirmaur and Solan. The lesser Himalayas includes the parts of Mandi, Sirmaur and parts of Chamba, Kangra and Shimla. The Alpine zone is at an altitude of 4500 m and beyond, includes Kinnaur and parts of Lahaul and Spiti, Chamba districts. The principal peaks are Mount Parbati (6633 m), Gaya (6794 m), Mulkila (6517 m), Indrasan (6220 m), Sikarveh (6200 m), Jeopango (5870 m), Sarcho peak (5741 m) in Lahaul and Spiti district, Leo-Pargial (6791m),Mani Range (6554 m), Kinnaur Kailash (6050 m), Shrikhand Mahadev (5227 m), in Kinnaur district, Mukarbeh (6060 m), DeoTibba (6001 m), Hanuman Tibba (5932 m) in Kullu district, Churdhar peak (3647 m), Hatu peak (3631 m) in Shimla district and Kailash peak (5656 m) in Chamba district.

44. The principal rivers of the state are Chenab, Ravi, Beas, Sutlej and Yamuna. These perennial rivers are fed by snow and rainfall.

45. **Climate.** The climate of the state varies from place to place depending on the altitude. It varies from hot and sub-humid tropical (450- 900 m) in the southern low tracts, warm and temperate (900-1800 m), cool and temperate (1900-2400 m) and cold alpine and glacial (2400-4800 m) in the northern and eastern high mountain ranges. The state as a whole mainly comes under the climatic type as stated below: -

Area	Climate Type
The districts Bilaspur, Kangra, Mandi, Sirmaur and Una	Sub-tropical monsoon, mild and dry winter and hot summer.
Shimla	Tropical upland, mild and dry winter, short warm summer
Chamba	Humid subtropical, mild winter, moist all seasons, long hot summer
Kullu	Humid sub-tropical, mild winter, moist all seasons, long hot summer and marine.

46. The year may be divided into four seasons. The winter season from January to February is followed by the pre-monsoon or hot weather season from March to May. The period from June to September constitutes the southwest monsoon season and the period from October to December is the post monsoon season.

47. During the period from January to February, generally low temperature prevails over the entire state and is generally very unpleasant due to biting cold. In this season, a series of western disturbances affect the climate of the state. Heavy snowfall occurs during this season.

48. In the summer months from March to May, weather is very dry. In the hilly regions, due to lower temperature, the climate of the state is comfortable.

49. Weather tends to be humid during July to September due to rise in moisture content of the atmosphere. These monsoon months are fairly comfortable due to reduced day temperature, although humidity continues to be high in comparison with the other months.

50. **Atmospheric sea level pressure and winds.** The seasonal variation of atmospheric pressure over the state takes place in a systematic manner with a maximum in winter and minimum in the southwest monsoon season. The pressure gradient over the state generally remains weak. During the winter season the higher pressure is to the north or northwest. In April, the pressure is maximum towards west or northwest and decreases towards south or southeast. During January to April, winds are mainly from south, southwest and southeast directions. In July, winds are mainly from south, southwest directions. October is the month of transition with the weakest pressure gradient. From October onwards, the changeover of the pressure pattern to winter type commences.

51. **Temperature.** The temperatures of hilly districts with deep valleys vary considerably from place to place depending on elevation. June is the hottest month with the mean daily maximum temperature at 35.5ºC in the plains and 28.7ºC in hilly places. The highest values are observed over the extreme southwestern region. With the onset of monsoon, the day temperatures fall appreciably. During August, an appreciable drop in the mean daily maximum temperature is observed with the values ranging between 20.4ºC and 32.6ºC. The value of mean daily maximum temperature in October ranges between 18.8ºC to 30.6ºC. The mean daily maximum temperature of January ranges between 9.3ºC and 20.3ºC.

52. The highest maximum temperature ever recorded in the state is 49.9ºC on 08 May 1958 at Gondla and the lowest minimum temperature ever recorded in the state is -25.9ºC at Keylong on 21 February.

53. **Rainfall.** The total annual rainfall for the state is 1490 mm and the total annual number of rainy days is 65. Kangra district receives the maximum amount of rainfall (1850 mm) in a year, whereas Una receives the minimum amount of rainfall (1210 mm).

54. The southwest monsoon season is the principal rainy season over the state.

    About 73% of the annual rainfall is received in the monsoon season (June to September), about 9% is received in the winter season (January and February), about 11% is received in the pre-monsoon season (March to May) and about 6% is received in the post-monsoon season (October to December).

55. During the southwest monsoon season, the state receives rainfall mainly due to low pressure areas and monsoon depressions originating from the Bay of Bengal.

    July and August are the rainiest months accounting individually to about 27% of the annual rainfall. By about 20 June Monsoon covers the entire state and the withdraws by 3rd week of September.

56. The most common rain-giving systems over the state during post-monsoon season are the depressions and cyclonic storms originating from the Bay of Bengal. The storms and depressions cause heavy to very heavy rainfall and contribute substantially to the season's total rainfall.

57. During winter, the state receives about 14 cm of rainfall. This rainfall generally occurs in association with induced low-pressure areas over the surface due to western disturbances moving from west to east, across the northern parts of the country.

58. **Snowfall.** Hills of Himachal Pradesh receive considerable amount of precipitation in the form of snowfall. Snowfall is experienced throughout the year except in southwest monsoon season. In winter season, amount of snowfall

received varies from 20 to 1300 mm and number of days of snowfall varies from 2 to 23. In pre-monsoon season, amount of snowfall received varies from 10 to 420 mm and number of snowfall days varies from 1 to 30. In post monsoon season amount of snowfall received varies from 20 to 390 mm and the number of snowfall days varies from 1 to 17. Annual amount of snowfall varies from 250 to 2040 and number of snowfall days varies from 6 to 77. In hills of Himachal Pradesh, north western sector receives maximum snowfall.

59. **Cyclonic Storms & Depressions.** Himachal Pradesh situated far inland does not experience the full fury of the cyclonic storms or depressions like the coastal regions. However, the monsoon lows/depression affect the state during August and September.

60. **Climatological Data: Temperature.**

Place	Temp (°C)	Jan	Feb	Mar	Apr	May	Jun	Jul	Aug	Sep	Oct	Nov	Dec
Nahan	Max Temp	18	21	25	31	34	36	30	29	29	27	24	19
	Min Temp	7	9	12	18	21	23	22	22	21	17	13	9
Shimla	Max Temp	9	10	15	19	23	24	21	20	20	19	15	12
	Min Temp	2	3	6	11	14	15	15	14	13	11	7	4
Manali	Max Temp	11	12	16	22	25	27	26	25	25	23	19	14
	Min Temp	-2	-1	3	6	9	12	15	15	11	5	1	0
Chamba	Max Temp	18	21	23	29	33	32	33	31	29	28	23	20
	Min Temp	4	6	7	13	17	21	21	21	19	14	9	6
Mandi	Max Temp	19	21	26	31	35	36	32	31	31	29	25	20
	Min Temp	2	4	9	14	17	19	21	20	18	12	7	3
Dalhousie	Max Temp	11	13	17	21	25	27	24	23	23	22	18	14
	Min Temp	2	3	6	11	14	16	16	15	14	11	7	4

61. **Climatological Data: Rainfall.**

Place	Rainfall	Jan	Feb	Mar	Apr	May	Jun	Jul	Aug	Sep	Oct	Nov	Dec
Shimla	Rainfall (mm)	77	49	64	38	58	132	435	378	178	62	9	19
	Rainy Days	6	4	5	4	5	9	20	18	9	3	1	2
	Annual rainfall	1,497 mm											
Manali	Rainfall (mm)	114	134	202	116	80	83	216	233	105	52	40	57
	Rainy Days	7	8	9	6	6	7	15	15	9	4	3	3
	Annual rainfall	1,431 mm											
Chamba	Rainfall (mm)	91	121	145	93	78	96	268	230	120	36	26	52
	Rainy Days	5	6	7	6	5	6	12	11	6	2	2	3
	Annual rainfall	1355 mm											
Mandi	Rainfall (mm)	68	66	80	43	66	169	457	372	156	40	17	29
	Rainy Days	5	5	5	4	4	8	16	15	8	2	1	2
	Annual rainfall	1,565 mm											
Dalhousie	Rainfall (mm)	155	122	156	96	63	131	588	573	246	73	38	71
	Rainy Days	7	6	7	5	4	6	18	20	9	3	2	3
	Annual rainfall	2,311 mm											

Uttarakhand

62. Uttarakhand state is mostly hilly and has international boundary with China (Tibet) in the north and Nepal in the east and state boundary with Himachal Pradesh in the northwest. The state has foothills areas in the south and southwest which are bounded by Uttar Pradesh. The state is rich in natural resources especially water and forests with many glaciers, perennial rivers, dense forests and snow-capped mountain peaks. The most of the northern parts of state are part of greater Himalaya ranges covered by the high mountain peaks and glaciers.

63. Uttarakhand lies on the south slope of Himalaya ranges and the climate varies from sub-tropical forests at lower elevation to glaciers at higher elevation. The altitude in the state varies from 200 to 7817 metre above mean sea level. Within this altitudinal variation state comprises five litho tectonically and physiographical distinct sub-divisions namely:

    a)   Outer Himalaya comprising Tarai and Bhabhar

    b)   Sub-Himalayan belt of Siwalik

    c)   The Lesser Himalaya

    d)   The Great Himalaya

    e)   The Trans-Himalaya or Tethys.

64. Important mountain peaks in northern part of Uttarakhand are Nanda Devi (7817 m), Mt. Kamet (7756 m), Abi Gamin (7355 m), Mana (7272 m), Mukut Parvat (7242 m), Hardeol (7151 m), Chaukhamba (7140 m), Trishul (7122 m), Dunagiri (7066 m), Kedarmath (6942 m), Gangotri (6674 m), Neelkanth (6597 m), Shivling (6501m), Nilgiri (6474 m), Bandarpoonch (6320 m). Major Glaciers in Uttarakhand are Maiktoli, Kaphini, Ralam, Sunderdhunga, Chorbani, Gangotri, Khatling and Nandadevi.

65. Uttarakhand state has a large number of rivers as its northern part is a home of glaciers to melt the ice and flow the water. Hence, there are almost perennial rivers in the state. The state is mostly drained by Ganga, Yamuna, Bhagirathi, Alaknanda, Kosi, Mandakini, Kali, Gori-Ganga, Ramganga and Pindar rivers, and other major rivers with their tributaries.

66. **Climate**. The climatic conditions experienced in Uttarakhand vary from hot and moist sub-tropical in the southern part to cold alpine in the upper reaches of the Himalayan mountain in the northern parts. Warm and cool temperate climate persists over the areas between southern and northern parts of the state.

67. The climate of Uttarakhand is sharply demarcated in case of two distinct divisions the predominant hilly terrain and the smaller plain region. The areas of high hills even become inaccessible in winter due to extreme cold of climate causing prolonged snowfall. The mountain range itself exerts an appreciable extent of influence on monsoon and rainfall patterns. Cold alpine climate is experienced at higher reaches where summers are cool and winters are harsh. At altitudes over 4800 m, the climate is bitterly cold with temperatures consistently below the freezing point of water and the area is perennially shrouded in snow and ice.

68. In general, the year may be divided into four seasons. The winter season from December to February is followed by pre-monsoon or hot weather season from March to May. June to September constitutes the southwest monsoon season and the period of October and November is of post monsoon season.

69. **Climate Type**. The state is divided in to two regions viz. western part is known as Garhwal region and eastern part is Kumaun region. Areas in the state under each climate pattern are as follows: -

Area	Climate
Hardwar and Udham Singh Nagar districts and some parts of Nainital, Bageshwar, Champawat, Almora, Dehradun, Pithoragarh, Tehri Garhwal, Pauri Garhwal, Rudraprayag and Uttarkashi districts	Subtropical monsoon, mild and dry winter, hot summer
Some parts of Chamoli, Almora, Dehradun, Nainital, Uttarkashi, Pithoragarh, Rudraprayag, Bageshwar, Champawat, Pauri Garhwal and Tehri Garhwal districts	Tropical upland, mild winter, dry winter, short warm summer
High altitudinal areas of Almora, Bageshwar, Chamoli, Dehradun, Pauri Garhwal, Pithoragarh, Rudraprayag, Tehri Garhwal and Uttarkashi districts	Humid continental, severe winter, moist all season, short warm summer
High altitudinal areas (peaks) of Bageshwar, Chamoli, Pithoragarh, Rudraprayag, Tehri Garhwal and Uttarkashi districts	Tundra and very short summer

70. **Sea Level Pressure and Winds**. The seasonal variation in atmospheric pressure over the state occurs in a systematic way with a maximum in the winter and minimum in the southwest monsoon season. The pressure gradient over the state is generally weak except during latter part of summer and south west monsoon season.

71. During winter season winds are generally variable as the state has mostly hilly terrain with lofty mountains and valleys. Owing to the nature of the terrain, local effects on winds are predominant. When the general prevailing winds are not too strong, there is a tendency for diurnal reversal of winds, blowing down the slopes at night (katabatic). With the progress of the monsoon, south-easterly, easterly and northerly components of the wind appear. October onwards, wind pattern slightly changes as sometimes north-westerly and westerly components are seen. In the wake of western disturbances and in association with thunderstorms the winds become quite strong.

72. **Temperature**. The state lies in the mountainous region except for some plain areas in the districts along the southern boundary of the state. The temperatures in the state therefore, vary considerably from place to place. The temperature starts to rise from March and steadily rises till it reaches its peak in May to the middle of June, when the mean maximum temperature in southern parts and valleys of the state is at about 34 ^{0}C to 38 ^{0}C. At places at about 2 km altitude mean maximum temperatures are around 23 - 24 ^{0}C. On individual days maximum temperature rises to 42 ^{0}C in the valleys and southern part of the state and 30 ^{0}C. The highest maximum temperature on record at any individual station was 47.4 ^{0}C at Roorkee on 22 May 1978.

73. With the onset of southwest monsoon, the maximum temperatures start falling. Temperature reches lowest values in January and early February. January is the coldest month with mean maximum temperature in southern part and river valleys about 20 ^{0}C and mean minimum temperature about 6 ^{0}C. A much lower temperature is experienced in the wake of western disturbances during winter. On such occasions minimum temperature falls below 0 ^{0}C in southern part and lss than -10 ^{0}C at high elevated areas in northern part of the state. The lowest minimum temperature on record at any individual station was -15.1 ^{0}C at Joshimath observatory on 15 January 1974.

74. **Rainfall**. The total annual rainfall for the state as a whole is about 1330 mm and total annual number of rainy days are about 63. Snowfall occurs mostly in winter months from December to February associated with western disturbances. January is the month with the heaviest snowfall. The rainfall in the state varies from place to place due to its rugged topography. Pithoragarh district in Kumaon division received the maximum amount of precipitation i.e. at about 1770 in a year, whereas Almora district in Kumaon division received the minimum amount of precipitation i.e. at about 970 mm in a year.

75. The southwest monsoon season is the main rainy season over the state. Of the total amount of annual rainfall, about 78% is received in the southwest monsoon season (June to September), 9% is received in the winter season (December to February), 10% in pre- monsoon season (March to May) and 3% in post monsoon season (October to November).

76. The state receives rainfall mainly due to low pressure areas and monsoon depressions originating in the Bay of Bengal during the southwest monsoon season. During this season most of the depressions originating in the Bay of Bengal cross inland and move north-westwards or westwards over the state, some of them reach near the hilly districts. Sometimes heavy rain occurs with the interaction of the monsoon system and extra-tropical system. The rest of the rainfall occurs in winter and early summer in association with the passage of western disturbances across north India. Winter precipitation over the northern parts of the districts is in form of snowfall.

77. July and August are the rainiest months and in these two months nearly 53% of the annual rainfall is received. The southwest monsoon sets in over the state by the last week of June. The monsoon starts to withdraw from the state by about third week of September and completely withdraws by 01 October.

Cyclonic storms and depressions.

77. In general, cyclonic storm does not reach towards the state and even though depressions become very weak as they rarely reach on the land of Uttarakhand state is about 1300 km away from the east coast of India. The state therefore does not experience the full fury of depressions like the coastal regions. However, in association with these systems, heavy rainfall occurs over the affected districts. During the course of movement, the disturbances sometimes turn or recurve towards north or northeast under the influence of deep westerly system moving across Pakistan and Northwest India.

78. The maximum number of storms / depressions originating from the Bay of Bengal affects the state during August and September. The monsoon disturbances during June to September generally form over the north Bay of Bengal and travelling west or north-westwards, they move across Odisha, Chattisgarh, Madhya Pradesh, Uttar Pradesh and move towards Uttarakhand. During the period 1891-2013, 34 storms/depressions affected the Uttarakhand state.

79. **Western Disturbances**. Western disturbances mostly affect Indian subcontinent to the north of Latitude 25°N. The state is strongly affected by WDs in winter. On an average 3 to 6 WDs per month move across the state in winter months. They are less frequent in post monsoon season. Its frequency is 2 in November and about 4 in March and April, and 2 to 3 in May. The weather associated with western disturbances over the state are of precipitation, cold wave condition and fog. The high elevated areas of the state often get heavy snowfall during the winter season. Hails also occur over the state in association with the WDs. The chilly weather in the state is experienced with the movement of WDs. Cold wave conditions generally prevail during November to March after moving away of the WDs. The minimum temperature in cold waves drops as much as 5°C to 8°C below normal. A spell of cold wave usually lasts for about 4 to 5 days. The state is most affected by severe cold wave approximately 4 spells per year.

80. **Climatological Data: Temperature**.

Place	Temp	Jan	Feb	Mar	Apr	May	Jun	Jul	Aug	Sep	Oct	Nov	Dec
Dehradun	Max Temp	20	22	27	32	35	34	31	30	30	29	35	21
	Min Temp	6	8	13	17	21	23	23	23	21	16	11	7
Musoorie	Max Temp	11	12	16	21	23	23	21	20	20	19	15	13
	Min Temp	3	3	7	12	14	16	15	15	14	11	7	4
Joshimath	Max Temp	11	12	18	22	24	26	24	23	23	21	17	13
	Min Temp	2	3	7	11	14	17	17	17	15	10	6	4
Nainital	Max Temp	11	12	16	21	23	23	22	21	21	19	15	13
	Min Temp	2	3	7	12	15	16	17	16	14	10	6	3
Roorkee	Max Temp	20	23	29	35	38	38	34	33	33	31	27	23
	Min Temp	6	8	13	18	23	24	25	25	23	17	11	7

81. **Climatological Data: Rainfall**.

Place	Rainfall	Jan	Feb	Mar	Apr	May	Jun	Jul	Aug	Sep	Oct	Nov	Dec
Dehradun	Rainfall (mm)	41	57	55	35	51	162	490	458	217	35	10	21
	Rainy Days	3	3	3	2	3	7	15	16	9	2	1	1
	Annual rainfall	1,632 mm											
Nainital	Rainfall (mm)	33	36	28	17	42	199	454	416	239	41	4	12
	Rainy Days	2	2	2	1	3	8	15	15	8	2	1	1
	Annual rainfall	1,521 mm											
Tehri Garhwal	Rainfall (mm)	47	59	49	32	51	128	321	326	147	25	8	24
	Rainy Days	3	4	3	3	4	7	14	14	7	1	1	1
	Annual rainfall	1,216 mm											
Udham Singh Nagar	Rainfall (mm)	21	21	10	9	29	129	324	307	178	33	2	8
	Rainy Days	1	2	1	1	2	5	11	11	6	1	0	1
	Annual rainfall	1,071 mm											
Chamoli	Rainfall (mm)	76	104	101	51	73	93	217	224	109	40	10	39
	Rainy Days	4	6	6	4	5	7	15	15	8	3	1	2
	Annual rainfall	1,136 mm											

Uttar Pradesh

West UP

82. Western Uttar Pradesh shares borders with the states of Uttarakhand, Himachal Pradesh, Haryana, Delhi, Rajasthan and Madhya Pradesh, as well as a brief international border with Nepal in Pilibhit district. Western Uttar Pradesh receives rain through the Indian Monsoon and the Western Disturbances. During monsoon season, the location and orientation of the Axis of Monsoon Trough (AMT) plays the most vital role in the rainfall activity. When the AMT passes along the normal positions, the region receives rainfall from both Arabian Sea and Bay of Bengal Branch. When the western end of the AMT is south its normal position, the region is affected only by the Bay of Bengal Branch. When AMT is north of its normal position, that is, close to the foothills, the areas close to the hills experience heavy fall. Shifting of the AMT to further south may cause Break Monsoon condition when West UP may experience scanty rainfall. The region is also affected by severe thunderstorm and Andhis during pre-monsoon season.

83. The annual cycle of weather at a place is guided by both climatic and seasonal variations. The cycle of weather over west UP can be broadly divided into four seasons. There are two main seasons viz. Winter (Cold) and Monsoon (Wet), the first extending from December to second week of March and the second from last week of June to end of September. The change over periods from winter to monsoon and vice-versa are transitional pre-monsoon and post-monsoon seasons.

84. **Winter season (December - March)**. It is characterized by low temperatures, high pressure, dry winds, stable atmosphere and strong zonal westerly regime. During the season a well-marked high-pressure ridge extends over the Indian Sub-continent, Iran and adjoining Afghanistan. Cold and dry north-westerlies sweep across entire NW India except during the passage of a western disturbance. A system moving across northern latitudes, on occasions has an induce low over Baluchistan, Central Pak or West Rajasthan. At times the induced low tags along the main system and moves /across Gulf, North Arabian Sea, Iran and adjoining Pakistan, Gujarat and Rajasthan region.

Poor to very poor visibility due to fog, sometimes smog, is one of the main characteristics of the season. Fog /Mist also occur after the passage of a WD. The frequency of Rain/Thunder shower accompanied with hail progressively increases from Dec to Mar. The other important characteristic is the very low temperature. Cold wave condition with near to zero-degree temperature under the influence of persistent dry and cold north-westerly continental winds is common from end of December to early February.

85. **Pre-Monsoon Season (April - June).** In this season the stable atmosphere of winter months progressively yields to unstable conditions prior to the onset of summer monsoon season and is marked by violent convective activity. There are various synoptic features that trigger convective activities. One important feature is the passage of WD/ Induced low which leads to Andhis over West Rajasthan, Haryana, Punjab, West UP, NW MP & adjoining areas. This season is characterized by high temperatures, dust haze, dust raising winds, dust storms and thunder storms generally accompanied with gale winds. At times these thunderstorms result in hail storms. Heat wave condition commonly prevails over the region during this season.

86. **Monsoon Season (July – September).** It is the main rainy season and has telling effect over the economy of the region. The weak pressure gradient of pre-monsoon months gradually changes over to well-marked seasonal low over West Rajasthan and strong gradient along west coast. The inter tropical convergence zone (ITCZ) shifts as far north as 27°N Lat and is referred to as Axis of Monsoon Trough (AMT). Winds south of AMT are strong south-westerlies (Arabian Sea branch) and to the North, south-easterlies (Bay branch).

The rainfall to a larger extent is guided by:

a)   Oscillation of Axis of Monsoon Trough (AMT) and its depth.

b)   Development and movement of Monsoon Low / Depression.

c)   Intensification of seasonal low either due to moving Western

Disturbance or merging of Low / Depression over Rajasthan or Haryana and Western Uttar Pradesh / Western Madhya Pradesh region.

d)   Prevalence of Break Monsoon Condition.

87. **Post-Monsoon Season (October – November).** This is the pleasant season when temperature starts falling gradually. Morning mist increases as the season progresses. Clear sky and pleasantly cool temperatures are the main characteristics of the season. Occasionally the dry spells are broken when induced westerly system affects the region.

88. **Rainfall.** About 75 % of annual rainfall is received during south west monsoon season. Annual average collection of rainfall in west UP is about 1115 mm. About 840 mm of rainfall is received in the monsoon months. 15% rainfall is received from pre-monsoon months and about 07% is received from the WDs in the winter season. July is the wettest month with 329 mm of rainfall followed by August with 319 mm. Monsoon rain drops to 192 mm in September. November is the driest month with just about 05 mm of rainfall.

89. **Temperature and Rainfall Data: Lucknow.**

Month	Jan	Feb	Mar	Apr	May	Jun	Jul	Aug	Sep	Oct	Nov	Dec
Max Temp (⁰C)	23	26	32	38	40	38	34	33	33	33	29	25
Highest Max (⁰C) Date/Yr	32 01/92	34 27/89	41 31/73	45 30/99	47 31/95	48 09/66	44 06/82	40 03/72	40 18/58	38 15/99	38 08/63	30 04/76
Minimum Temp (⁰C)	07	10	15	21	25	27	26	26	24	19	13	08
Lowest Min (⁰C)	-01 31/64	00 02/64	05 10/79	11 01/68	17 03/87	20 12/79	22 01/50	22 23/96	17 29/50	10 31/84	04 29/52	01 30/73
Rainfall (mm)	20	16	10	5	18	123	270	255	211	41	07	13

Heaviest in 24 Hr (mm) Date/Yr	85 02/80	47 01/82	43 25/82	20 01/65	57 28/59	162 26/71	272 09/60	250 30/2k	177 14/85	200 10/85	39 26/97	49 09/97
Highest in a Month (mm) Yr	94 1996	61 1970	62 1982	36 1965	115 2000	412 1975	761 1960	654 1967	698 1985	346 1985	61 1997	96 1997
Total Annual rainfall	989 mm											

East UP

90. East UP is bounded by Nepal to the north, Bihar to the east and Bagelkhand region of MP to the south.

91. East UP, lying in Indo-Gangetic plains, experiences tropical as well as sub-tropical weather. While the winter herald cold sub-tropical weather, the summer months usher in the vanguards of tropical weather with transitory seasons in between. A humid subtropical climate is a zone of characterized by hot, usually humid summers and mild to cool winters. This essentially means that during summers one can expect intense perspiration over the region.

92. **Winter Season (December – Mid-March)**. East UP experiences generally dry and cold weather conditions during the season. Generally, induced lows affect the weather of the region with rain or thundershowers. But at times, strong westerly troughs give rise to widespread cloudiness. The surface temperature starts falling by the middle of November and reaches the lowest value by the end of January and beginning of February. Often after the passage of WD, cold wave condition prevails over the region. Beside cold condition, Occurrence of dense and persistent fog is the other important characteristic of the season.

93. **Pre Monsoon Season (April to May)**. This is the hot and dry season. The temperature continues to rise as the season progresses. The weather which is dry and hot becomes humid too after low level easterlies set in. Passage of induced low, leads to cloudy skies accompanied by rain/thundershowers. The frequency of these systems increases from 2 - 3 to 3 – 4 as the season progresses. Dust raising winds are quite common in the month of May and June over UP and Bihar. Thunderstorm and strong surface winds are important weather features during this season. Temperature rises to very high scale and often remains above 40 ^{0}C in the month of May and June. Occasionally, the region including Bihar becomes the hottest one in the country during May.

94. **Monsoon Season (June to September)**. Monsoon brings in plenty of rain over the region and sets in over the region by middle of June. Violent Thundershower activity is witnessed during the onset phase. Rain, drizzle, thundershower, low clouds, high humidity and sultry weather are the main characteristics of the monsoon season. Occurrence and diurnal variation of significant weather over the region is governed by the position of the Axis of Monsoon Trough (AMT), the monsoon systems (Low/Depression, Cyclonic circulations forming over the Bay of Bengal) and their movement and the in-situ systems, forming over East UP. The oppressive heat of summer season is driven away by welcome monsoon showers with the onset of monsoon and temperature drops by as much as 10°C. Thereafter it remains moderate throughout the season.

95. **Post Monsoon Season (October to November)**. The monsoon easterlies are gradually replaced by dry north-westerlies/westerlies. Weather improves considerably after the withdrawal of monsoon. Violent thunderstorm may still be experienced during the withdrawal of monsoon, which takes place by middle of of October. One or two induced systems may affect this region in the later part of November. After mid-October, sky remains almost free of clouds. Temperature continues to fall gradually throughout this season. Cold continental winds start sweeping the Gangetic plains during November.

96. **Rainfall**. There is progressive increase in the amount of mean rainfall from May (50 mm) to July (365 mm) and then starts decreasing from 355 mm in August to 65 mm in October. In the other months, base receives very less amount

of rainfall and no pronounced winter rainfall is experienced. The base is lying close to the foothills of Himalayas and receives copious amount of rainfall during weak/break monsoon conditions. This could be the reason for July and August recording maximum amount of rainfall over base. The frequency of rainfall days for these two months (23 days) is also high for the said reason. November is the driest month with less than 3 mm of monthly average rainfall. Average annual rainfall for the region is 1,325 mm.

Bihar

97. Bihar is an entirely land-locked state, having an average elevation of about 150 meters above mean sea level. The state shares its boundary with Nepal to the north, the states of West Bengal to the east, Jharkhand to the south and Uttar Pradesh to the west.

Topographically, Bihar state can be divided into three regions.

   a) **The Sub-Himalayan foot hills**. This region lies in the northern part of the state. There are some small hills like Someshwar and the Dun hills in the extreme north of West Champaran district. These hills are off-shoots of the Himalayan system. South of it lies the Tarai region, a belt of marshy and sparsely populated region.

   b) (b) **The Indo Gangetic Plains**. The Bihar state has a number of rivers, the most important of which is the Ganga. The Gangetic plain of Bihar is divided into north and south by the Ganga river which flows through it from west to east. The Ganga is the most dominant river of Bihar state and is joined by the rivers Ghaghra, Gandak, Burhi Gandak, Bagmati, Kamla-Balan, Kosi and Mahananda flowing southward from Himalayas in northern part of the Gangetic Plain. In the southern part of the plain, there are some rivers like Sone, Uttari Koel, Pinpun, Panchane and Karmnaska which flow towards north from the plateau region. In the central part of the state, there exist small hills like Rajgir Hills and Kharugpur Hills which are two parallel ridges extending around 65 kms. These hills are around 300 meters high.

   c) (c) **The Southern Plateau Region**. To the further south of Bihar plains lie the plateau region which consists of Kaimur plateau in the west and Chhota Nagpur plateau in the east.

98. **Climate**. The year may be divided into four seasons. The winter season from December to February is followed by the pre-monsoon or hot weather season from March to May. The period from June to September constitutes the southwest monsoon season. The period of October and November is the post monsoon season. The state mainly has subtropical monsoon climate which is, mild and dry winter, hot summer except the districts viz. Jamui, Banka, Munger, Lakhisarai, Khagaria, Shekhpura and some parts of Bhagalpur, Saharsa and Begusarai located in the extreme southeastern part of the state which come under Tropical Savanna type of climate which is hot, and seasonally dry (usually winter).

99. **Surface Winds**. Surface wind pattern over Bihar are as described below: -

   a) In winter season winds are mostly light or calm. Generally, winds blow from south, southwest, west and sometimes northwest direction in winter.

   b) By April the winds veer clockwise and mainly blow from east.

   c) With increase of pressure gradient with advance of monsoon, the winds from east direction also strengthen, reaching the maximum value in June. In July, pressure gradient increases further and and correspondingly the winds become mainly easterly. Easterly component of the wind becomes increasingly predominant with the progress of the monsoon.

   d) October onwards, the changeover of the pressure and wind pattern to winter pattern commences.

100.　**Temperature**. Characteristics of temperature over the state are as follows: -

　　a)　Pre-monsoon is the hottest season while winter is the coldest season of the year. May is the hottest month with mean maximum temperature of about 37°C in the plains, while the plateau region and elevated places record about 3°C lower temperature.

　　b)　The mean maximum temperature ranges from 34°C to 41°C over the state during May and the values progressively increase southwestwards. The highest values observed over extreme southwestern region.

　　c)　With the arrival of monsoon, there is an appreciable drop in the mean maximum temperature during July which ranges between 32°C and 33°C.

　　d)　The maximum temperature pattern of October is quite similar to that of July. The mean maximum temperature in October ranges between 31°C and 32°C.

　　e)　The mean maximum temperature of January ranges between 22°C and 25°C. In this month, the minima of the mean minimum temperature is observed over the eastern region of the state. The values range between 8°C and 12°C. The temperature higher than 10°C is observed over the southeastern region of the state.

　　f)　The gradient of the mean minimum temperature increases in the month of April. The values range between 20°C and 23°C. The temperature is lower than 20°C over the extreme northwestern and eastern regions of the state.

　　g)　The gradient of mean minimum temperature is observed to decrease during the month of July. The values of minimum temperature range between 23°C and 26°C.

　　h)　The values of mean minimum temperature ranges between 19°C and 23°C during the month of October. The temperature value is less than 20°C over extreme southwestern parts of the state.

　　i)　The extreme maximum temperature increases from 43°C in the north to 49°C in southwestern parts of the state.

　　j)　The values of extreme minimum temperature range from 1°C to 4°C. The lowest temperature is experienced over southwestern part of the state.

　　k)　The highest maximum temperature and the lowest minimum temperature ever recorded in the state of Bihar are 49.5°C and -1.0°C respectively on 11th May 1988 and 18th January 1977 both at Dehri in Rohtas district.

101.　**Rainfall**. In general, rainfall is maximum over the northeastern part of the state. The normal rainfall pattern in Bihar state are brought out below: -

　　a)　The total annual normal rainfall for the state is about 1160 mm and the state receives on an average rainfall exceeding 2.5 mm for about 50 days.

　　b)　Kishanganj district in northeast sector receives maximum amount of rainfall of about 2210 mm in a year, whereas Arwal and Jahanabad districts in southwest sector receive the minimum with about 860 mm.

　　c)　The southwest monsoon season is the main rainy season over the state and the total amount of rainfall of about 86% is received in the southwest monsoon season (June to September), about 2% in the winter season (December, January and February) and about 6% each in the pre-monsoon (March-May) and post monsoon season (October and November).

　　d)　The state receives rainfall mainly due to low pressure areas or monsoon depressions originating in the Bay of Bengal during the southwest monsoon. The southwest monsoon sets in over the eastern parts of the state by second week of June and covers the entire state by the end of the second week of June. July and August are the rainiest months, each accounting individually to about 28% and 24% the annual rainfall respectively.

e) The most common rain giving systems over the state during the post monsoon season are the depressions and cyclonic storms originating in the Bay of Bengal.

f) The state receives about 30 mm of rainfall during winter. This rainfall generally occurs in association with induced low- pressure areas over the surface due to western disturbances moving from west to east, across the northern parts of the country.

g) The state in all receives about 75 mm of rainfall during the pre-monsoon season. This rainfall generally occurs in association with thunderstorms.

102. **Climatological Data.**

Patna

Month	Jan	Feb	Mar	Apr	May	Jun	Jul	Aug	Sep	Oct	Nov	Dec
Max Temp (^{0}C)	23	26	32	37	38	37	33	33	32	32	29	25
Minimum Temp (^{0}C)	09	11	16	22	25	27	26	26	32	31	29	25
Rainfall (mm)	11	08	07	09	28	116	305	238	201	53	05	03
No. of Rainy Days	1	1	1	1	2	5	13	11	9	2	1	1
Total Annual rainfall	9 8 5 mm											

Gaya

Month	Jan	Feb	Mar	Apr	May	Jun	Jul	Aug	Sep	Oct	Nov	Dec
Max Temp (^{0}C)	23	27	33	39	41	38	33	33	33	32	29	25
Minimum Temp (^{0}C)	09	12	16	23	26	27	26	26	25	21	14	09
Rainfall (mm)	12	10	12	05	17	133	268	249	179	41	08	07
No. of Rainy Days	1	1	1	1	1	6	12	12	9	2	1	1
Total Annual rainfall	941 mm											

Darbhanga

Month	Jan	Feb	Mar	Apr	May	Jun	Jul	Aug	Sep	Oct	Nov	Dec
Max Temp (^{0}C)	23	26	31	35	36	35	33	33	33	32	29	25
Minimum Temp (^{0}C)	09	11	16	20	22	24	24	25	24	22	16	11
Rainfall (mm)	11	08	07	19	57	160	314	261	184	60	05	05
No. of Rainy Days	1	1	1	1	3	7	13	11	8	2	1	1
Total Annual rainfall	1,093 mm											

Purnea

Month	Jan	Feb	Mar	Apr	May	Jun	Jul	Aug	Sep	Oct	Nov	Dec
Max Temp (^{0}C)	24	27	32	35	35	34	32	32	32	31	29	25
Minimum Temp (^{0}C)	08	10	15	20	22	24	25	25	24	21	14	09
Rainfall (mm)	09	09	12	37	127	277	479	358	322	85	08	09
No. of Rainy Days	1	1	1	2	6	10	17	14	11	3	1	1
Total Annual rainfall	1,733 mm											

Jharkhand

103. The Jharkhand state is located in the eastern part of India. The total area of Jharkhand is 79,714 square Km. It is bounded by 21°59'N to 25°18'N latitude and 82°52'E to 87°54'E longitude. The state shares its boundary with Bihar to the north, Uttar Pradesh and Chattisgarh to the west, Odisha to the south and West Bengal to the east. It is a land locked state with coastline around 150 km away. The tropic of cancer passes through Kanke, few Km away from Ranchi.

104. Jharkhand state is blessed with small hills and rivers and form it's topography. Major parts of the Jharkhand state lie on the Chhota Nagpur plateau. This plateau plateau is a series of hills ranges and flat-topped plateau with dissecting river valleys. The upper watershed of the rivers, the Koel, Damodar, Brahmani, Kharkai and Subarnarekha lies within Jharkhand.

105. **Climate**. The year may be divided into four seasons. The winter season from December to February is followed by the pre-monsoon or hot weather season from March to May. The period from June to September constitutes the southwest monsoon season and the period of October and November is the post monsoon season. This broad classification is based on temperature and rainfall. The districts located near the southern and western boundary of the state mainly comes under the climate type of Tropical Savanna, hot, seasonally dry (usually winter). Climate of Ranchi district situated in the centre of the state comes under subtropical monsoon which is characterized by mild and dry winter, hot summer and districts in the north-western part of the state come under the type of Interior Mediterranean climate with mild winter, dry and hot summer. Whereas, the region of Bokaro district and neighbourhood has marginally varying climate between Tropical Savanna, hot seasonal dry (usually winter) and subtropical monsoon. Mild winter and hot summer are its characteristics.

106. **Temperature**. Day temperatures are more or less uniform over the plains except during summer when temperatures rise westwards. In general, the temperatures at night are low in higher latitudes except during the southwest monsoon season when they are lower in south-west. Day and night temperatures are both lower towards the plateau and at high level stations than over the plains.

107. May is the hottest month with the mean maximum temperature at about 39°C in the plains, with the plateau regions and elevated places recording 2°C to 3°C lower temperatures. Sometimes heat waves are also experienced in some part of the state in later part of the pre-monsoon season. The highest maximum temperature recorded at any individual station in the plains is 48.5°C at Dumka in Santhal Parganas on 6th May 1989, which is about 10°C higher than the normal of the warmest month.

108. January is the coldest month when the mean minimum temperatures for the state as a whole is 10.3 °C, varying from 9°C in the north to 12°C in the south. A much lower temperature may be experienced in the wake of western disturbances during winter. On such occasions minimum temperature may fall below the freezing point and are recorded at a few stations in the northern parts of Jharkhand. Sometimes cold waves prevail in some part of the

state during the winter season. The lowest minimum temperature on record at any individual station (other than hill station) was –0.3ºC at Bokraro observatory on 24 December 1966.

109. The maximum temperature rises rapidly from February onwards till June. The increase in maximum temperature in the period from February to May ranges from 9ºC to 13ºC at individual stations whereas minimum temperature rises by 10ºC to 14ºC as we proceed from East to West of the state. The night temperatures start falling rapidly after September while the day temperature follow this trend after October and both attain the lowest values by January.

110. **Rainfall**. The total annual rainfall in the state is maximum over the region of Pakur, Simdega, Sahebganj and Dumka district, its neighbourhood and the extreme North-eastern part of the state. The total annual rainfall for the state is 1280 mm. Pakur district receives the maximum amount of rainfall (1500 mm) in a year whereas Garhwa receives the minimum amount of rainfall (1080 mm) in a year.

111. During the post monsoon season, the rainfall over the state decreases north-westwards. The southwest monsoon season is the main rainy season over the state. Of the total amount of rainfall, about 84% is received in the southwest monsoon season (June to September), about 3% is received in the winter season (December to February) and about 6% is received in the pre-monsoon (March-May) and 7% in post monsoon season (October and November).

112. The state receives rainfall mainly due to low pressure areas and monsoon depressions originating in the Bay of Bengal during the southwest monsoon. The southwest monsoon sets in over the southern parts of the state by about the beginning of second week of June and covers the entire state by the end of the second week of June. In the month of July and August, the state receives about 50% of the annual rainfall. The withdrawal of the southwest monsoon begins from the northern parts of the state by the end of first week of October and the monsoon completely withdraws from the state by about the end of second week of October. The most common rain giving systems over the state during the post monsoon season are the depressions and cyclonic storms originating in the Bay of Bengal. The storms and depressions cause heavy to very heavy rainfall and contribute substantially to the season's total rainfall.

113. The state receives about 90 mm of rainfall during post monsoon season. During winter, the state receives about 40 mm of rainfall. This rainfall occurs in association with induced westerly systems over the region and its neighbourhood due to western disturbances moving from west to east, across the northern parts of the country.

114. **Cyclonic storms and depressions**.The cyclonic storms and depressions which affect India, mostly originate over the Bay of Bengal, mainly during the months of May to November. They usually travel north-westwards or westwards and cross the east coast of India. In general, storms and depressions weaken on entering land. Jharkhand though an inland state, the east coast is not far away (about 150 km) and hence, remains under the domain of effect of cyclones. However, the state does not experience the full severity of the storms but the associated heavy to very heavy rainfall caused by the storms affect the state.

115. During the course of their movement, the storms sometimes re-curve towards north or northeast. This point of re-curving progressively shifts westwards till September. In May, these disturbances recurve while still out in Bay of Bengal. Hence, exceptionally few of them cross the coast and travel inland, affecting the weather of the state.

116. During the months from December to April, the state has not been affected by Bay but during the month of November, it has been affected four times since 1891. The number of storms/depressions that affected the state in September and October was 89 and 29 respectively, while the maximum number being 143 in the month of August (Monsoon depression). The monsoon depressions during June to September, generally form over the Bay of Bengal and traveling westwards or north-westwards, across Odisha, Jharkhand, Chattisgarh, Bihar, Madhya Pradesh and Uttar Pradesh. During the period 1891-2010, total 450 storms /depressions influenced the weather of Jharkhand state.

117. The Bay of Bengal storms/depressions progressively form in the lower latitudes, with the advance of the year. The tracks of the Bay cyclones are observed in more southerly latitudes in October and November, influencing the weather of Jharkhand during the period. Majority of storms and depressions that make landfall over Odisha coast across north Puri, mostly affect Jharkhand.

118. **Climatological Data**

Ranchi

Month	Jan	Feb	Mar	Apr	May	Jun	Jul	Aug	Sep	Oct	Nov	Dec
Max Temp (^{0}C)	23	25	31	35	37	33	29	29	29	28	26	23
Minimum Temp (^{0}C)	10	12	17	21	23	23	23	22	22	19	14	10
Rainfall (mm)	16	21	19	21	43	208	330	315	242	73	11	09
No. of Rainy Days	1	2	2	2	3	10	16	16	12	4	1	1
Total Annual rainfall	1,308 mm											

Dhanbad

Month	Jan	Feb	Mar	Apr	May	Jun	Jul	Aug	Sep	Oct	Nov	Dec
Max Temp (^{0}C)	25	28	34	38	38	35	32	31	31	31	29	26
Minimum Temp (^{0}C)	11	13	18	22	23	24	23	23	23	20	15	11
Rainfall (mm)	12	17	17	17	46	195	323	309	264	87	08	05
No. of Rainy Days	1	1	1	1	3	10	16	16	12	4	1	1
Total Annual rainfall	1,301 mm											

Hazaribagh

Month	Jan	Feb	Mar	Apr	May	Jun	Jul	Aug	Sep	Oct	Nov	Dec
Max Temp (^{0}C)	22	25	30	35	37	34	29	29	29	28	25	22
Minimum Temp (^{0}C)	09	12	16	21	23	24	23	23	22	19	13	09
Rainfall (mm)	13	13	12	13	33	182	316	303	239	84	09	07
No. of Rainy Days	1	1	1	1	2	8	15	15	11	3	1	1
Total Annual rainfall	1,226 mm											

West Bengal

Sub-Himalayan West Bengal (SHWB)

119. Because of the high altitude, Darjeeling and Jalpaiguri experience a cool temperate climate. The average temperature in summer is about 15 °C, and winter temperature is about 2 °C. Snowfall occurs in some parts of this region. Being obstructed by the Himalayas, the region receives heavy rainfall. Due to the scenic beauty and temperate climate of the region, a huge number of tourists visit the areas. Here Kalimpong is another hill station that is visited by many tourists in all seasons for its scenic beauty and the average cool temperature throughout the year.

120. Darjeeling has a temperate subtropical highland climate. The average annual precipitation in Darjeeling is approximately 3,100 mm. Eighty percent of the annual rainfall takes place between the months during the monsoon season. In contrast, just 03% of the annual rainfall takes place between December and March. Darjeeling's altitude—which is greater than some other regions of the Eastern Himalayas at the same latitude months of May, June, and July.

121. **Winter Season (December – February).** Siberian high is the most prominent feature during winter season. NE India and SHWB & Sikkim lies at its southern periphery. By the end of November, a ridge of high establishes over NE India, extending well into Gangetic plains. Another ridge from the northern part of the country penetrates into Myanmar and adjoining Bay. The pressure gradient over SHWB & Sikkim remains diffused during the entire season. The region, especially the hilly areas of Darjeeling and Sikkim receives good amount of rainfall during this season from westerly disturbance which move east wards in form of Induced Low or trough in westerly at middle or higher tropospheric levels across Gangetic plains or along the foothills of Himalaya. Sometimes, the westerly disturbances moving across north of the Himalayas fail to influence the weather over north India but affect sub-Himalayan West Bengal after the systems reach central Tibet region.

122. **Pre-Monson Season (March – May).** The surface winds during this season over SHWB & Sikkim are predominantly north-easterlies/easterlies. By middle of March the temperature starts rising. Over the hilly areas temperatures are still in comfortable zone. Over plains, maximum temperature rarely crosses 40°C. This is mainly due to occurrence of frequent thunderstorm activity over the region. The thunderstorms are triggered by both local heating and orographic lifting. Both anabatic and katabatic winds cause thunderstorm activities over the region. The eastern zone of the region, which is close to Assam and Bhutan, is known for very violent thunderstorms during April and May.

123. Monsoon Season (June – September):- Date of arrival of monsoon, more or less coincides with that of NE India. Normally onset of monsoon takes place between 05 and 10 June which is a few days after the arrival of monsoon over central Bay and parts of Myanmar. Monsoon mostly remain active over the region. However, during the formation period of a monsoon low/depression over north Bay, rainfall activity decreases over the region. It again get activated when the monsoon systems enters land mass. The monsoon systems generally affect the region indirectly as the track of movement of the system is away from the region. The region also experience heavy rainfall during Break Monsoon condition when most parts of the country remains practically dry. Position of the AMT to north of it's normal position is also favourable for the region for good monsoon activity. Besides, movement of WD across higher latitudes also enhances monsoon activity over the region.

124. **Post-Monsoon Season (October – November.** Withdrawal of SW Monsoon starts by first week of October over SHWB & Sikkim and by the end of the first fortnight it withdraws from the entire NE India. With the withdrawal of SW monsoon, the moist maritime air mass is gradually replaced by dry continental air mass. Like most of the regions in the country, SHWB also experience pleasant weather condition during this season. A few western disturbances may affect the region causing light rainfall and thunderstorm.

125. **Rainfall over Plain (Jalpaiguri) Area.** The rainfall gradually increases from Jan to Jul and there after decreases. July is the rainiest month with about 917 mm of rainfall followed by June (628 mm), August (624 mm) and September (511 mm). Rainfall of July accounts for 27% of the annual rainfall. The monsoon season accounts for 80% of annual rainfall. During October, the area receives about 154 mm of monsoon rain. Monthly rainfall during November to February remains less than 35 mm. A few pre-monsoon shower/thunder showers occur in march (mainly under the influence of WD moving across higher latitudes). During April to may there is a rapid rise in convective activities. Monthly rainfall in March, April and May are 35, 119 and 332 mm respectively. Average annual rainfall for the region is about 3357 mm.

Month	Jan	Feb	Mar	Apr	May	Jun	Jul	Aug	Sep	Oct	Nov	Dec
Max Temp (^{0}C)	27	30	34	36	36	37	36	36	36	34	33	29
Highest Max (^{0}C)	35	35	37	40	40	39	40	38	39	38	38	35
Minimum Temp (^{0}C)	08	09	13	16	20	21	23	21	18	14	09	08
Lowest Min (^{0}C)	04	02	08	10	11	15	18	18	15	09	06	02
Rainfall (mm)	09	16	35	119	332	628	917	624	511	154	07	03
No. of Rainy Days	1	1	2	7	13	17	21	17	14	05	07	03
Total Annual rainfall	3357 mm											

Gangetic West Bengal

126. Gangetic West Bengal (GWB) is the mainly plain region that stretches from SHWB (South Jalpaiguri) in the north to the coast and Odisha in the south. The western border is bound by Bihar and Jharkhand, while to the east is Bangladesh. The whole of GWB is a vast plain region except for the the south-western part of (Purulia, Bankura, Birbhum, Bardhaman, and Paschim Medinipur districts) lies in the eastern fringes of the Chota Nagpur Plateau. There are some isolated small hills over this region. Some of the important hills in the area include Ajodhya Hills (677 m), Panchet (643 m), and Baghmundi in Purulia, and Biharinath (452 m) and Susunia (442 m) in Bankura. Gorgaburu in the Ayodhya Hills (677 m) is the highest point in the region. The Chota Nagpur plateau/hills plays an important role in causing thunder storm in south GWB during the pre-monsoon season. The thunderstorm (Kal Baisakhi) built-up in the chota Nagpur hill often are of violent nature.

127. In general, GWB experiences Tropical wet-dry climate. Rainfall is mainly received in the monsoon season. Thunderstorm activity with showers also takes place during pre-monsoon season. Post monsoon season, is practically short and winter season in the southern part is mild. However, temperature in pre-monsoon season shoots up to the range of 38-39 0C. Such temperature in the prevailing humid condition make the weather very sultry and highly uncomfortable.

128. The region is prone to ill effects of Tropical Cyclones in both pre and post- monsoon seasons. During monsoon season the formation and subsequent movement of the monsoon depression from the head Bay sometimes causes torrential rainfall in the southern part of GWB.

129. **Winter Season (December - February).** GWB experiences mild winter during the because of its tropical location and proximity to sea. The southward shifting of ITCZ, in the post Monsoon period, changes the wind pattern over this part of Indian continent from SW'ly to NE'ly. The NW'ly wind sets over GWB in the month of December. In this season Western Disturbances move across North India at lower latitudes, which subsequently induces system and affect weather over NE India. Reduced sun shine period and cold N'ly wind decreases day time temperature and lowest minimum temperature is reported in January. Clear skies, poor visibility in Fog/Mist/Haze, till noon, are regular features of this season. Occasional Thunderstorm with light precipitation occurs during the passage of induced systems across SHWB. Fog occurs for a day or two after passage of the Induced WDs. Often cold wave condition prevails over northern parts of the region for 2-4 days in January.

130. **Pre-monsoon Season (March - May).** Pre-monsoon sets over this region by end of March. After a long dry and cool winter season the region gets heated up by the northern oscillation of sun's position, which increases day time insolation. The weather over this region during the season is controlled by the formation of heat low over east UP and adjoining Bihar- Jharkhand and formation of lower level anticyclone over the Bay of Bengal. These are responsible for establishing a zone of convergence over Orissa-Jharkhand-GWB where hot dry air mass from west-northwest converges with moist air mass from Bay. The frequent convergence is causing the formation of convective clouds over Chotanagpur Plateau during day time. The clouds move eastward, gaining strength and

causing severe Thunderstorm, often with Gale speed wind, over GWB in the evening-night. These weather activities are quite destructive and known as Nor-wester, which is locally known as Kalbaisakhi. The season (April to May) is also characterized by dusty atmosphere due to migration of dust from central and north India with the dry WNW/NW'ly lower level winds. The sea surface temperature over Bay of Bengal rises during April to May and wind pattern changes with the advancement of ITCZ. Few Cyclonic Storm forms over Indian Ocean or South Bay of Bengal and travels NW/N wards. Most of the storms make landfall over AP Coast and a few of them may travel up to West Bengal/Bangladesh coast.

131. **Southwest Monsoon Season (June - September).** During SW monsoon, spanning over four months of Jun to Sep, this region receives highest rainfall of the year. The south-westerly wind flow over Indian Ocean and Arabian Sea gains strength and wind discontinuity forms over peninsular India in the month of May. Though monsoon sets over GWB around 10 June. On set of monsoon depends upon the strength and advancement of monsoon front (ITCZ), movement and strength of higher- level Anticyclone over NE India. The easterly waves start appearing over Andaman Sea and Bay islands, which sometimes intensifies into a Low/Depression over Bay of Bengal and move N / NW wards. Once monsoon is established over the country in the first week of July, mean position of AMT mostly passes across south GWB (Kolkata) for about a month. The monsoon current occasionally oscillates, weakens due to passage of WD, which moves across Himalayan region or due to formation of Low/Depression over north/head Bay. At times in the months of Aug the ITCZ shift to extreme North, along the foothills of Himalaya causing heavy rainfall activity over the region. In this period (Break Monsoon) the monsoon rainfall activity decreases over other parts of the country India. In the months of July to September, occasionally Low/Depression forms over North Bay and move NW wards along the Monsoon Trough and may reach up to SW UP and east MP. In this period GWB receives maximum amount of rainfall, which often leads to flood in the catchment area.

132. **Post Monsoon Season (October - November).** The period of post monsoon season is the best period for the region. By the middle of October southwest monsoon withdraws from this region and northeast monsoon sets in. weather improves with onset of NE monsoon, clouding decreases and temperature starts falling. Western disturbances start appearing at low latitude over Jammu and Kashmir. The WDs move in W'ly direction and induce LOPAR and low-level circulation over SHWB region. Few Cyclonic storm forms over Bay along 10-20 ^{0}N, which move in WNW/NW'ly direction and crosses Tamil Nadu coast. At times the storm re-curves and travels North, and crosses north Orissa/GWB coast wrecking extensive damages and devastations. However, if the system moves NE wards from south of 20° N, only extensive clouding and strong surface wind are experienced by his region.

133. **Rainfall.** Maximum rainfall is received in the month of July (335 mm) followed by August and September with 310 mm each. Rainfall received in the month of June is 249 mm, which is a combination of both pre-monsoon and monsoon rain. Total average annual rainfall received by GWB is about 1,645 mm of which about 73% of annual rainfall is received during the monsoon months. Rainfall received in October and November is 152 and 27 mm respectively, which is about 11% of the annual rainfall. During March, April and May, which are the pre-monsoon months, rainfall received are 38, 51 and 123 mm respectively which is about 13% of the annual rainfall. During three winter months, GWB receives just 67 mm of rainfall which is about 04 %. December is the driest month with about 5 mm of rainfall and in January and February months rainfall received is 15 and 23 mm respectively.

134. **Climatological Data: Kolkata (Alipore).**

Month	Jan	Feb	Mar	Apr	May	Jun	Jul	Aug	Sep	Oct	Nov	Dec
Max Temp (^{0}C)	30	34	37	39	39	38	36	36	35	35	33	30
Highest Max (^{0}C)	33	38	41	43	44	44	40	38	39	39	35	33
Minimum Temp (^{0}C)	11	12	18	21	22	24	24	25	24	21	17	12
Lowest Min (^{0}C)	07	07	10	16	18	20	21	23	21	17	11	07
Rainfall (mm)	15	25	37	35	119	277	371	372	325	180	33	06
No. of Rainy Days	1	1	2	3	6	13	17	17	14	7	1	1
Total Annual rainfall	1,795 mm											

135. **Climatological Data: Shanti Niketan.**

Month	Jan	Feb	Mar	Apr	May	Jun	Jul	Aug	Sep	Oct	Nov	Dec
Max Temp (^{0}C)	29	33	38	41	42	40	36	35	35	34	32	29
Highest Max (^{0}C)	33	37	42	45	47	47	42	37	39	37	34	31
Minimum Temp (^{0}C)	12	15	19	23	25	26	26	26	25	23	17	13
Lowest Min (^{0}C)	05	06	11	15	18	19	20	22	18	16	10	06
Rainfall (mm)	12	25	33	52	113	221	343	297	267	88	11	09
No. of Rainy Days	1	2	2	3	6	12	16	15	12	5	1	1
Total Annual rainfall	1,471 mm											

Sikkim

136. Sikkim's climate ranges from sub-tropical in the south to tundra in the north. Most of the inhabited regions of Sikkim experience a temperate climate, with temperatures seldom exceeding 28 °C in summer. The average annual temperature for most of Sikkim is around 18 °C. Sikkim's geographical location with its altitudinal variation allows it to have tropical, temperate and alpine climatic conditions. Sikkim is also the most humid region in the whole range of the Himalayas, because of its proximity to the Bay of Bengal and direct exposure to SW monsoon.

137. **Winter (December to March).** The winter season is charming and perfect for vacation as the whole area is enveloped in snow. Most areas of the state receive snowfall and get engulfed in the blanket of thick snow, thus luring tourists. During November, even though it does not snow, the temperature is quite appropriate for the tourists. Precipitation during the season comes mainly under the influence of the WDs and it's induced systems. Most of the WDs that move across Tibet cause increase in clouding and precipitation mainly in form of snow.

138. **Summer (Pre-Monsoon Season) (April to May).** Summers in Sikkim aren't as harsh as everywhere else in India. During this season, the weather remains pleasant over most part of the state. Summer is pleasantly warm and moderate and is the ideal season for trekking where you can see blooming rhododendrons and orchids all around in this beautiful Himalayan kingdom. Apart from this adventure activity, summer time in Sikkim is ideal for other tourism activities like Mountaineering, Kayaking, Mountain biking, Canoeing, River rafting and Hang gliding. The season is characterized by pre-monsoon thunderstorm activities during this season. Because of the hilly terrain, thunderstorms are rather violent in nature mainly over the southern parts. An appreciable amount of rainfall is received by the region during this season.

139. **Monsoon Season (June to September).** Onset of monsoon takes place between 05 and 10 June with other north-eastern states and SHWB. But the monsoon in Sikkim takes the form of incessant rain during the months from June till September while July be the wettest of all. Around 600-700 mm of rainfall is received each month

during the season and the temperature remains between 17- 22 degree Celsius. Due to the loft hill ranges, monsoon remains more active over the region. Apart from the normal southerly or south-easterly monsoon current, influence of WDs moving along the higher latitudes also enhances rainfall activities.

140. **Post-Monsoon (October to November).** Monsoon rains mostly continue till mid of October. After withdrawal of monsoon, sky condition improves rapidly and temperature starts falling. By November, the weather condition starts exhibiting winter characteristics with very low temperatures and fog at isolated places.

141. **Rainfall.** The southern part of Sikkim receives more summer rains and the northern part receives more winter rains (including snowfall). Rainfalls are heavy and well distributed from May to October. July is the wettest month. The intensity of rainfall decreases from south to north during monsoon season. The annual rainfall at some places may exceed 5000 mm. Number of rainy days (Rainfall 2.5 mm or more in a day) may range from 100 at Thangu to 184 days at Gangtok.

142. **Snowfall.** Sikkim is one of the coldest regions of the country. Temperature in Kanchenjunga falls to as low as -40 ^{0}C in winter. Sikkim is one of the few states to receive snowfall on a regular basis. Most of the peaks in Sikkim remains snowbound throughout the year, particularly over north-west part.

143. **Climatological Data: Gangtok.**

Month	Jan	Feb	Mar	Apr	May	Jun	Jul	Aug	Sep	Oct	Nov	Dec
Max Temp (^{0}C)	16	18	23	24	25	25	25	26	25	24	21	17
Highest Max (^{0}C)	20	22	28	27	29	28	28	30	29	27	25	24
Minimum Temp (^{0}C)	05	06	09	12	14	16	17	17	16	13	09	06
Lowest Min (^{0}C)	-02	-01	01	03	07	10	11	11	08	04	02	-02
Rainfall (mm)	27	72	126	206	406	610	626	567	439	173	37	19
No. of Rainy Days	02	05	09	15	20	24	27	25	21	08	02	02
Total Annual rainfall	3,308 mm											

Arunachal Pradesh

144. Arunachal Pradesh is primarily a hilly tract nestled in the foothills of the Himalayas in northeast India. Elevation ranges from mountains that are above 7,000 m to the towns in the plains with an elevation of less than 300 m. Arunachal shares international borders with Bhutan, Tibet (China) and Burma (Myanmar). Internally, Arunachal borders the states of Assam and Nagaland.

145. Relief range varies between plains that are a few hundred meters in height and mountains above 7,000 metres (23,000 ft).[5] The elevation of the towns of Naharlagun, Pasighat and Tezu in the south are 290 m, 155 m and 210 m respectively, while Kangto, Nyegi Kangsang and the Gorichen group of mountains are some of the highest peaks in this region of the Himalayas. The southern borders of Arunachal Pradesh are encompassed by the Shivalik ranges which merge into plains. The hills and mountains have associated features such as valleys and intermontane plateaus. Parts of the Lohit district, Changlang district and Tirap district are covered by the Patkai hills. The hills extend towards Nagaland, and form a natural boundary between India and Burma.

146. River systems in the region, including those from the higher Himalayas and Patkoi and Arakan Ranges, eventually drain into the Brahmaputra River.

147. **Climate.** Climate of the state is influenced greatly by the Himalayan Mountains and large variations in altitude across the state. Areas that are at a very high elevation in upper Himalayas close to the Tibetan border experience

alpine and tundra climates. In the middle Himalayas temperate climate is experienced. Areas at the sub Himalayan generally experience humid sub-tropical climate with hot summers and mild winters. The rainfall of Arunachal Pradesh is amongst the heaviest in the country receiving more than 3500 mm in a year. The state receives rainfall over a period of 8 to 9 months excepting in winter, however, most of rainfall is between May and September. Higher regions experience snowfall during winter. The average annual rainfall is 1000 mm in the higher elevations and 5750 mm in the foot hill areas.

148. **Temperatures**. Winter months have average temperatures in the range 15°C to 21°C, and the monsoon month temperatures are in the range of 22 °C – 33 °C, and the summer months temperatures sometimes are higher well over 37 °C. The foot hills experience maximum temperatures around 40 °C during Summer

149. **Rainfall**. The mean south-west monsoon season (June to September) rainfall (1815 mm) contributes 64% of annual rainfall (2818 mm). Mean monthly rainfall during July (581 mm) is highest and contributes about 21% of annual rainfall. The mean rainfall during June is slightly lower and contributes about 18% of annual rainfall. August and September rainfall contribute 15% and 13% of annual rainfall, respectively. Contribution of pre-monsoon (March to May) rainfall and post-monsoon (October to December November & December) rainfall in annual rainfall is 23% and 8% respectively.

150. Average number of rainy days in the state during the south west monsoon is about 78 days and varies from 60 days to 88 days. Days when there is high rainfall events range from 1 to 9 days and similarly the extreme rainfall days are less and is about 1 to 2 days. Average number of rainy days in the state during the post monsoon (October to December) is about 15 days and varies from 12 days to 17 days. Days when there is high and extreme rainfall events are negligible.

151. **Synoptic Features Affecting Weather over Arunachal Pradesh**.

    a)   **Winter Season**.

        i.   Mainly northerly to north-easterly dry and light winds prevail over the state during winter season.

        ii.   The dry spells are broken when WDs move across eastern Himalayas or Tibet/China. The state experience light to moderate rainfall and over the areas of lower altitudes and most of the valley and snowfall over the high hills to the north.

        iii.   After the passage of the WDs the sky clears rapidly and temperatures fall sharply. Dense fog also occurs during morning hours.

    b)   **Pre-monsoon Season**.

        i.   Pre-monsoon season is characterized by rise in temperature. The foot-hill areas adjoining Assam are the hottest places. From second half of March the state experiences thunderstorm activities during afternoon and evening due to uplift of winds (Anabatic wind) along the hilly slopes. Katabatic winds also cause thunderstorm during late night/early morning hours. May is the month with maximum thunderstorm activities.

        ii.   Moisture incursion to the state from the Bay of Bengal due to location of a low-level anticyclone over north bay or Myanmar enhances the thunderstorm activity over Arunachal.

        iii.   The WDs, though follow a more northerly track, continues to affect the state in form of trough in the middle or upper tropospheric level. Occasionally, an induced CYCIR may form in and around the region which further enhances rainfall/thunderstorm activities.

c) **Monsoon Season.**

    i.   During monsoon season the south-westerly monsoon current brings in rain to the state. The hilly barrier of the state often converts the nature of the rains to convective type due to 'orographic lifting' effect. However, thunderstorm activity is much less than the pre-monsoon season.

    ii.   A few WDs may still affect the area in indirect manner. Passage of WDs along higher latitudes. Such episodes enhance the effect of monsoon current.

    iii.   Extension of the AMT up to NE region or formation of a secondary trough of monsoon and passing up to NE region increase rainfall activity over the state. During Break Monsoon condition when the AMT passes along the foot-hills of Himalaya, Arunachal Pradesh experience heavy rainfall.

d) **Post-monsoon Season.**

    i.   Monsoon withdraws from the state during mid-October. Hence, the monsoon rain continues till this period. However, the intensity is lesser than the previous month.

    ii.   Withdrawal of monsoon results in rapid improvement in weather and gradual drop in temperature. Towards the end of the season, the higher ridges experience chilling cold weather.

    iii.   The track of WDs gradually shifts southward and a few of them starts affecting the state from November.

    iv.   Generally northerly to north-easterly dry winds prevail over the region except during the period of passage of WDs across China/Tibet.

152. **Climatological Data of Arunachal: Temperature and Rainfall.**

	Jan	Feb	Mar	Apr	May	Jun	Jul	Aug	Sep	Oct	Nov	Dec
Average Max (^{0}C)	24	26	30	31	32	32	30	33	31	30	27	24
Average Min (^{0}C)	11	13	16	19	21	24	24	25	23	20	15	12
Highest Max (^{0}C)	29	34	37	41	40	40	39	39	39	32	32	28
Lowest Min (^{0}C)	6	7	10	15	14	20	18	22	17	13	9	5
Rainfall (mm)	15	42	61	135	189	326	492	409	276	82	16	6
No. of Rainy Days	4	6	12	20	24	26	27	27	22	13	4	2
Annual Rainfall	2,049 mm											

153. **Climatological Data: Pasighat.**

	Jan	Feb	Mar	Apr	May	Jun	Jul	Aug	Sep	Oct	Nov	Dec
Average Max (^{0}C)	27	29	32	33	35	36	36	36	36	33	31	28
Average Min (^{0}C)	10	11	14	16	18	21	22	22	21	18	14	11
Highest Max (^{0}C)	30	34	34	37	37	39	39	39	38	36	33	30
Lowest Min (^{0}C)	7	7	11	13	11	19	19	20	17	13	8	7
Rainfall (mm)	47	101	163	291	386	786	1135	667	601	191	33	26
No. of Rainy Days	4	7	11	14	13	19	22	15	15	8	2	2
Annual Rainfall	4,427 mm											

154. **Climatological Data: Ziro.**

	Jan	Feb	Mar	Apr	May	Jun	Jul	Aug	Sep	Oct	Nov	Dec
Average Max (^{0}C)	13	13	17	19	21	23	25	25	24	23	18	16
Average Min (^{0}C)	0	2	5	8	12	15	16	15	14	10	4	0
Highest Max (^{0}C)	21	20	26	28	30	33	32	32	32	30	29	25
Lowest Min (^{0}C)	-7	-3	0	1	3	8	9	10	9	2	-4	-6
Rainfall (mm)	89	106	79	169	199	203	249	232	200	64	51	14
No. of Rainy Days	5	7	7	11	13	16	15	15	13	5	3	1
Annual Rainfall	1,744 mm											

155. **Climatological Data: Itanagar.**

	Jan	Feb	Mar	Apr	May	Jun	Jul	Aug	Sep	Oct	Nov	Dec
Average Max (^{0}C)	23	25	28	29	31	31	32	32	31	30	28	25
Average Min (^{0}C)	11	11	14	18	21	23	23	24	23	20	14	11
Highest Max (^{0}C)	27	30	34	34	36	36	38	37	35	36	31	28
Lowest Min (^{0}C)	7	5	9	14	19	20	21	21	17	16	10	7
Rainfall (mm)	34	68	106	227	469	633	578	503	492	242	19	11
No. of Rainy Days	3	5	7	12	14	20	21	16	16	9	2	1
Annual Rainfall	3,382 mm											

156 **Climatological Data: Tawang.**

	Jan	Feb	Mar	Apr	May	Jun	Jul	Aug	Sep	Oct	Nov	Dec
Average Max (^{0}C)	4	4	6	10	13	15	15	16	15	12	8	5
Average Min (^{0}C)	-9	-7	-3	0	3	7	9	9	7	2	-2	-5
Rainfall (mm)	56	102	195	249	331	526	643	466	305	121	45	41
Annual Rainfall	3,080 mm											

Assam

157. Assam is located in the northeastern part of the country and is bounded to the north by the kingdom of Bhutan and the state of Arunachal Pradesh, to the east by the states of Nagaland and Manipur, to the south by the states of Mizoram and Tripura, and to the west by Bangladesh and the states of Meghalaya and West Bengal. The state of Assam comprises three physiographical divisions, namely, the Brahmaputra Valley, the Barak Valley and the Karbi-Anglong and the North-Cachar hills. The Brahmaputra Valley in Assam is approximately 80 to 100 km wide and almost 1000 km long. The width of the river itself is 16 km at many places within the valley. The hills of Karbi Anglong and Dima Hasao district and those in and around Guwahati (along with the Khasi and Garo Hills) are originally parts of the Meghalaya Plateau to the south. These are eroded and dissected by the numerous rivers in the region. The average height of these hills in Assam varies from 300m to 400m to a maximum of about 2000 m. There are also hills in the North Bank bordering Bhutan and Arunachal Pradesh as well as those near Burma. The southern Barak Valley is separated by the Karbi Anglong and North Cachar Hills from the Brahmaputra Valley in Assam. The Barak originates from the Barail Range in the border areas of Assam, Nagaland, and Manipur and flowing through the district of Cachar, it converges with the Brahmaputra in Bangladesh.

158. With the tropical monsoon climate, Assam is temperate (summer maximum temperature at 35–38 °C and winter minimum temperature at 6–8 °C) and experiences heavy rainfall and high humidity. The climate is characterized by heavy monsoon downpours reducing summer temperatures. Pre-monsoon (March–May) and post monsoon (October to November–October) are usually pleasant with moderate rainfall and temperature. The pre-monsoon season, though, most of the days are characterized by moderately high temperature, is moderated by the frequent afternoon/evening thunderstorms. Unlike the other regions of the country, July and August are the hottest months of the state. Though Assam is mainly influenced by the south west monsoon current, the topography and geographical locations of various locations experience diverse climatic condition. Besides, the orographic lifting of air by the hills in the summer months, the availability of abundance of moisture from the Brahmaputra and its tributaries, the anabatic and katabatic effect of the nearby hills, etc. play vital role in the climate of Assam. These factors are responsible for the prolonged rainy season of the state which starts from mid of march to early October.

159. Every year, flooding from the Brahmaputra and other rivers such as Barak River etc. deluges places in Assam. The water levels of the rivers rise because of rainfall resulting in the rivers overflowing their banks and engulfing nearby areas. Apart from houses and livestock being washed away by flood water, bridges, railway tracks, and roads are also damaged by the calamity, which causes communication breakdown in many places. Fatalities are also caused by the natural disaster in many places of the State. Assam is perhaps the only state which experiences flood from overflowing rivers during the pre-monsoon season also. Two to three episodes of flood in a year in various parts of the state is common in the state.

160. The main synoptic features responsible for rainfall in the state are as stated below: -

   a) During the pre-monsoon season, the main features are the instability of atmosphere caused due to unequal heating of surface and triggering through anabatic and katabatic process. Incursion of moisture from the Bay due to location of a low level anti-cyclone over north/head bay in combination with inflow of dry and hot continental hot winds from Gangetic Plains enhances the instability causing severe thunderstorm. Occasionally, Tropical Cyclones moving northwest ward over Bay of Bengal and re-curving towards Bangladesh/Myanmar also affect the state, though, severity of the systems are not experienced.

   b) During monsoon season the main features are: -

      i.   The position of the AMT plays an important role. Whenever the AMT passes through the Brahmaputra valley or its neighbourhood, the state of Assam experience good monsoon activity.

      ii. Prevalence of strong southerly or south-westerly monsoon current (Bay branch) enhances monsoon activity in the state.

      iii. During 'Break Monsoon' condition the AMT passes along the foothills of Himalaya. In this condition when most parts of the country experiences dry weather, Assam gets plenty of rainfall.

      iv. During the formation of a monsoon system (Low/Depression) over north/head Bay, there is a decrease in rainfall over Assam.

   c) During post monsoon season, monsoon activity persists up to mid of October. After the withdrawal of monsoon, as the winds and pressure pattern tend to change over, there is a long spell of clear sky condition. However, towards the end of the season, due to shifting of the track of WDs, the state starts being affected by westerly trough originating from the parent system moving across Tibet. Such westerly trough at middle or higher levels causes increase in clouding and light to moderate rainfall. Occasionally, Tropical Cyclones moving northwest ward over Bay of Bengal and re-curving towards Bangladesh/Myanmar also affect the state, though, severity of the systems are not experienced.

There are four seasons in Assam, namely winter season (December to February), Pre-monsoon season (March to May), Monsoon season (from June to September) and the post-monsoon season (from October to November).

161. **West (Lower) Assam.** Pre-monsoon season is characterized by increased thunderstorm activity due to advection of positive vorticity influenced by the east ward moving westerly systems. Katabatic winds play a major role for more than 70% of thunderstorms after 2000 hours. During four months of SW Monsoon season, this region experiences approximately 3300 mm of rainfall which is about 80% of annual rainfall. Out of this, 60% of precipitation is between sunset and sunrise. Post monsoon is the most pleasant season. The winter is generally mild in compared to the north Indian regions.

162. **Rainfall.** Almost 75- 80% of total annual rainfall takes place during the monsoon months i.e. from Jun to Sep. Frequency of rainy days (2.5 mm of rain in 24 hrs) 20 days or more from May to Sep with the highest of about 29 days in July. Amount of rainfall and number of rainy days start declining rapidly from the month of October to January. Then there is gradual rise from February to march and the rise rapidly increases from the month of march. Average monthly rainfall along with number of rainy days for the various months are as follows.

Month	Jan	Feb	Mar	Apr	May	Jun	Jul	Aug	Sep	Oct	Nov	Dec
Rainfall (mm)	12	25	48	187	381	796	1114	811	577	181	17	06
No. of Rainy Days	3	5	8	17	22	26	29	27	23	11	4	5
Total	4,155 mm annual rainfall											

163. **Middle Assam.** The mid-Brahmaputra valley region has a moderate climate characterized by heavy rainfall during the monsoons and slight rainfall with isolated thunderstorms from October to March. The city does not experience very high temperatures at any time of the year, though the moderately high temperature during summer make life uncomfortable due to prevailing high humidity. The daily maximum temperature is 36 °C in August, the hottest month and the daily minimum temperature is 8.9 °C in January, the coldest month.

164. The region to the north of Brahmaputra is not far from the Himalayas and hence, the weather is affected by the hills in most of the months in some or other ways. The southern region is also bound by the Karbi hills to the south. The combination of the hills to the either sides and the presence of the large water body (Brahmaputra) make the region highly prone to weather phenomena like thunderstorm during summer months and fog during winter.

165. **Winter Season (December-February).** During winter season, the weather over the region generally remains good on most of the days except the incidence of atmospheric obscurities like fog, mist and haze in the morning hours. It is the coldest and driest season of the year. The average max temp is 23-25°C and minimum temperature is of the order of 9-12°C. On a clear weather day, one can often enjoy the panoramic view of snow-clad peaks of Eastern Himalayas.

166. **Pre-Monsoon Season (March-May).** The temperature shows continuous rise during this season. The average max temperature during the season is 29°C and minimum temperature rises gradually from 14°C to 23°C. This is the season of high convective activity. The rainfall generally occurs from convective cloud and starts at late night and early morning hrs.

167. **Monsoon Season (June-September).** The middle Assam region experiences copious rainfall in the monsoon season. The monsoon season accounts for about two third of the annual rainfall. On an average it rains on 80% of the days in this season. The rainfall generally occurs during night and morning hours. Unlike the other parts of the country, and despite the monsoon rains, the temperatures are the highest in this season with record highest temperature of 39.2°C. This is the most uncomfortable season of the year because of high temperature and high

humidity conditions. The average maximum and minimum temperature do not show much of variation and remain between 31-32°C and 24-25°C respectively. However, unlike most places in India, maximum temperature in this region is attained during July/August.

168. **Post Monsoon Season (Oct-Nov).** The monsoon spell continues till 15 Oct after which there is contrasting change to almost dry weather for rest of the period. On an average it rains on 40% of the days in Oct and 13% in Nov. The max temp gradually drops from 30°C to 26°C and min temp from 22°C to 13°C.

169. **Precipitation.** Maximum rainfall is received during the month of June in the region when in most of the areas over north India the hot and dry pre-monsoon season prevails. Most of the monsoon rains in this month are convective in nature due to the prevailing hill topography, availability of moisture from the large water bodies like the rivers and lakes and the prevailing high temperature. Normal monthly rainfall for June is 374 mm. The lowest amount of rainfall is recorded in the months of December and January (08 and 11 mm respectively). There is a sharp rise in the monthly rainfall after March and a sharp fall after October. The average monthly rainfall with number of rainy days (with rainfall 2.5 mm or more) are given in the following table.

170. **Climatological Data.**

Month	Jan	Feb	Mar	Apr	May	Jun	Jul	Aug	Sep	Oct	Nov	Dec
Rainfall (mm)	11	30	55	190	286	374	340	301	243	121	21	08
No. of Rainy Days	5	7	13	20	21	25	27	24	20	11	4	3
Total	1,980 mm annual rainfall											

The records of extremes observed over Tezpur are as follows: -

Highest maximum temperature was 39.2 °C (16 August 2006)

Lowest Temperature was 4.2°C (14 January 1989)

Highest Monthly Rainfall 1564.2 mm (July 1999)

Heaviest Rainfall in 24 hrs 220 mm (11 Jul 1999)

Maximum wind recorded was 55 KT from WSW direction (23 April 1986 &13 Jul 1986)

171. **Upper (East) Assam.** The climate over upper Assam is typically tropical. It varies from hot and humid in the summer (May to August) to cold nights and warm days in the winter. Being more the interior of land, the region is characterized by more mild winters. Rainfall occurs 12 months in a year. Prevailing rainfall, availability of a number of large water bodies including the perineal might Brahmaputra and existence of thick forest and large tea gardens keep the region warm and humid throughout the year. The region is bound by the hills of Arunachal from the north and east and Nagaland hills to the south and plains of Brahmaputra valley to the west. The main factors affecting the climate of the region are similar to that over the middle Assam with some variation due to binding of the region by hills from three sides and existence of more forest vegetations.

172. **Winter Season (December – February).** A ridge of high with Northwest-Southeast orientation establishes over NE India and sometimes extending well into the Gangetic plains. Another ridge from Northeast Arunachal Pradesh penetrates into Myanmar & adjoining Bay. The pressure gradient over NE India gets diffused. Due to low temperatures, this season is also known as Cold Season. January is the coldest month of the year. Weather over the region remains generally good due to dry North/Northeasterly flow of winds except during the passage of Western disturbances (WD). Generally, mist/ fog occurs in early morning hours. Frequency of thunderstorms and precipitation is less. Clear air turbulence (CAT) and mountain waves are also experienced over adjoining Arunachal hills mainly during afternoon hours.

173. **Pre-Monsoon Season (March – May)**. A low pressure establishes over Myanmar which protrudes Northwestwards up to extreme parts of NE India. As the Sun crosses the equator, temperatures start rising rapidly in the Indian sub- continent. That is why, this season is also known as Hot Season. However, marked rise in temperature is not observed over Northeast India. Due to the rising temperatures and incursion of moisture, thunderstorms occur over the region. Western disturbances moving across the region cause thunderstorms, which are more severe and prominent. Frequency of severe thunderstorms is at its peak in the month of April. When cyclone/ depression forms over Bay of Bengal, weather over NE region becomes P' Cloudy. In rare cases, when these systems move in North/ Northeast ward direction, after crossing Bangladesh and Myanmar coast causes heavy rain/ showers over this region. Over Assam, 26 % of the total annual rainfall occurs in this season. Clear Air Turbulence (CAT) and mountain waves are also experienced over high ridges of hills.

174. **Monsoon Season (June – September)**. During this period 64% of the annual rainfall is received at this region. Southwesterly wind flow sets in and onset of monsoon takes place over NE India by first week of June. It withdraws by 10 October from this region. The diurnal variation of temperature is small as compared to Indian mainland due to cloud cover and precipitation. High humidity and high temperatures give rise to sultry weather over the region. The main characteristics of the season are moderate to heavy rain, high humidity & uncomfortable weather. At times, NE India is inundated due to heavy and prolonged spells of precipitation. Secondary trough, which is generally marked from Bihar to East Assam, is a semi-permanent feature. During break monsoon conditions, rainfall over this region increases like other parts of NE India. Depressions / lows, which form over-head Bay of Bengal, influence the weather, especially during starting and end of monsoon season. The majority of rainfall occurs at night/ early morning hours. Generally, the weather in afternoon remains good with isolated spells of passing showers.

175. **Post-Monsoon Season (October – November)**. During this season the moist maritime air mass is gradually replaced by dry continental air mass over the region. Temperatures start decreasing and by late October it becomes cool over upper Assam. The nights are cool and pleasant. During this season, the summer pattern starts giving way to winter pattern. Pressure pattern is ill defined and rather diffused. The katabatic winds which remain subdued in the SW monsoon season gradually become prominent again. Zonal westerly makes its appearance over the region. Cyclones forming over Bay of Bengal rarely recurve East/ Northeastwards and disturb the seasonal weather over the region. WDs start making appearance over North Arunachal Pradesh. The frequency of mist/ fog in early morning hours increases as the season progresses. Weather in this season is generally good and is in fact the best weather period of the year.

176. **Rainfall**. The rainfall gradually increases from January to July and thereafter decreases in the same manner. July is the rainiest month, which accounts for about 18% of the annual rainfall and December is the month of least rainfall which accounts for less than 1% of the annual rainfall. The monsoon season (June – September), accounts for about 64% of the annual rainfall. The maximum numbers of rainy days are around 27 in the month of July followed by 24 and 23 rainy days in August and June respectively. The least number of rainy days are in the month of December (around 4 days). The average temperature and rainfall data in respect of the region is as follows: -

Month	Jan	Feb	Mar	Apr	May	Jun	Jul	Aug	Sep	Oct	Nov	Dec
Mean Max Temp (^{0}C)	23	25	27	28	30	32	32	32	31	30	28	25
Mean Min Temp (^{0}C)	9	12	15	19	22	24	25	25	24	21	14	10
No. of Rainy Days	7	11	17	20	21	24	27	22	20	12	9	4
Rainfall (mm)	24	66	118	259	321	412	503	423	358	81	23	12
Total Annual Rainfall	2,600 mm											

177. **Barak Valley**. Barak Valley is situated in the south Assam. In the north there is Cachar Hills. It is bounded by Manipur in the east, Cachar hills in the north, Meghalaya and Bangladesh in the west States of Tripura and Mizoram in the south. The region is actually bound by hills from all sides except for the Bangladesh and Tripura plains. The region is mainly drained by the Barak river. This river originates in the hills of Manipur and enters Bangladesh before it joins the Ganges and Brahmaputra delta.

178. **Winter Season (December to February)**. This is the coldest season and is characterized by mild cold temperature, mainly clear sky and light winds. The region experiences very less rainfall in this season. Rainfall occurs under the influence of WD which affect about 3-4 days in a month.

179. **Pre-Monsoon Season (March to May)**.   There is gradual increase in temperature in this season. Severe thunderstorm with shower occurs during the season. Rainfall increases progressively.

180. **Monsoon Season (June to September)**. As the rainfall activity continues from the pre-monsoon season, much difference is not felt when monsoon sets in. However, frequency of thunderstorm and heavy rainfall episode decreases in this season.

181. **Post Monsoon Season (October to November)**. The monsoon withdraws from the region by second week of October. Temperature starts falling gradually and rainfall activity reduces significantly after withdrawal of monsoon. Mainly northerly to north-easterly winds prevail over the region during this season.

182. **Rainfall**. There is gradual increase in rainfall from January to July. July is the wettest month with above 600 mm of rainfall which accounts for about 20% of the annual rainfall. The monsoon season accounts for 60% of the annual rainfall. With about 3,545 mm of annual rainfall and about 212 rainfall days the region is one of the wettest regions of the country. There is gradual decrease in rainfall from August and the decrease significant from October. November to January is the period of least rainfall with 35, 18 and 17 mm of rainfall respectively.

Month	Jan	Feb	Mar	Apr	May	Jun	Jul	Aug	Sep	Oct	Nov	Dec
Rainfall (mm)	17	61	172	362	473	586	604	571	452	204	35	18
No. of Rainy Days	4	6	12	19	23	26	28	30	27	22	12	3
Total Annual Rainfall												

Meghalaya

183. Meghalaya is one of the Seven Sister States of northeast India. It is a mountainous state, with stretches of valley and highland plateaus. Meghalaya has many rivers. Most of these are rainfed and seasonal. The important rivers in the Garo Hills region are Ganol, Daring, Sanda, Bandra, Bugai, Dareng, Simsang, Nitai and the Bhupai. In the central and eastern sections of the plateau, the important rivers are Khri, Umtrew, Digaru, Umiam or Barapani, Kynshi (Jadukata), Umngi, Mawpa, Umiam Khwan, Umngot, Umkhen, Myntdu and Myntang. In the southern Khasi Hills region, these rivers have created deep gorges and several waterfalls.

184. The elevation of the plateau ranges between 150 m (490 ft) to 1,961 m (6,434 ft). The central part of the plateau comprising the Khasi Hills has the highest elevations, followed by the eastern section comprising the Jaintia Hills region. The highest point in Meghalaya is Shillong Peak in the Khasi Hills overlooking the city of Shillong. It has an altitude of 1961 m. The Garo Hills region in the western section of the plateau is nearly plain. The highest point in the Garo Hills is Nokrek Peak with an altitude of 1515 m.

185. **Climate**. The varied physiological features of the state and the altitudinal differences gives rise to varied types of climate ranging from near tropical to temperate and alpine. With the average annual rainfall as high as 12,000 mm

in some areas, Meghalaya is the wettest place on Earth. The western part of the plateau, comprising the Garo Hills region with lower elevations, experiences high temperatures for most of the year. The Shillong area, with the highest elevations, experiences generally low temperatures. The maximum temperature in this region rarely goes beyond 28 °C, whereas sub-zero winter temperatures are common. the town of Sohra (Cherrapunji) in the Khasi Hills south of capital Shillong holds the world record for most rain in a calendar month, while the village of Mawsynram, near Sohra (Cherrapunji), holds the record for the most rain in a year.

186. Synoptic situations that affect weather over Meghalaya are:-

   a) The southerly or south-westerly monsoon current in the monsoon season. These monsoon current after travelling through the plains of Bangladesh suddenly encounters the high hills of Meghalaya and get orographically lifted over the state, which results in magnification of the effect of monsoon current.

   b) During the period of formation of a monsoon low/depression over north bay, the region experiences south-easterly to southerly monsoon current which are somewhat less effective than the normal south-westerly current.

   c) The other important feature is the AMT. On many occasions the AMT extends eastward upto the NE states or a secondary trough forms and extend towards NE states. On both such occasions Meghalaya experiences enhanced monsoon activity. Also, during the Break Monsoon condition when the AMT passes along the foot hills of Himalaya, rainfall activity over Meghalaya increases.

   d) During the pre-monsoon season there is thunderstorm activity on most of the days due to prevailing temperature, availability of moisture and orographic lifting of the warm moist air. Location of a low-level anticyclone over north Bay, especially over north-east bay enhances the thunderstorm and rainfall activity over the state.

   e) During both, post-monsoon and monsoon seasons, the movement of tropical system across the vicinity of Bangladesh or West Bengal result in increase in clouding and rainfall over Maghalaya. The cyclones entering into Bangladesh, invariably have more impact.

   f) During winter season Meghalaya mostly experience north-easterly or northerly winds. The anabatic and katabatic winds play important role in weather formation during winter season. In fact, anabatic and katabatic wind flows effect the region in all the seasons, though they are subdued during the monsoon season.

   g) Passage of WD across Arunachal or Tibet pla important role in the weather in this hilly state during winter season.

187. **Winter Season (Dec to Feb).** The NE Monsoon sets in over the country by November. The weather is in general good except during the passage of WD when weather activity takes place over NE India. The frequency of active WDs is 3 - 4 per month. The normal seasonal rainfall is 39.2 mm only. January is the coldest month of the season. The average maximum temperature is 15-18°C and average minimum temperature ranges from -0.4 to 4.7°C. The absolute minimum temperature of -7.5°C was recorded in Shillong on 12 January 89.

188. **Pre-Monsoon Season (March to May).** The surface temperatures show continuous rise during this season. In March and April months, surface winds are generally 10-15 Kt often gusting to 20-25 Kt. By end of May, the seasonal heat low gets established over NW India and adjoining Pak. Rain/Thundershower activities increase as the season progresses. Pre- monsoon season has most intense thundery activity over NE India. Average rainfall during the season is 500.4 mm. The average maximum temperatures range from 18°C to 22°C and average minimum temperatures range from 5.4°C to 13.9°C. The highest maximum temperature of 28.4°C was recorded on 02 May 66. In this season significant convective activity and precipitation generally occur during Night / Afternoon hours.

189. **Monsoon Season (June to September).** In monsoon season Shillong experiences copious rainfall. The monsoon season lasts from June to September, often extending up to mid-October. On an average 4-5 low / depressions per month form over Bay of Bengal which move west-north-west ward. Over NE India Bay systems affect changes in the upper air circulation, moisture feed and alignment of monsoon trough causing weather activity over the region. Surface winds are generally south-southwesterly 10-15 Kt. This season accounts for about two third of the annual rainfall. On an average, it rains on 80% of the days and normal seasonal rainfall is 1790.1 mm. The rainfall generally occurs during morning and night. Despite of the monsoon clouding, temperatures remain fairly high during this season. Highest maximum temperature of 28.4 °C was recorded on 16 Sep 1996.

190. **Post-Monsoon Season (October to November).** Weather becomes pleasant during this season. The monsoon season may spill over till mid-October and after its withdrawal there is a contrasting dry weather condition up to November. The surface winds are light E/NE'lies. Rainfall reduces significantly by November month. The average rainfall during the season is 264.7 mm. The minimum temperatures drop to about 05°C in November. The lowest minimum temperature ever recorded in this season was -3.0°C on 27 November 1971.

191. **Climatological data of Shillong: Extremes.**

Highest Maximum Temp	: 28.4°C 02 May 66
Lowest Minimum Temp	: -07.5°C 12 Jan 1989
Highest Rainfall in a Month	: 1430.3 mm Jul 1974
Lowest Rainfall in a Month	: 0.0 mm December 1999,2009,2010
Highest Rainfall in 24 hrs	: 319.8 25 Jun 1988
Highest wind speed	: 250/70 Kt 28 Feb 1974

192. **Rainfall.** The maximum rainfall occurs in Shillong in the month of

July with the mean of 551.6 mm. Lowest rainfall of 5.5 mm occurs in December month. The frequency of rainy days is highest in July with 29 days. The lowest frequency of rainy day is 03 in Jan & Dec months.

Month	Jan	Feb	Mar	Apr	May	Jun	Jul	Aug	Sep	Oct	Nov	Dec
Rainfall (mm)	13	21	57	141	301	534	552	397	308	222	42	05
No. of Rainy Days	3	6	9	14	22	26	29	27	24	16	5	3
Total	2,593 mm annual rainfall											

Nagaland

193. Nagaland lies in the hills and mountains of north-eastern part of the country. It is one of the smaller states of India. Nagaland is bounded by Arunachal Pradesh to the northeast, Manipur to the south, and Assam to the west and northwest and Myanmar to the east. Nearly all of Nagaland is mountainous. In the north the Naga Hills rise abruptly from the Brahmaputra valley to about 610 m and then increase in elevation toward the southeast to about 1,830 m. The mountains merge with the Patkai Range, part of Arakan system, along the Myanmar border, reaching a maximum height of 3,826 m at Mount Saramati.

194. The region is deeply dissected by rivers; the Doyang and Dikhu in the north, the Barak in the southwest, and the tributaries of the Chindwin River (in Myanmar) in the southeast. Nagaland has a monsoonal (wet-dry) climate. Annual rainfall averages between 1,800 and 2,500 mm and is concentrated in the months of the southwest monsoon (June to September).

195. Average temperatures decrease with greater elevation. In the summer temperatures range from the about 21-23 ^{0}C to about 38-40 ^{0}C, while in the winter they rarely drop below 4 ^{0}C, though frost is common at higher elevations. The temperature during the summer season remains between 16 and 31 ^{0}C. Winter often arrives early, with bitter cold and dry weather striking certain regions of the state. The maximum average temperature recorded in the winter season is 24 ^{0}C.

196. Nagaland receives most of the annual rainfall during the monsoon season (June to September). However, the state receives rainfall in all the months. July is the wettest month and December is the driest month. There is a gradual rise in rainfall from January to July/Aug. The rise is rapid after March and fall is rapid after September. The arrival of monsoon over the state does not make much difference in the weather pattern as abundant of rainfall continues from the pre-monsoon season (March to May). The rainfall during the pre-monsoon season is mainly convective in nature and generally affect during evening, late night and early morning hours due to local insolation and Katabatic & Anabatic effect.

197. The post-monsoon season is pleasant with cool temperature and decreased rainfall. However, towards higher elevation, temperatures are too cold. Winter season (December to February) is a dry season with least rainfall and very low temperature. However, a few western disturbances moving across Arunachal or Tibet cause rainfall activity over the state. Snowfall occurs in the higher elevations, but it is rare and most of the state does not witness any snow.

198. **Climatological Data.**

Description	Data
Average Annual Maximum Temperature	28.9ºC
Average Annual Minimum Temperature	19.0 ºC
Average Monthly Rainfall	171.2 mm
Hottest Month	June (32.3ºC)
Coldest Month	January (11.3ºC)
Wettest Month	July (381 mm)
Driest Month	December (11 mm)
Number of Days with Rainfall (≥ 1.0 mm)	208 days (57%)
No. of Dry Days	157 days (43%)

199. **Climatological Data: Kohima.**

	Jan	Feb	Mar	Apr	May	Jun	Jul	Aug	Sep	Oct	Nov	Dec
Average Max (^{0}C)	18	20	24	25	25	26	26	26	26	24	21	19
Average Min (^{0}C)	6	8	11	14	16	18	19	19	18	15	10	7
No. of Rainy Days	4	6	12	21	27	28	26	29	26	18	3	2

Manipur

200. The state lies at a latitude of 23°83'N – 25°68'N and a longitude of 93°03'E – 94°78'E. The capital (Imphal) lies in an oval-shaped valley of approximately 700 sq mi (2,000 km²), surrounded by blue mountains, at an elevation of 790 m ASL. The slope of the valley is from north to south. The mountain ranges create a moderate climate, preventing the cold winds from the north from reaching the valley and barring cyclonic storms.

201. Manipur is bordered by Nagaland to its north, Mizoram to its south, Assam to its west, and shares an international border with Myanmar to its east.   The state has four major river basins: the Barak River Basin (Barak Valley) to

the west, the Manipur River Basin in central Manipur, the Yu River Basin in the east, and a portion of the Lanye River Basin in the north.

202. Manipur may be divided into two distinct physical regions: an outlying area of rugged hills and narrow valleys, and the inner area of flat plain. These two areas are distinct in physical features. The valley region has hills and mounds rising above the flat surface. The Loktak Lake is an important feature of the central plain. The altitude ranges from 40 m at Jiribam to 2,994 m at Mount Tempü peak along the border with Nagaland.

203. Climate. Manipur's climate is largely influenced by the topography of the region. The state is wedged among hills on all sides. It generally has an amiable climate, though the winters are chilly. The maximum temperature in the summer months is 32 ˚C. The coldest month is January, and the warmest is July.

204. The state receives an average annual rainfall of 1,467 mm between April and mid-October. Precipitation ranges from light drizzle to heavy downpour. The capital city Imphal receives an annual average of 933 mm. Rainfall in this region is caused by The South-west Monsoon current picking up moisture from the Bay of Bengal.

205. Like other states of North-east, Manipur also experience increase in rainfall from March to May which are the pre-monsoon months. Rainfall activities during these months are generally convective in nature that is triggered by local insolation and Katabatic and Anabatic effect. A few of the rainfall episodes can be attributed to the westerly systems that move across Tibet. Besides, location of an anticyclone during the pre-monsoon months over north bay or north-east bay supply moisture to the state, which eventually causes convective activities.

206. Most of the annual rainfall in the state is received during the monsoon season (June to September), though the pre-monsoon season also contributes substantially. There is a decrease in rainfall in the post-monsoon months (October-November) and the season is characterized by pleasant temperature conditions, though, during November the upper ridges experience very cold condition.

207. **Climatological Data**.

Description	Data
Average Annual Maximum Temperature	26.3ºC
Average Annual Minimum Temperature	14.2ºC
Average Monthly Rainfall	111 mm
Hottest Month	May (29.4ºC)
Coldest Month	January (5.7ºC)
Wettest Month	July (277 mm)
Driest Month	December (5.8 mm)
Number of Days with Rainfall (≥ 1.0 mm)	161 days (44 %)
No. of Dry Days	204 days (55.84%)

208. **Month Wise Average Data**.

	Jan	Feb	Mar	Apr	May	Jun	Jul	Aug	Sep	Oct	Nov	Dec
Average Max (ºC)	22	25	28	29	29	28	27	28	28	27	25	22
Average Min (ºC)	6	8	12	16	17	20	21	20	19	16	10	8
Highest Max (ºC)	27	32	33	35	36	35	34	34	33	32	29	25
Lowest Min (ºC)	-1	0	5	9	12	16	18	16	12	6	4	0
Rainfall (mm)	6	7	17	74	159	232	278	217	171	151	16	6
No. of Rainy Days	2	1	3	14	21	26	27	25	21	17	3	1
Annual Rainfall	1,334 mm											

Mizoram

209. Mizoram is a landlocked state in North East India. The southern portion shares 722 Km long international borders with Myanmar and Bangladesh. The Northern part share borders with Manipur, Assam and Tripura. It extends from 21°56'N to 24°31'N, and 92°16'E to 93°26'E. The tropic of cancer runs through the state nearly at its middle.

210. Mizoram is a land of rolling hills, valleys, rivers and lakes. As many as 21 major hill ranges or peaks of different heights run through the length and breadth of the state, with plains scattered here and there. The average height of the hills to the west of the state is about 1,000 m. These gradually rise up to 1,300 m to the east. Some areas, however, have higher ranges which go up to a height of over 2,000 m. Phawngpui Tlang also known as the Blue Mountain, situated in the south-eastern part of the state, is the highest peak in Mizoram at 2,210 m.

211. The biggest river in Mizoram is Chhimtuipui, also known as Kaladan (or Kolodyne). It originates in Chin state in Myanmar and passes through Saiha and Lawngtlai districts in the southern tip of Mizoram and then goes back to Myanmar's Rakhine state. Although many more rivers and streams drain the hill ranges, the most important and useful rivers are the Tlawng, Tut, Tuirial and Tuivawl which flow through the northern territory and eventually join the Barak River in Cachar District.

212. Climate. Mizoram has a mild climate, being relatively cool in summer 20 to 29 °C but progressively warmer, with summer temperatures crossing 30 degrees Celsius and winter temperatures ranging from 7 to 22 °C. The region is influenced by monsoons (June to September) raining heavily from June to September with little rain in the dry (cold) season. Good amount of rainfall is also received during the pre-monsoon season (March to May). The climate pattern is moist tropical to moist sub-tropical, with average state rainfall 2540 mm per annum. In the capital Aizawl, rainfall is about 2150 mm and in Lunglei, another major centre, about 3500 mm.

213. During the monsoon season, the main synoptic features that affect the state are: -

   a) The strength of south-westerly monsoon current: – Strong monsoon surge from the Bay of Bengal enhances monsoon activity over the region

   b) Formation of monsoon low/depression over Head bay: - In general, during the period when monsoon systems form over the head Bay/north Bay, rainfall decreases over the state. However, as the systems strengthens before they start moving north-west ward, monsoon activity may revive over the state.

   c) Location of the AMT: - AMT or a secondary trough passing through NE region is favourable for good monsoon activity over the region.

214. The state being close to Bangladesh are often affected by the tropical cyclones that hit Bangladesh during pre and post monsoon season.

215. **Climatological Data.**

Description	Data
Average Annual Maximum Temperature	29.9°C
Average Annual Minimum Temperature	16.6 °C
Average Monthly Rainfall	129 mm
Hottest Month	April (34.2 °C)
Coldest Month	January (7.8 °C)
Wettest Month	July (287 mm)
Driest Month	December (4.5 mm)

Number of Days with Rainfall (≥ 1.0 mm)	173 days (47 %)
No. of Dry Days	192 days (53 %)

216. **Month Wise Average Data.**

	Jan	Feb	Mar	Apr	May	Jun	Jul	Aug	Sep	Oct	Nov	Dec
Average Max (^{0}C)	24	29	32	34	33	30	29	30	30	30	29	26
Average Min (^{0}C)	8	10	14	19	20	22	22	21	21	18	13	10
Highest Max (^{0}C)	31	34	39	40	40	38	35	35	36	36	34	30
Lowest Min (^{0}C)	-1	3	6	14	17	18	19	18	16	12	4	2
Rainfall (mm)	12	7	18	76	181	281	286	202	224	102	21	5
No. of Rainy Days	1	1	4	14	20	26	28	28	26	21	3	1
Annual Rainfall	1,415 mm											

217. **Climatological Data for Aizawal.**

	Jan	Feb	Mar	Apr	May	Jun	Jul	Aug	Sep	Oct	Nov	Dec
Average Max (^{0}C)	20	22	25	27	26	25	25	25	26	25	23	21
Average Min (^{0}C)	11	13	16	17	18	19	19	19	19	18	15	12
Rainfall (mm)	13	23	73	168	289	406	320	321	305	184	43	15
Annual Rainfall	2,160 mm											

Tripura

218. The state is located between 22°56′ & 24°32′ north latitudes and 91°09′ & 92°20′ east longitudes. It is bordered by Bangladesh to the north, south, and west, and Assam and Mizoram to the east. It consists of eight districts, namely West Tripura, Khowai, Sipahijala, Dhalai, North Tripura, Unakoti, Gomati and South Tripura. About 60% of its land is hilly, while the remaining 40% is plain land. Five hill ranges - Boromura, Atharamura, Longtharai, Shakhan and Jampui Hills, run almost parallel from north to south. The capital Agartala is located on a plain to the west with a height of 15 meters above mean sea level.

219. **Climate of Tripura.**   The state has a tropical savanna climate. The undulating topography leads to local variations, particularly in the hill ranges. The state observes moderately warm temperatures during summer and moderately cold temperatures during winter. Due to the presence of Bay of Bengal to its south, the humidity in the state is fairly high during summer.

220. As per IMD's classification of meteorological seasons in India, the period of the year has been classified into four seasons, namely, Winter Season (January to February) Pre-monsoon Season (March to May) Monsoon Season (June to September) Post-monsoon Season (October to December)

221. **Winter.**   The winters in Tripura are generally moderately cold, but on one or two occasions, temperatures as low as 2-3 ^{0}C have also been observed in the two observatories of IMD. Winter conditions commence in the state from December itself. During this period, moderate to dense fog and sometimes very dense fog is observed during morning hours. January is the coldest month of the year. Average minimum temperatures are around 10 ^{0}C in this month. The days are generally dry, cloud free and light northerly surface winds are observed.

222. **Pre-monsoon.** Temperatures start rising from March which also brings thunderstorms accompanied with rain to the state. These thunderstorm events in the pre-monsoon season are known as 'Norwesters' or 'Ka Baisakhi' in local language. They generally move from northwest to southeast direction. Their duration may be from a few

minutes to a few hours. Sometimes thundershowers are accompanied with squall, with wind speed of more than 150 km per hour or hail. The activity begins in March and progressively increases with the advance of the season reaching to its peak in May. It is the secondary rainy season of the state, with nearly 30% of the annual rainfall. On an average, thunderstorms occur more than 30 days in Tripura during this season. The average maximum temperatures are around 32-33 ^{0}C during this season, with April being the hottest month of the year.

223. **Monsoon**. Normally south-west monsoon enters in the season in the first week of June. With this the wind direction changes from northerly/northwesterly to southerly which brings humid air from the Bay of Bengal to the state. Cloudiness increases over the state in this season. Severe thunderstorm activity decreases but rainfall increases during this season. Although some thunderstorms are observed in this season, but they are of lower intensity as compared to pre-monsoon season. It is the main rainy season for the state with an average of more than 1300mm. It is about 60% of the annual rainfall. June is the rainiest month of the year with more than 400mm of average rainfall. The average maximum temperatures are around 31-32 ^{0}C and minimum temperatures are around 24-25 ^{0}C. The south-west monsoon normally withdraws from the state by mid-October.

224. **Post-monsoon**. Rainfall decreases in the state from October. The temperatures also start decreasing. The average maximum and minimum temperatures fall from 31 and 22 ^{0}C in October to 26 and 11 ^{0}C respectively in December. From November itself the weather becomes dry, but sometimes one or two cyclonic circulations in the Bay of Bengal bring some rainfall for 2-3 days during this season. Mornings start becoming foggy from December and winter conditions are setup. The surface winds change direction once again to northerly/north-westerly from November.

225. The maximum temperatures keep on increasing from January to April, but due to significant increase in rainfall from May, it decreases slightly during that period. Again, due to decrease in monsoon rainfall from August, temperatures increase slightly. From November onwards, both temperatures and rainfall decrease significantly.

226. **Extremes**. The highest ever temperature recorded at Agartala is 42.2 °C on 01 May 1960, whereas the lowest ever temperature is 02.0 °C on 30 December 1972. The highest one-day rainfall recorded was 257.2 mm on 22 May 1993.

227. **Rainfall Data**. Month and season wise average rainfall data for Tripura are as follows: -

Month	Monthly Rainfall (mm)	Season	Seasonal Rainfall (mm)
January	8	Winter	35
February	28		
March	79	Pre-Monsoon	678
April	211		
May	388		
June	460	Monsoon	1458
July	403		
August	340		
September	254		
October	171	Post-Monsoon	222
November	39		
December	12		
Annual Rainfall	2393 mm		

Kutch

228. Climate of Kutch district of Gujarat can be classified as semi-arid. Some weather elements vary largely from normal from day to day and on yearly basis. The region receives very less rainfall and is ideal for the salt industry. As a whole climatic condition of the region is pleasant. Its mainly due to the fresh breeze that mostly prevails over the region in most of the month. Winds are even stronger from June to September when pressure gradient over northeast Arabian Sea remains mostly steep.

229. **Winter Season (December – February).** Mainly clear sky condition prevails over Kutch during winter season. Clouding increases whenever any Induced Low move across south Pakistan and Rajasthan (4-5 per month). When the systems move south of 26°N latitude fog/mist occurs during morning hours. Persistant prevalence of North-westerly continental winds causes cold wave condition over the region.

230. **Pre – Monsoon Season (March – May).** During this season also, mainly clear sky condition prevails over the region. Whenever any western system moves across south Pakistan and Rajasthan, Kutch experiences thunderstorm activity. Surface winds are mainly in westerly to south-westerly with average speed of 10-20 Kt and gradually increasing to 25-30 Kt towards end of April.

231. Tropical Storms which form over Arabian Sea move generally towards west-north-west ward. Occasionally the cyclones recurve towards north or north-east ward and affect Kutch region. Occurrence of Hot day or heat wave condition is rare. It may occur about 2-3 days in the month of May. About 2-3 days prevails on just 2-3 days in the month of May.

232. **Southwest Monsoon Season (June – September).** The SW-Monsoon advances into Kutch region during the last week of June and covers the entire area by first week of July. The withdrawal of Monsoon over Kutch region takes place by mid-September. The region being semi-arid, monsoon condition is generally weak over the region. The main features which affect the weather over Kutch region during this season are Monsoon Low / Depression / Mid-tropospheric Cyclonic Circulation (MTCC) and trough of low off west coast. The SW-monsoon becomes active over Kutch under the influence of the E-W trough & MTCC, the frequency being 3-4 per month with each spell persisting for 2-3 days. Most of the annual rainfall occurs during this season. Major rainfall activity takes place in the month of Jul and Aug. The year to year variability of precipitation over Kutch is very high. When monsoon is active/vigorous, very heavy rainfalls have been recorded on a single day which is nearly equal to monthly average. This region experiences heavy rainfall when W/WNW ward moving monsoon Low / Depression is close to W/NW MP and adjoining Gujarat.

233. **Post Monsoon Season (Oct – Nov).** Generally, this is a good weather period. However, under the influence of induced lows which form over South Pakistan and adjoining South Rajasthan and rarely due to a cyclonic storm which cross Saurashtra coast, this region experiences weather in the form of Rain/Thunderstorm. Surface Wind remains W/NW'ly with average speed of 10-15 kt.

234. This season is favourable period for formation of tropical storms over the Indian Seas. Tropical Storm/Depression which forms in the neighborhood of Andaman Sea move further westward and strike Tamil Nadu/Andhra Coast and move further westwards and revive in the Arabian Sea. Some of these systems intensify and move northwestwards. At times, some cyclone may re-curve, move NW/NNE wards and cross Gujarat and Saurastra Coast. Cyclonic storm forming over Arabian Sea is a rare occurrence.

235. **Rainfall.** Annual rainfall over Kutch is about 390 mm which indicates that this is one of the driest regions of the country. Major amount of rainfall is received in the months of July, August and September, highest being in July which is about 150 mm. The months from December to April are very dry and receives 1 mm or less rainfall

each month. Pre-monsoon showers give 34.5 mm rainfall in June and about 8 mm in May. During post-monsoon Kutch receives about 18 mm of rainfall in October which drastically reduces to less than 5 mm in November.

Saurashtra

236. Saurashtra is a large peninsular region of Gujarat which is bound on the south and south-west by the Arabian sea, on the north-west by the Gulf of Kutch and on the east by the Gulf of Khambhat. Various parts of the region exhibit different weather condition. In spite of being surrounded by sea, most of the places of Saurashtra are semi-arid. However, weather conditions remain comparatively comfortable than many other semi-arid regions. The coastal areas enjoy sea breeze almost throughout the year. As a whole the region has hot semi-arid climate. Rainfall received over the region varies largely from the normal and year to year. Many synoptic situations individually or in combination influence the rainfall activity over the region. The prominent weather systems that affect the region are Monsoon Depression, Mid Tropospheric Cyclonic Circulation, East-West shear Zone and Deep Amplitude Trough in Westerly. The other important system is Tropical Cyclone that form over south-east Arabian Sea during pre-monsoon season and recurve north or north-east ward from the normal north-west ward direction. Occasionally one odd cyclone originating over Bay of Bengal after crossing north Konkan get revived over Arabian Sea and recurve towards Saurashtra or Gujarat.

237. **Winter Season (December – February).** This is the coldest part of the year. Moderate cold wave conditions are experienced over the region after the passage of westerly systems. Sometimes advection fog occurs over coastal areas during morning hours

238. **Pre-Monsoon Season (March – May).** Prevailing of dry weather, weak and diffused pressure pattern are the main features at beginning of the season. As the season progresses gradual and steady rise in temperature is experienced and this finally leads to the build-up of pressure gradient over the region. Again, due to the movement of westerly systems such as Induced Low/WDs, pressure gradient over Saurashtra becomes steep, causing strong surface winds, at times reaching to 30 Kt or more in afternoon/evening over the coastal areas.

239. **Southwest Monsoon Season (June – September).**The onset of monsoon over Saurashtra usually by 20 June. During this season, a Heat Low develops over Central Pakistan region. Steep pressure gradient is generally seen over Saurashtra. Most of the annual rainfall occurs during this season. The occurrence of rainfall is normally in the form of passing Rain / Showers.

240. **Post Monsoon Season (Oct – Nov).** The withdrawal of monsoon takes place by 20 September in this region. Depression/Cyclonic Storms over Arabian Sea at times recurve and move North/North-Eastwards to cross Saurashtra and Kutch coast causing widespread heavy thundershower with hurricane winds.

241. **Rainfall.** Western coast of Saurashtra receives about 570 mm of rainfall annually of which about 83% is received in the months of July to September. 15% of annual rainfall is received in the months of June and October. Rest of the months practically remain dry.

242. Rajkot, which is located in the centre of Saurashtra, experiences hot semi-arid climate. with hot, dry summers from mid-March to mid-June and the wet monsoon season from mid-June to October. Rajkot city receives 670 mm annually. However, this rainfall varies greatly from year to year. For instance less than 160 mm was received in 1911 and 1939 but more than 1,300 mm was received in 1878 and over 1,450 mm in 1959.

Gujarat

243 Gujarat's climate is greatly affected by its location near the Arabian Sea. It has a wide range of weather conditions and temperatures throughout the year. Practically, Gujarat has three distinct seasons: summer (Pre-monsoon), SW monsoon, and winter. Summers prevail from March to May with temperatures reaching as high as 45 0C in some regions. The monsoon season (June to September) brings relief from the scorching summer heat and is characterized by rainfall of varied nature (Light to heavy fall). The winter season, spanning from October to February, is pleasant with temperatures ranging from 15 to 25 0C (October to November is actually the post monsoon season).

244. The monsoon season is of utmost importance for Gujarat's economy and agriculture. The region receives an average rainfall of around 900 mm during this period, which is essential for the growth of crops such as cotton, groundnut, and sugarcane. However, heavy rainfall can also result in flooding and waterlogging in some areas, causing significant damage to crops and infrastructure.

245. Gujarat's climatic zones are classified into two broad categories, coastal, central. The coastal zone, encompassing cities such as Ahmedabad and Surat, experiences a tropical monsoon climate with hot and humid summers and mild winters. The central zone, which includes Vadodara has a semi-arid climate with hot summers and cool winters.

246. Gujarat, like the rest of the world, is experiencing the impact of climate change, including rising temperatures, changing precipitation patterns, and extreme weather events. The region has witnessed several severe heat waves in recent years, with temperatures surpassing 45 0C in some places. Heavy to very heavy rainfalls have also become more frequent, leading to flooding and damage to infrastructure.

Madhya Pradesh

247. Madhya Pradesh is located in a sub-tropical climate region and has a tropical monsoon climate. The climate of Madhya Pradesh is governed by a monsoon weather pattern. The distinct seasons are summer (March to May), winter (November to February), and the intervening rainy months of the south-west monsoon (June to September). Presence of Tropic of Cancer is responsible for placing Madhya Pradesh in Tropical Climate region. The Tropic of Cancer passes through 14 districts located in the central part of Madhya Pradesh which are Ratlam, Ujjain, Agar-Malwa, Rajgarh, Sehore, Bhopal, Vidisha, Raisen, Sagar, Damoh, Katni, Jabalpur, Umaria and Shahdol.

248. The state is very large (latitudinal spread of 870 km from east to west) and hence, there is large variations in the climatic conditions over different parts.

    a) **Regional variation.** Madhya Pradesh weather is markedly different in the following geographical zones: -

        i. The Northern Plains

        ii. The Hilly Region of the Vindhyanchal

        iii. The Narmada Valley

        iv. The Malwa Plateau.

    b) **Temperature variation.** The maximum temperature in the state is in May and the lowest in January.

    c) **Unequal distribution of the Rainfall.** There is a wide regional variation in the rainfall. Annual average rainfall for the State is 1160 mm, with the heaviest rains in the south eastern parts and decreasing towards the north-west. Most of the rainfall is received from the South Asian monsoon during June to September which, in Madhya Pradesh, accounts for more than 90% of the total annual rainfall. Pachmarhi has average rainfall of 2120 mm while Bhind gets only 620 mm.

249. The important factors that affect the climate of Madhya Pradesh are: -

    a) **Geographical location**. The region is land locked one and is away from both Arabian Sea and Bay of Bengal. Hence, there is no moderating effect of sea over the region.

    b) The Tropic of Cancer passes almost through the centre of the region. The region has a high temperature range and undergoes large diurnal and seasonal variations.

    c) Presence of the mountain ranges (Vindhyachal mountain, Satpura mountain and Rajpipla mountain series) in the region changes the effect of the prevailing winds in different ways in different seasons.

250. **Summer Season (Pre-monsoon) (March to May).**Summer starts in Madhya Pradesh from mid-March and lasts till mid-June, In this period, after March 21, due to the northward movement of sun, the temperature increase and the air pressure decreases, as a result of which there is a severe heat in the month of May. The average May temperature in Madhya Pradesh is around 40° Celsius. In Madhya Pradesh summer, season is locally known as Unala. During this time, the temperature in Gwalior, Morena, Datia and Southern Balaghat is around 47°C. Ganj Basoda (Vidisha district) is the hottest place in the state which has recorded temperature of up to 48.7 °C in summer.

251. This season is normally dry with very low humidity. The sky is generally clear across the state. In May and June, high temperature in north-west India builds steep pressure gradient, Hot, dusty laden and strong wind known as Loo blows. Loo normally starts blowing by 9.00 A.M. increases gradually and reaches maximum intensity in the afternoon. It blows with an average speed of 30-40 km per hour and persists for days. The Sun being perpendicular to the Tropic of cancer, MP receives the maximum insolation during May to mid of June.

252. Monsoon Season (June to September).   The south-west Monsoon usually breaks out in mid-June and the entire state receive a major share of its rainfall between June and September. It comes from both the southwest monsoon branches of Arabian Sea and Bay of Bengal branch. Rainfall first starts from the south-eastern regions and then progresses towards north-west through east and north-east parts.

253. Madhya Pradesh receives about 85 % of the annual rainfall i.e. 100 to 125 cm from the southwest monsoon which is about 1000 to 1250 mm. There is large variation from the average in the actual rainfall received year to year. The main synoptic features that influence the strength of monsoon and intensity of rainfall over Madhya Pradesh are movement of monsoon depression / Low and the location, orientation and depyh of the AMT. Besides, the east-west shear zone at mid tropospheric levels also plays important role in enhanced rainfall activity over south MP. Above all, it is only the west ward moving monsoon systems that causes good rainfall, and sometime in access which causes flood in Madhya Pradesh.

254. The south and south-east regions receives higher rainfall whereas the parts of north-west receive less. Mandla, Balaghat, Sidhi, Jabalpur and other extreme eastern parts receive more than 1500 mm rainfall. The south-eastern part of Madhya Pradesh receives the highest rainfall as the Arabian branch from Keral and Bay branch of monsoon hit the Satpura Maikal Range, the highest mountain range of the state. The amount of rainfall in Madhya Pradesh decreases from east to west and south to north. In the eastern and southeastern regions of Madhya Pradesh, most of the rainfall is received from Bay of Bengal branch, while in the northwest and intermediate regions, both branches Arabian Sea and Bay of Bengal cause rainfall. Major contribution, however, is from the Arabian Sea branch as the AMT mostly remains north of MP or along north MP.

255. The highest rainfall occurs in the state during June to September. In Madhya Pradesh, the rainy season is locally known as Chaumasa. This season leads to a sudden fall in the day temperature due to the clouds and rains.

Till July, the complete Eastern Madhya Pradesh comes under the impact of these monsoon winds. The hilly and plateau regions receive more rain as compared to the plains. For example, the Mahadeo hills of Satpura range receive an average monthly rainfall of 230 mm while, the average rainfall m plain areas of Jhabua, Dhar and Khargone is 40-50 mm.

256. **Annual Rainfall**. As mentioned earlier, The South-Eastern part of the state receives an excess rainfall i.e. more than 1500 mm. Such places in Madhya Pradesh are Pachmarhi, Mahadeo hills, Mandla, Sidhi and Balaghat. This part is called Cherrapunji of Madhya Pradesh due to being the rainiest place. Pachmarhi in Madhya Pradesh receive the highest rainfall of 2150 mm. Betul, Chhindwara, Seoni, Narsinghpur districts receives above normal rainfall. The average rainfall in these areas is 1250 mm to 1500 mm. The North-Eastern places of the state generally receives close to normal rainfall. Central highlands, plateau of Bundelkhand, Rewa-Panna plateau are the areas which receive lesser rainfall. These areas include places with 750 mm to 1000 mm of rainfall. The Western areas of the state receive the minimum rainfall i.e. of the order of 500 to 750 mm. Neemuch, Mandsaur, Ratlam, Dhar, Jhabua, etc comes under this part of the state. Bhind (Gohad) is the least rainy place in Madhya Pradesh.

257. **Post Monsoon Season (October to November).**   Like most parts of the country, this is the most pleasant season of Madhya Pradesh. Withdrawal of monsoon by end of September or beginning of October is characterized by clear sky and gradual fall in temperature. Rainfall received during this season is the minimum.

258. **Winter Season (December to Mid-March).** This season is locally known as Siyala in the state. The season is characterized by sharp decline in temperature. Temperature often falls below 10 0C. The northern parts of the state come under cold wave condition after passage of an western disturbance or due to persistent prevalence of dry and cold north-westerly winds.

259. The Induced WDs passing across south Rajasthan sometimes affect north and central Madhya Pradesh region. Also, occasionally, a deep amplitude trough extending from a parent WD at higher latitude, extends up to Madhya Pradesh causing increase in clouding and light to moderate rainfall.

260. **Temperature.**   In winter, the mean minimum temperature is 10°C, though it can drop as low as 1°C and rise to a mean maximum temperature of 25°C. Summer temperatures fluctuate between a minimum of 22°C and a maximum of 38°C, though it can get very hot with temperatures up to 48°C.

261. **Average Climatological data: Madhya Pradesh.**

Month	Jan	Feb	Mar	Apr	May	Jun	Jul	Aug	Sep	Oct	Nov	Dec
Daily Mean Temp (^{0}C)	20	25	31	36	39	35	29	28	28	28	25	21
Max Temp (^{0}C)	26	30	35	40	43	38	32	30	31	32	30	27
Highest Max (^{0}C)	32	37	43	46	47	47	41	38	37	37	34	32
Minimum Temp (^{0}C)	12	16	21	27	30	29	25	23	23	21	16	12
Lowest Min (^{0}C)	4	7	13	19	24	18	20	21	16	13	11	6
Rainfall (mm)	8	7	10	4	2	108	220	246	160	26	1	2
No. of Rainy Days	1	2	2	1	1	12	23	21	14	4	1	1
Total Annual rainfall	794 mm											

Chhattisgarh

262. Chhattisgarh is a landlocked state with Maharashtra and Madhya Pradesh on the west, Uttar Pradesh on the north, Jharkhand on the north-east, Odisha on the east, and Telangana & Andhra Pradesh on the south. The northern and southern parts of the state are hilly, while the central part is a fertile plain. The highest point in the

state is the Gaurlata near Samri, Balrampur-Ramanujganj district. In the north lies the edge of the great Indo-Gangetic plain. The Rihand River, a tributary of the Ganges, drains this area. The eastern end of the Satpura Range and the western edge of the Chota Nagpur Plateau form an east–west belt of hills that divide the Mahanadi River basin from the Indo-Gangetic plain.

263. **Climate.** Chhattisgarh has a tropical climate. It is hot and humid in the summer because of its proximity to the Tropic of Cancer and its dependence on the monsoons for rains. Summer temperatures in Chhattisgarh can reach up to 49 °C. Late in the month of June, Monsoon (July-September) arrives in the state as a respite from the scorching heat. Chhattisgarh receives good amount of rainfall with an average of 1,292mm. Since it falls under the rice-agro-climatic zone, rainfall proves to be the main source of irrigation. A significant variation in the annual rainfall adversely affects the harvest. The elevated regions in the north and south observe moderate climate round the year.

264. During Pre-monsoon season the state experience thunderstorm activities due to local instability and synoptic features like north-south wind discontinuity at lower tropospheric levels, which generally extends from interior Tamil Nadu or north/south Interior Karnataka, at times extending up to Chhattisgarh. Another important synoptic features that triggers thunderstorm activity in the state is the nort-south trough at lower levels which extends south ward from Bihar/SHWB under the influence of east ward moving westerly systems along higher latitudes. The hilly region towards the north which is located in the western periphery of the Chota Nagpur plateau, also experience thunderstorm activity due to orographic lifting of the westerly winds and simultaneous inflow of maritime winds from the Bay of Bengal due to location of a low- level anticyclone over north Bay of Bengal.

265. During the monsoon months, Chhattisgarh receives rainfall from both Arabian Sea and Bay of Bengal branches of monsoon. However, the main synoptic feature for good monsoon activity is the west ward movement of the monsoon low/depression from north Bay of Bengal.

266. During October the state gets some rainfall from the retreating monsoon before monsoon withdraws from the state. During the months of November to February, the state receives very less rainfall. However, during the post monsoon season (October to November), the state may receive appreciable amount of rainfall from the tropical cyclones that make land fall over Odisha/AP coast. Winter season generally receives less rainfall, though, occasionally, under the influence of moving induced westerly system across north India, there is increase in clouding associated with light rainfall.

267. In October to November, cool breeze envelops the entire state as if heralding the arrival of winters. The winter season (December-February) doesn't necessarily mean wearing loads of woolens in Chhattisgarh. At this time, the temperature drops down to 10 ^{0}C or less. However, hilly areas are much colder.

268. **Climatological Data.**

Raipur

Month	Jan	Feb	Mar	Apr	May	Jun	Jul	Aug	Sep	Oct	Nov	Dec
Max Temp (^{0}C)	28	31	36	40	42	37	32	31	32	32	30	28
Highest Max (^{0}C)	37	38	43	46	48	47	41	37	37	38	36	32
Minimum Temp (^{0}C)	13	16	20	25	27	26	24	24	24	21	16	12
Lowest Min (^{0}C)	05	05	08	15	14	16	17	20	18	14	08	04
Rainfall (mm)	14	14	12	09	30	221	327	300	201	50	10	07
No. of Rainy Days	1	1	1	1	2	9	14	14	9	3	1	1
Total Annual rainfall	1,193 mm											

Bilaspur

Month	Jan	Feb	Mar	Apr	May	Jun	Jul	Aug	Sep	Oct	Nov	Dec
Max Temp (^{0}C)	23	25	30	35	40	38	28	27	28	28	25	23
Minimum Temp (^{0}C)	10	12	16	21	30	26	26	22	21	17	12	10
Rainfall (mm)	20	30	20	20	20	200	370	360	200	70	10	0
Total Annual rainfall	1,320 mm											

Ambikapur

Month	Jan	Feb	Mar	Apr	May	Jun	Jul	Aug	Sep	Oct	Nov	Dec
Max Temp (^{0}C)	23	27	32	37	39	35	30	29	30	29	26	23
Highest Max (^{0}C)	31	35	40	44	45	45	38	36	36	35	32	29
Minimum Temp (^{0}C)	09	12	16	21	25	25	23	23	22	18	13	09
Lowest Min (^{0}C)	01	03	08	11	16	17	16	20	16	09	04	02
Rainfall (mm)	26	20	19	14	21	235	411	352	227	48	14	11
No. of Rainy Days	2	2	2	1	3	10	17	16	11	3	1	1
Total Annual rainfall	1,399 mm											

Odisha

269. The state of Orissa lies in the northeastern part of the Indian Peninsula, roughly between latitudes 22036'N and 17049'N and longitudes 81036'E and 87018'E. The state is bound by the districts Ranchi, Singbhum (of Bihar) and Medinipur (of West Bengal), on the north; by the districts Raigarh, Raipur, Bastar (of Madhya Pradesh) on the west; by the districts Khammam, East Godavari, Vishakhapatnam, and Srikakulam (of Andhra Pradesh) on the south and by the Bay of Bengal, on the east.

270. The climate of the state is characterized by hot summer and cold winter in the interior and climate of the coastal region near the Bay of Bengal is moist and equable.

271. The state may be broadly divided into four geographical regions, viz. the northern plateau, central river basins, eastern hills and coastal plains. The northern plateau region includes mainly Mayurbhanj, Keonjhar and Sundargarh districts. This is an undulating upland frequently intersected by hill ranges and sloping from north to south. The average elevation in central section of the plateau is 900 m ASL. The central river basins lie between the northern plateau and eastern hills and include Bolangir, Sonepur, Sambalpur, Deogarh, Bargarh, Jharsuguda, Dhenkanal and Angul districts, and a part of Cuttack district.

272. The eastern hills which constitute the last portion of the eastern ghats, lie to the south and southwest of central river basins through the districts of Koraput, Rayagada, Nawarangpur, Malkangiri, Kalahandi, Nawapara, Gajapati and a part of Ganjam district. The eastern hills are elevated and are generally 900 m ASL. The coastal plains comprise mostly of Balasore, Bhadrak, Kendrapara, Jagatsinghpur, Jajpur, Puri, Khurda, Nayagarh and a portion of Ganjam and Cuttack districts.

273. **Climate.** Climate of Odisha is Tropical Savanna Hot which is seasonally dry during winter. However, the climate of Rayagada district and neighbourhood marginally varyies between Tropical Savanna Hot; seasonal dry (winter) and subtropical monsoon, mild and dry winter; hot summer. Whereas, Kandhamal and Koraput districts come under the types of subtropical monsoon, mild winter; hot summer. The year may be divided into four seasons. The

winter season from December to February is followed by the pre-monsoon or hot weather season from March to May. The period from June to September constitutes the southwest monsoon season and the period of October and November is the post-monsoon season.

274. During the period from December to February, generally low temperatures prevail over the state except in the coastal belt. In the hot weather season from March to May, weather is generally dry and uncomfortable in the interior, while due to lower temperatures, the plateau regions are, comparatively less uncomfortable. Weather tends to be unbearable during July due to high humidity and high temperature. The rest of the period of the monsoon is fairly comfortable due to reduced day temperatures.

275. **Temperature**. The variation of maximum temperature in the coastal districts is not very pronounced during the southwest monsoon and post-monsoon seasons. May is the hottest month with the mean daily maximum temperature of 36.6 ^{0}C in the plains. During this month, the mean daily maximum temperature ranges from 32.3 ^{0}C to 43.2 ^{0}C over the state, the values increasing northwestwards. During July, an appreciable drop in the mean daily maximum temperature is observed, with the values ranging between 25.5 ^{0}C and 33.5 ^{0}C. The temperature pattern of October is quite similar to that of July. The values of mean daily maximum temperature in October ranges between 27.2 ^{0}C to 33.5 ^{0}C. The mean daily maximum temperature of January ranges between 24.6 ^{0}C and 30 ^{0}C.

276. In the month of December, the minima of the mean daily minimum temperature is observed over the central region of the state. The values range between 8.2 ^{0}C to 17.2 ^{0}C. In April, the values range between 19.9 ^{0}C to 26.4 ^{0}C.

277. The highest maximum temperature ever recorded in the state is 49.6 ^{0}C on 28 May 1988 at Jharsuguda and 26 April 1976 at Titlagarh respectively. The lowest minimum temperature ever recorded in the state is 0 ^{0}C at Phulbani observatory on 06 January 1986 and 27 December 1990.

278. **Rainfall**. The total annual rainfall in the state is maximum over the region of Baudh district and neighbourhood, the coastal districts and the south-western part of the state. The total annual rainfall for the state is 1450 mm and the total annual number of rainy days is 69. Malkangiri district receives the maximum amount of rainfall (1670 mm) in a year, whereas Ganjam receives the minimum amount of rainfall (1280 mm) in a year. The southwest monsoon season is the principal rainy season and the state receives about 79% of the annual rainfall in this season. About 3% is received in the winter season (December to February), about 8% is received in the pre-monsoon season (March to May) and about 10% is received in the post-monsoon season (October-November).

279. During the southwest monsoon season, the state receives rainfall mainly due to low pressure areas and monsoon depressions originating in the Bay of Bengal. The southwest monsoon sets in over the southern parts of the state by about first week of June and covers the entire state by the second week of June. July and August are the rainiest months. The number of rainy days ranges from 10 to 16 in the southwest monsoon Season. The withdrawal of the southwest monsoon begins from the northern parts of the state in the second week of October and the monsoon withdraws from the state completely by about 15 October.

280. The most common rain-giving systems over the state during post-monsoon season are the depressions and cyclonic storms originating in the Bay of Bengal. The storms and depressions cause heavy to very heavy rainfall and contribute substantially to the season's total rainfall.

281. During winter, the state receives about 40 mm of rainfall. This rainfall generally occurs in association with induced low-pressure areas over the surface due to Western Disturbances moving from west to east, across the northern parts of the country.

282. **Climatological Data**

Angul

Month	Jan	Feb	Mar	Apr	May	Jun	Jul	Aug	Sep	Oct	Nov	Dec
Max Temp (^{0}C)	27	31	35	39	40	36	32	31	32	32	29	27
Minimum Temp (^{0}C)	14	17	20	23	26	26	24	24	24	22	17	14
Rainfall (mm)	13	27	24	27	53	225	348	358	217	86	20	03
No. of Rainy Days	1	2	2	2	4	10	16	15	11	5	1	1
Total Annual rainfall	1,402 mm											

Balasore

Month	Jan	Feb	Mar	Apr	May	Jun	Jul	Aug	Sep	Oct	Nov	Dec
Max Temp (^{0}C)	27	30	34	36	36	34	32	32	32	32	30	27
Minimum Temp (^{0}C)	14	17	21	24	25	26	25	25	25	23	18	14
Rainfall (mm)	15	32	34	62	108	221	309	332	268	170	35	05
No. of Rainy Days	1	2	2	3	6	10	14	15	12	6	1	1
Total Annual rainfall	1,592 mm											

Cuttack

Month	Jan	Feb	Mar	Apr	May	Jun	Jul	Aug	Sep	Oct	Nov	Dec
Max Temp (^{0}C)	29	32	35	38	38	36	33	32	33	33	31	29
Minimum Temp (^{0}C)	16	19	22	25	26	26	25	25	25	23	19	15
Rainfall (mm)	10	29	25	18	71	210	308	339	229	126	45	04
No. of Rainy Days	1	2	2	2	4	10	14	15	12	6	2	1
Total Annual rainfall	1,424 mm											

Jharsuguda

Month	Jan	Feb	Mar	Apr	May	Jun	Jul	Aug	Sep	Oct	Nov	Dec
Max Temp (^{0}C)	28	31	36	40	41	37	31	31	32	32	30	28
Minimum Temp (^{0}C)	12	15	19	24	27	26	25	25	24	21	16	12
Rainfall (mm)	14	23	18	15	28	219	386	383	210	55	08	04
No. of Rainy Days	1	2	1	1	2	9	17	16	11	3	1	1
Total Annual rainfall	1,363 mm											

Koraput.

Month	Jan	Feb	Mar	Apr	May	Jun	Jul	Aug	Sep	Oct	Nov	Dec
Max Temp (^{0}C)	25	28	32	33	34	31	25	26	27	27	25	24
Minimum Temp (^{0}C)	10	13	17	20	21	21	19	18	19	18	14	10
Rainfall (mm)	06	09	18	55	82	207	376	394	256	126	33	07
No. of Rainy Days	1	1	1	4	5	11	19	19	14	7	2	1
Total Annual rainfall	1,226 mm											

Maharashtra

283. Maharashtra is bounded by Arabian sea on its western side. The state of Gujarat, lies north of it, while Madhya Pradesh and Chattisgarh lie on the northern and eastern sides of Maharashtra respectively. On the southern side, it is bounded by Karnataka and Andhra Pradesh. The Western Ghats (Sahyadri) run north to south separating the coastal districts of Thane, Mumbai City, Mumbai Suburban, Raigad, Ratnagiri and Sindhudurg from the rest of Maharashtra. The average height of the range is about 1 km. West to east the region stretches across a distance of 800 kms.

284. As the ridge runs across at right angles to the monsoon stream, it forms an important climatic divide. The western slopes and the coastal districts get very heavy monsoon rains, while to the east of the Ghats rainfall drops to less than a tenth within a short distance from the Ghats. The state receives rainfall mainly during the southwest monsoon season (June to September). There is heavy rainfall in coastal region (about 2000 mm), scanty rains in the rain shadow areas in central parts (about 500 mm) and medium rains in eastern parts (about 1000 mm) of the state. Fortunately, all the important rivers like Godavari, Bhima and Krishna which originate from the water sheds of the Ghats flow east across these drier regions and contribute to their economic benefit. The influence of the smaller and less cohesive east west oriented ranges of Satpura and Ajanta is not so marked.

285. There are four meteorological subdivisions, viz. Konkan, Madhya Maharashtra, Marathwada and Vidarbha in the state.

286. **Climate**. The state mainly comes under the climatic type Tropical Savanna, hot, seasonally dry, usually winter. Based on temperature and rainfall variation, the climate of Maharashtra can be classified under the following main types: -

   a) **Monsoon**. This type characterized by an annual rainfall of more than 1000 mm is confined to the coastal belt and the adjoining Ghats region covering the districts of Thane, Raigad, Ratnagiri, Sindhudurg and the western hilly parts of Pune, Satara and Kolhapur districts. The coastal region experiences very small annual range of temperature, not exceeding 5 °C. The mean daily temperature is above 22 °C throughout the year. The coastal belt is thus hot and humid with plentiful rain during the southwest monsoon season. Over the remaining portion covered by this climate type, the mean daily temperature of the coldest month is between 18 °C and 22 °C.

   b) **Dry climate**. This type covers the semi-arid portions of Jalgaon, Nashik, Aurangabad, Pune, Beed, Satara, Osmanabad and Kolhapur and almost the whole of Dhule, Nandurbar, Ahmednagar, Solapur and Sangli districts. Mean daily temperature is between 18 °C and 22 °C during winter and above 22 °C during remaining months. Annual rainfall is low being 600 to 900 mm and is confined mainly to southwest monsoon season.

   c) **Tropical Rainy**. Parts of Nashik, Jalgaon districts, eastern portions of Aurangabad, Jalna, Beed and Osmanabad, as well as the remaining districts of Marathwada viz. Hingoli, Latur, Parbhani and Nanded and entire Vidarbha have tropical rainy climate. The precipitation is confined to the monsoon season and is above 700 mm. Mean daily temperature is between 18 °C and 22 °C during winter and above 22 °C during the remaining months.

287. **Sea Level Pressure and Winds**. During January, atmospheric pressure is higher over India and decreases southwards. But the gradient is weaker. Over Maharashtra, the pressure is uniform along the coast. Over the interior, pressure gradient is weak and is oriented in a north-south direction. Winds are light and mainly from north or northeast.

288. Pressure begins to decrease after January. During March, pressure is uniform throughout the state. A reversal of the pressure gradient occurs in April with the establishment of a low over north India. Winds are north-westerly to westerly. With the advance of summer, the isobars gradually orient themselves in a west to east direction. The winds strengthen and the pressure gradient remains strong throughout the monsoon season till September. Winds are from the westerly direction.

289. October is the month of transition when the pressure gradient weakens considerably and reverses gradually to the winter pattern. Winds are weak and mostly north-north-easterly over the state. Pressure continues to rise thereafter till January, winds remaining northerly. Due to the sea breeze effect, winds are from a westerly/north-westerly direction over the coastal belt throughout the year during evening.

290. **Temperature**. May is generally the hottest month when the mean maximum temperature for the coastal belt is 33 $^\circ$C while for the interior it increases from about 38 $^\circ$C in the western parts to 43 $^\circ$C in the extreme east. Plains of Vidarbha and adjoining eastern parts of Madhya Maharashtra experience severe summers with mean maximum temperature of 42 to 43 $^\circ$C in May. As there are gaps in the Western Ghats, sea breeze can penetrate through them into the parts of Madhya Maharashtra. In this region, therefore, high temperatures persist for lesser duration and evenings tend to become sufficiently pleasant. Temperatures begin to fall with the onset of southwest monsoon in June.

291. During winter, cold waves in the wake of western disturbances passing across north India affect the northern parts of the state particularly in north Madhya Maharashtra resulting in a rapid and appreciable fall in temperatures. December is the coldest month for the interior regions with the mean minimum temperature of 13 $^\circ$C and January for Konkan with the mean minimum of 18 $^\circ$C. Temperatures then begin to rise rapidly from February to April.

292. The lowest minimum temperature ever recorded in the state was −0.6 $^\circ$C at Malegaon on 01 February 1929 and the highest maximum temperature ever recorded was48.4 $^\circ$C at Jalgaon on 28 May 1989 and at Wardha on 17 May 1989.

293. **Rainfall**. Maharashtra state experiences extremes of rainfall ranging from 6000 mm over the Ghats to less than 600 mm in Madhya Maharashtra. The coastal strip and the adjoining Western Ghats exposed to the southwest monsoon receive the heaviest rains exceeding 2000 mm. Rainfall over the Ghats may exceed 5000 mm annually. Rainfall decreases very rapidly, towards the eastern slopes and the plateau area. It again increases as we proceed further eastwards. Thus, Madhya Maharashtra is the region of lowest rainfall. In this region, the annual rainfall is less than 600 mm at most places.

294. July is generally the rainiest month except in the central region of the state where September is the rainiest. The state gets its rainfall chiefly in the southwest monsoon season (June to September). The rainfall becomes heavy to vigorous on occasions in association with the cyclonic storms and depressions. The monsoon advances from south to north and reaches Mumbai by about 10 June. The withdrawal of the monsoon normally takes place by 01 October.

295. **Storms and Depressions**. In June, a few of the depressions which originate in the Bay of Bengal and travel westward affect the weather in the east or northeast Vidarbha. Also, in June, troughs in the Arabian sea off Konkan coast cause heavy rain over the coastal region and sometimes even in the interior. The Bay depressions progressively take more southerly track till September, when the course of depressions is across the state, through Marathwada. This is one of the reasons why this part of the state has the rainiest weather in September. These depressions, if and when they recurve towards north or northeast, cause very heavy and prolonged spells of rain.

296. During October, some depressions from the Bay of Bengal cross the Peninsula through the southern states and emerge into the Arabian sea. Some of the depressions intensify into cyclonic storms and move north-eastward towards inland. They cause heavy rainfall and gales leading to damage to life and property in the coastal region.

297. Climatological Data: Temperature.

Madhya Maharashtra

Place	Temp	Jan	Feb	Mar	Apr	May	Jun	Jul	Aug	Sep	Oct	Nov	Dec
Kolhapur	Max Temp	31	33	36	37	36	30	27	27	29	31	31	30
	Min Temp	15	16	19	21	22	22	21	21	21	20	17	15
Satara	Max Temp	28	31	34	35	36	30	27	25	27	29	28	28
	Min Temp	14	15	19	21	22	22	21	21	20	19	15	14
Pune	Max Temp	30	32	36	38	37	32	28	27	29	32	30	29
	Min Temp	11	12	16	20	22	23	22	21	21	18	14	12
Ahmednagar	Max Temp	31	33	37	39	40	34	30	30	31	33	31	30
	Min Temp	12	14	18	22	23	22	22	21	20	19	15	13
Nasik	Max Temp	29	31	35	38	38	33	29	28	30	32	30	29
	Min Temp	10	12	16	19	21	22	22	21	20	17	13	11

Marathwada

Place	Temp	Jan	Feb	Mar	Apr	May	Jun	Jul	Aug	Sep	Oct	Nov	Dec
Osmanabad	Max Temp	29	32	36	38	39	34	30	29	29	30	29	29
	Min Temp	15	17	21	23	25	23	21	21	21	19	15	15
Beed	Max Temp	30	33	37	40	40	36	32	31	31	33	31	29
	Min Temp	13	14	18	23	25	24	23	22	22	19	15	12
Aurangabad	Max Temp	30	32	36	39	40	35	31	29	31	33	31	29
	Min Temp	14	16	20	23	24	23	22	21	21	20	17	14
Nanded	Max Temp	31	34	38	41	42	37	33	31	32	33	31	30
	Min Temp	13	16	19	23	26	24	23	22	22	19	15	13

Vidarbha

Place	Temp	Jan	Feb	Mar	Apr	May	Jun	Jul	Aug	Sep	Oct	Nov	Dec
Amravati	Max Temp	29	32	37	41	42	37	32	31	32	34	31	29
	Min Temp	15	18	21	25	28	26	24	23	23	21	18	15
Akola	Max Temp	30	33	37	41	42	37	32	31	32	34	31	29
	Min Temp	14	16	20	25	28	26	24	23	23	20	16	13
Yavatmal	Max Temp	29	32	37	40	42	37	31	29	31	32	30	28
	Min Temp	15	17	22	25	28	25	23	22	22	20	17	15
Nagpur	Max Temp	29	31	36	40	42	38	32	30	32	33	31	28
	Min Temp	13	15	19	24	28	26	24	24	23	20	15	12
Chandrapur	Max Temp	30	33	37	41	43	38	32	31	32	33	31	29
	Min Temp	14	17	21	26	28	27	25	24	24	21	16	13

298. **Climatological Data: Rainfall**.

Madhya Maharashtra

Place	Rainfall	Jan	Feb	Mar	Apr	May	Jun	Jul	Aug	Sep	Oct	Nov	Dec
Kolhapur	Rainfall (mm)	1	1	7	27	66	279	619	392	172	111	35	8
	Rainy Days	0	0	1	2	4	12	20	18	10	6	2	1
	Annual rainfall	1719 mm											
Satara	Rainfall (mm)	3	1	4	17	41	121	240	142	132	85	35	7
	Rainy Days	0	0	1	1	3	7	12	10	7	5	2	1
	Annual rainfall	828 mm											
Pune	Rainfall (mm)	1	1	2	8	29	177	408	282	183	82	29	5
	Rainy Days	0	0	0	1	2	8	14	12	9	5	2	1
	Annual rainfall	1208 mm											
Ahmednagar	Rainfall (mm)	20	1	3	5	21	101	100	87	148	66	27	7
	Rainy Days	0	0	0	1	1	6	7	6	7	4	1	1
	Annual rainfall	567 mm											
Nasik	Rainfall (mm)	2	1	2	6	17	135	317	234	171	58	27	5
	Rainy Days	0	0	0	1	1	7	14	12	9	3	1	1
	Annual rainfall	976 mm											

Marathwada

Place	Rainfall	Jan	Feb	Mar	Apr	May	Jun	Jul	Aug	Sep	Oct	Nov	Dec
Osmanabad	Rainfall (mm)	3	3	8	9	30	140	148	152	187	78	28	4
	Rainy Days	0	1	1	1	2	8	10	10	10	4	1	0
	Annual rainfall	790 mm											
Beed	Rainfall (mm)	3	3	5	7	21	124	143	141	179	68	24	9
	Rainy Days	0	0	1	1	1	7	8	8	9	4	1	1
	Annual rainfall	726											
Aurangabad	Rainfall (mm)	3	2	4	3	19	125	160	162	156	53	24	9
	Rainy Days	0	0	1	1	1	7	10	10	8	3	1	1
	Annual rainfall	719 mm											
Nanded	Rainfall (mm)	7	6	8	7	17	156	256	252	193	69	17	5
	Rainy Days	1	1	1	1	1	8	13	12	9	3	1	1
	Annual rainfall	991 mm											

Vidarbha

Place	Rainfall	Jan	Feb	Mar	Apr	May	Jun	Jul	Aug	Sep	Oct	Nov	Dec
Amravati	Rainfall (mm)	13	12	7	4	8	133	249	226	151	43	17	7
	Rainy Days	1	1	1	1	1	7	13	11	8	2	1	1
	Annual rainfall	871 mm											

Akola	Rainfall (mm)	10	7	8	4	10	134	227	194	137	45	18	8
	Rainy Days	1	1	1	1	1	7	12	11	7	2	1	1
	Annual rainfall	801 mm											
Yavatmal	Rainfall (mm)	10	9	11	9	14	171	294	250	167	55	14	8
	Rainy Days	1	1	1	1	1	9	15	12	8	3	1	1
	Annual rainfall	1012 mm											
Nagpur	Rainfall (mm)	19	16	15	10	10	161	326	286	181	52	15	11
	Rainy Days	1	1	1	1	1	8	15	13	9	3	1	1
	Annual rainfall	1101 mm											
Chandrapur	Rainfall (mm)	11	12	16	11	12	179	396	374	206	60	11	7
	Rainy Days	1	1	1	1	1	9	17	15	9	3	1	1
	Annual rainfall	1295 mm											

Konkan

299. Konkan is a coastal strip occupies the entire west coast of Maharashtra. The land of Konkan is bounded by the Sahyadri Mountain range ("Western Ghats") on the East and the Arabian Sea on the West. The region consists of costal districts such as Raigad, Ratnagiri, Sindhudurg and Thane.

300. The climate of Konkan is characterized by an oppressive summer, dampness in the atmosphere nearly throughout the year and heavy southwest monsoon rainfall. The cold season from December to February is followed by the pre-monsoon season from March to beginning of June. The period from June to about the end of September constitutes the southwest monsoon season. October and November form the post monsoon season.

301. **Rainfall**. The average annual rainfall is 2145 mm. About 96% of the annual rainfall in Mumbai is received during the southwest monsoon months (June to September). July is the rainiest month when about one third of the annual rainfall is received. Some rainfall mostly as thundershowers is also received during May and the post monsoon months. During the period December to April there is very little rainfall. On an average there are 75 rainy days (i.e. days with rainfall of 2.5 mm or more) in a year. The heaviest rainfall in 24 hours recorded in Mumbai was 944.2 mm at Santacruz on 27 July 2005.

302. **Temperature**. During summer, due to prevailing high humidity the weather is very oppressive. On some days the maximum temperature goes up to 42°C. The afternoon sea breezes bring some welcome relief from heat. After the onset of the monsoon by about the beginning of June, the weather becomes progressively cooler. But, towards the end of the southwest monsoon season, day temperatures begin to increase slightly and a secondary maximum in day temperature is reached in November. Nights however, become progressively cooler after the withdrawal of the monsoon. After November, the day temperatures also begin to decrease. January is generally the coldest month when the mean daily maximum temperature is 29.4°C at Colaba and 30.4°C at Santacruz. The mean daily minimum temperature is 19.2°C at Colaba and 16.6°C at Santacruz. In the cold season, in association with passing western disturbances across north India, the minimum temperature occasionally drops down up to about 7°C. The highest maximum temperature recorded at Colaba was 40.6°C on 19 April 1955 and that at Santacruz was 42.2°C on 14 April 1952. The lowest minimum temperature was 11.7°C at Colaba on 15 th January 1935 and 01 February 1929 and that at Santacruz was 7.4°C on 22 January 1962.

303. Climatological Data: Temperature (^{0}C).

Place	Temperature	Jan	Feb	Mar	Apr	May	Jun	Jul	Aug	Sep	Oct	Nov	Dec
Dahanu	Max Temp	27	28	30	32	33	33	30	30	30	32	32	30
	Min Temp	17	17	21	24	27	27	25	25	24	23	21	19
Colaba	Max Temp	29	29	31	32	33	32	30	30	30	33	33	31
	Min Temp	19	20	23	25	27	26	25	25	25	25	23	21
Santacruz	Max Temp	30	31	33	33	33	32	30	29	30	33	33	32
	Min Temp	17	17	21	24	26	26	25	25	24	23	21	19
Ratnagiri	Max Temp	31	31	31	32	33	31	29	29	29	32	33	32
	Min Temp	19	19	22	25	26	25	24	24	24	23	21	20

304. Climatological Data: Rainfall.

Place	Rainfall	Jan	Feb	Mar	Apr	May	Jun	Jul	Aug	Sep	Oct	Nov	Dec
Mumbai	Rainfall (mm)	1	1	0	1	11	532	745	477	301	58	14	04
	Rainy Days	0	0	0	0	1	14	23	20	13	3	1	1
	Annual rainfall	2145 mm											
Thane	Rainfall (mm)	1	1	0	2	12	433	960	633	321	68	17	1
	Rainy Days	0	0	0	0	1	13	25	23	13	3	1	0
	Annual rainfall	2449 mm											
Raigad	Rainfall (mm)	1	0	0	3	23	655	1241	789	410	101	23	4
	Rainy Days	0	0	0	0	1	17	27	26	16	6	2	0
	Annual rainfall	3341 mm											
Ratnagiri	Rainfall (mm)	1	0	0	4	37	833	1201	847	370	129	32	5
	Rainy Days	0	0	0	0	2	20	28	26	16	6	2	0
	Annual rainfall	3461 mm											

Goa

305. Goa is state, located on the western coast (Konkan coastal belt) of the Indian peninsula. Goa state is located between the co-ordinates14°53'57" N -15°47'59" N latitudes and 73°40'54"E-74°20'11"E longitudes. The geographical area of the state is 3,702 Sq km and it is bounded by Sindhudurg district of Maharashtra in the North, Belgavi and Dharwar districts of Karnataka along the East, Uttar Kannada district in the south and Arabian sea in the west. Most of the Goa is a part of the coastal region known as the Konkan. The western Ghats range separates Goa from the Deccan Plateau. The highest point is the Sonsogor, with an altitude of 1167 metres. Goa has a coastline of 101 Km. Goa can be divided into four different regions:

a) The coastal plains with areas like Mormugao, Tiswadi, Salcete and Bardez.

b) The Area of the Western Ghats like Sanguem, Sattari, Canacona and Ponda forms the Eastern Hilly region.

c) The Flood Plains comprising of the rolling uplands.

d) The coastal plains as well as the central valley lands consisting of the areas like Eastern Sanguem, Bicholim, Pernem and Quepem.

306. There are nine major rivers in Goa flowing from east (Western Ghats) to west (Arabian Sea) except Sal River. Terekhol, Chapora, Baga, Mandovi, Zuari, Sal, Saleri, Talpona and Galgibag are the main nine rivers of Goa.

Among these rivers Mandovi and Zuari drain 2553 square km., about 70% of the total geographical area of Goa. Rivers in Goa are unique and are both tidal as well as rain fed.

307. **Climate**. Goa has a moderate, tropical monsoon climate, hot with seasonally excessive rainfall. with hot summers and chilly winters. Goa lies in the tropical zone and it is also bordered by the Arabian Sea on the west. It experiences hot and humid weather throughout the year. However, the temperature in Goa is always moderate with the summer being not too hot and winter being mild. Monsoon is the wet season and July is the wettest month. The year may be divided into four seasons. The months of January and February constitute the winter season. This is followed by the hot season which lasts till about the end of May. The period from June to September is the southwest monsoon season. The three post monsoon months October, November and December constitute a transition period from the monsoon to winter conditions.

308. **Sea Level Pressure and Winds**. Except during the summer months, the pressure gradient over Goa generally remains weak. The seasonal variation pressure takes pace in a progressive manner. Atmospheric pressure is maximum during winter and minimum during monsoon season. Being a coastal state, land breeze and Sea breeze blow throughout the year except during southwest monsoon season when winds are predominantly south-westerly or westerly. During summer (Pre-monsoon) season winds are moderate and easterly to north-easterly winds are observed in the morning and westerly to north-westerly in the afternoon. Winds strengthen by evening. With the advance of the summer, seasonal low over north India deepens and shifts towards the northwest. Winds strengthen and mostly westerly to south-westerly are seen in the state from June to August. From September winds become moderate and easterly to north-easterly are observed in the morning and westerly, north-westerly and South-westerly are seen in the afternoon. October is the month of transition when reversal of pressure gradient takes place. Winds weaken and take a northerly component. The seasonal low begins to establish itself over the Bay of Bengal and conditions revert to the winter pattern. Pressure thereafter, continue to rise till January.

309. **Temperature**. Pre-monsoon season is the hot period while winter is the cold period. April and May are the hottest months with mean maximum temperature about 33 ^{0}C and mean minimum temperature about 26 ^{0}C. Temperatures begin to fall with the onset of the southwest monsoon season in the first week of June. After the withdrawal of the monsoon there is a tendency for a rise in temperatures till November due to increased insolation. During winter period the night temperatures fall rapidly and day temperatures decrease slowly. January is the coldest month in the state with mean maximum temperature at about 32 ^{0}C and mean minimum temperature at about 21 ^{0}C.

310. Both day and night temperatures then begin to rise slowly from March to May. The highest maximum temperature ever recorded 39.8 ^{0}C at Panjim on 07 April 1989 which is about 7 ^{0}C higher than the normal. The lowest minimum temperature ever recorded in the state was 12.2 ^{0}C on 18 November 1946 at Marmugao which is about 10 ^{0}C lower than the normal. The night temperature starts to fall from October while the day temperature follows this trend from December and maximum temperature attains the lowest values in February while the minimum temperature attains the lowest values in January.

311. The climate is pleasant from December to February. Summer months of April and May are uncomfortable with oppressive heat. The period of June to October is warm and humid, and uncomfortable.

312. **Rainfall**. The normal annual rainfall for the state is about 3370 mm and normal annual number of rainy days is about 103. Central and southern parts of the state slightly more rainfall. Goa receives maximum rainfall in the monsoon season and the minimum during winter and pre-monsoon season. The percentage of the seasonal number of rainy days with respect to the annual number of rainy days shows that 86% during the southwest monsoon season, 3.6% during the pre-monsoon season, 0.5% during the winter season and 10% in post-monsoon season.

313. The state, is situated windward side of Western Ghats and hence, receives good amount of rainfall during southwest monsoon season. The heavy rainfall received during southwest monsoon season are known to be associated with off-shore trough along west coast of India, low pressure areas and monsoon depressions originating in the Bay of Bengal, Arabian Sea during the southwest monsoon season. July is the rainiest month with 33% of the annual rainfall received in this month. It decreases rapidly after withdrawal of southwest monsoon in September.

314. Precipitation during the pre-monsoon months is mostly associated with thunderstorms and constitutes about 3.6% of the annual rainfall. The southwest monsoon sets in over the state by about the first week of June. The monsoon withdraws from the state by about third week of September.

315. **Cyclonic storms and depressions.** The state of Goa, though is a coastal state, is not so much vulnerable to tropical storms. The cyclonic storms and depressions which affect India mostly originate and/or intensify over the Bay of Bengal, mainly during the months of May to December. But the west coast of India is less affected as less number of storms and depressions originate in the Arabian Sea. As per the data available from 1891 to 2017, the state has not been affected by tropical cyclone during the period from January to September. The number of storms / depressions that affected the state in October was 1 and in November it was 3.

316. **Climatological Data: Temperature.**

Places	Temp	Jan	Feb	Mar	Apr	May	Jun	Jul	Aug	Sep	Oct	Nov	Dec
Panjim	Max	32	32	32	33	34	31	29	29	30	32	33	32
	Min	20	21	23	25	26	25	24	24	24	24	23	21
Marmugao	Max	32	31	32	33	33	31	29	29	30	32	33	33
	Min	22	22	24	26	27	25	25	24	24	25	24	23

317. **Climatological Data-North Goa: Rainfall.**

Rainfall	Jan	Feb	Mar	Apr	May	Jun	Jul	Aug	Sep	Oct	Nov	Dec
Monthly (mm)	1	0	3	6	70	928	1109	706	315	160	40	10
Rainy Days	1	0	1	1	3	22	27	25	14	7	2	1
Annual rainfall	3347 mm											

318. **Climatological Data-South Goa: Rainfall.**

Rainfall	Jan	Feb	Mar	Apr	May	Jun	Jul	Aug	Sep	Oct	Nov	Dec
Monthly (mm)	2	0	3	8	92	889	1128	709	340	171	39	13
Rainy Days	1	0	1	1	3	22	27	25	15	8	2	1
Annual rainfall	3394 mm											

Kerala

319. Kerala is situated between the Lakshadweep Sea to the west and the Western Ghats to the east. Kerala's coast runs some 580 km in length, while the state itself varies between 35–120 km in width. The topography consists of a hot and wet coastal plain gradually rising in elevation to the high hills and mountains of the Western Ghats.

320. With around 120–140 rainy days per year, 80 Kerala has a wet and maritime tropical climate influenced by the seasonal heavy rains of the southwest summer monsoon (June to August) and northeast winter monsoon. Around 65% of the rainfall occurs from June to August corresponding to the Southwest monsoon, and the rest from September to December corresponding to Northeast monsoon. The influence of the Northeast monsoon is, however, mainly confined to the southern districts.

321. The average daytime temperature ranges from 28 ⁰C to 32 ⁰C on the plains, but drops to about 20 ⁰C in the highlands. During the summer months Kerala experiences gale force winds, storm surges, and cyclone-related torrential downpours; while the winter months are much cooler and calmer. However, tropical cyclones also affect Kerala occasionally during winter season. The storms originating over south Bay of Bengal moves west ward during winter season, follow a path across Tami Nadu or Sri Lanka and affects Kerala.

322. The state receives an average annual rainfall of 3107 mm against the all India average of 1,197 mm. Parts of Kerala's lowlands may average only 1250 mm annually while the cool mountainous eastern highlands of Idukki district receive in excess of 5,000 mm of rainfall annually.

323. **Month wise Average Temperature and Rainfall**.

Month	Jan	Feb	Mar	Apr	May	Jun	Jul	Aug	Sep	Oct	Nov	Dec
Max Temp (⁰C)	30	31	32	34	34	30	29	29	29	30	30	31
Min Temp (⁰C)	22	23	24	25	25	24	23	23	23	23	23	22
Rainfall (mm)	15	17	36	111	253	653	687	404	252	270	159	46

324. **Lightening**. Kerala is a place of high lightning incidence compared to most of the other parts in India because of its weather patterns and the location of the Western Ghats. Higher population density and vegetation density result in more casualties. Lack of awareness also aggravates the situation. Accidents caused by ground conduction from trees, which is a special feature of Kerala, add to the casualties and loss of property. The records show that the months April, May, October and November have the highest lightning rates. The most active time of the day is from 15:00 to 19:00. The severe impact of the hazard on the state and its people is seen from the very high average casualty rates of 71 deaths, 112 injuries and 188 accidents per annum. Losses to telephone communications, networked systems and electrical equipment are also very high.

Karnataka

325. The state of Karnataka is situated in the south western part of India and spreads over the Deccan Peninsular region. Karnataka is blessed with the bounty of nature that is revealed in the spectacular topography of the state. The state exhibits a splendid mix of lofty mountains, coastal plains, residual hills and plateaus. Karnataka is located in the region where the Western Ghats and the Eastern Ghats converge into the Nilgiri Hills. It is surrounded by chains of lush green mountains on three sides: east, west and south. There are many rivers that criss-cross through Karnataka. Karnataka starts with a narrow coastal plain that extends towards the east with short and small plateaus having different altitudes, and then suddenly rises up to great heights. This is followed by the gentle east and east-north-west sloping plateau.

326. Weather of Karnataka is pleasant especially in the hilly areas or high plateau. Karnataka weather is at its best in Bangalore which is known as an 'air-conditioned city'. Here, the weather is pleasant throughout the year. Moreover, Bangalore gets showers in both the seasons of summer and winter rejuvenating the city throughout the year. Karnataka's weather can be best described as dynamic and erratic as it changes from place to place owing to the changes in altitude, relief features and its distance from the sea. The hilly regions and plateau present a different climatic condition with the plains being comparatively warmer. Due to this climatic difference, Karnataka is divided into three Meteorological Divisions - Coastal Karnataka, North Interior Karnataka (NIK) and South Interior Karnataka (SIK).

a) **Coastal Karnataka.** This region stretches over the districts of Udupi, Uttara Kannada and Dakshina Kannada. The entire coastal belt and the adjoining areas have tropical monsoon. The area receives heavy rainfall. The average annual rainfall in Coastal Karnataka is about 3456 mm, which is much more than the rainfall received in the other parts of the state.

b) **North Interior Karnataka.** This region extends over the districts of Bagalkot, Belgaum, Bijapur, Bidar, Bellary, Dharwad, Haveri, Gadag, Gulbarga, Koppal and Raichur. This area is an arid zone. North Interior Karnataka receives the least amount of rainfall in the state and the average annual rainfall is just 731 mm.

c) **South Interior Karnataka.** This region spreads over the districts of Bangalore Rural, Bangalore Urban, Chitradurga, Chamrajnagar, Chikmagalur, Hassan, Kodagu, Kolar, Mysore, Shimoga and Tumkur. This zone experiences semi-arid type of climate. South Interior Karnataka receives an annual average of 1286 mm rainfall.

327. **Pre-Monsoon Season (March – May).** Summer starts in March and lasts till May. April and May are the hottest months in almost all places of Karnataka. It is during this season that humidity is quite less in the atmosphere. The average temperature is 34.5 ^{0}C. Towards NIK an average high temperature of 28-34 ^{0}C and a low temperature of 20-22 ^{0}C. While, SIK have the average high temperature which ranges from 25-34 ^{0}C and the low average temperature going up to 15-22 ^{0}C. The highest recorded temperature in the state of Karnataka was at Raichur with 45.6 ^{0}C in 1928. The temperature of Karnataka remains more or less the same during day and night in the coastal and the highly elevated regions of the state.

328. **Monsoon Season (June – September).** Monsoon begins in beginning of June and continues until September. The temperature dips very low and the humidity is quite high. Coastal Karnataka comprising experiences tropical monsoon climate. Here, rainfall is quite heavy and ranges up to 3638.5 mm per annum. The climate of this area is generally hot accompanied by excess rainfall till September. On the other hand, NIK mainly lie in comparatively arid zone and the average rainfall per annum is only 711.5 mm. SIK, receives an average rainfall of 1064.8 mm per annum while in the Western Ghats it is 2540 mm. The districts of Chitradurga, Bijapur, Koppal receives the least rainfall which is of the order of 570 mm. Udupi receives the highest average rainfall of 4119 mm.

329. **Post Monsoon Season (Oct – December).** The Post monsoon season of Karnataka begins from October and extends up to December. During this season, an intermittent shower of rain is associated with the north-eastern monsoon. These winds affect the south-eastern part of the state.

330. **Winter Season (January – February.** Winter season extends from January to February which is pleasant and not extreme in any of the cities of Karnataka. The state receives winter showers during October-November in the south eastern part. The temperature during this season ranges from 32 ^{0}C to 20 ^{0}C or even below. December and January are generally the coolest months in most of the cities in Karnataka.

331. **Average Annual Rainfall**.

Coastal Karnataka : 3456 mm

North Interior Karnataka : 731 mm

South Interior Karnataka : 1126 mm

Tamil Nadu

332. Tamil Nadu is bordered by Kerala to the west, Karnataka to the northwest, Andhra Pradesh to the north, the Bay of Bengal to the east and the Indian Ocean to the south. Cape Comorin (Kanyakumari), the southernmost tip of the Indian Peninsula which is the meeting point of the Arabian Sea, the Bay of Bengal, and the Indian Ocean is located in Tamil Nadu.

333. The western, southern and north-western parts are hilly and mix of vegetation and arid. Tamil Nadu is the only state in India that has both the Western Ghat and the Eastern Ghat mountain ranges which both meet at the Nilgiri Hills. The Western Ghats dominate the entire western border with Kerala, effectively blocking much of the rain-bearing clouds of the South West Monsoon from entering the state. The eastern parts are fertile coastal plains. The northern parts are a mix of hills and plains. The central and the south-central regions are arid plains.

334. The geographical location of Tamil Nadu is such that the climatic condition shows only slight seasonal variations. Due to close proximity to the Sea, the temperatures and humidity remain relatively high all the year round. The climate of Tamil Nadu is tropical in nature with little variation in summer and winter temperatures. While April-June is the hottest summer period (with the temperature rising up to the 40°C mark), November-February are the cool months with temperatures hovering around 20°C, making the weather condition quite pleasant.

335. Tamil Nadu gets most of its rains from the North-east Monsoons between October and December while the state remains largely dry during the South West Monsoon season. The average annual rainfall in Tamil Nadu ranges between 635 and 1900 mm a year. During summers (April to June), the coastal areas of Tamil Nadu become awkwardly warm and humid, but the nights become cool and pleasant due to sea breeze.

336. The state can be divided into two sub-divisions, Interior TN and Coastal TN as the proximity of the sea causes difference in the climatic conditions.

Interior Tamil Nadu

337. **Winter Season (December – February)**. A mild winter prevails over this region during mid-December to February. This is generally a period of uninterrupted good weather. Surface winds are generally ENE'ly or N'ly. Strong inflow of moisture associated with an easterly wave or a low-level trough in easterlies may sometimes result in mist/fog or low clouds for short duration in the morning hours.

338. Tamil Nadu is normally affected by 1 – 2 transitory easterly waves during the month of Dec which cause wide spread rain/thundershowers. On rare occasion, cyclone or depression which form over the SW Bay, cross TN and cause widespread precipitation in this region. Months of January and February are generally free from these systems and precipitation.

339. **Pre – Monsoon Season (March – May)**. The weather during this period is governed by the seasonal N-S oriented peninsular trough/discontinuity. The combination of increased heating and influx of moisture results in the development of thunderstorms in the afternoons and evenings. There is a rapid increase in thunderstorm frequency from 3.2 days in March to 10 – 12 days in April and May.

340. Pre-monsoon season is favourable for the formation of tropical systems over Bay. The depressions and cyclonic storms take north-westerly course over the SW Bay. Occasionally, they move in west-northwesterly direction and affect the region. About 23% of rainfall is received by the region during this season.

341. **Southwest Monsoon Season (Jun – September)**. SW Monsoon over this airfield normally sets between the first and second week of June and is preceded by marked thunderstorm activity followed by strong south-westerlies throughout the day and night. Interior TN receives only about 10% of the annual rainfall from June to August aggregating to 21% in the season in the form of light precipitation. Persistent, strong and gusty surface winds are the typical feature of the monsoon season over the region.

342. **Post – Monsoon Season (October – November)**. Post Monsoon season also known as the NE Monsoon season is the rainiest season for this region. It accounts for nearly 47% of the annual rainfall. The region experiences moderate to heavy rain during this season. The frequency for the precipitation is 18 to 19 days and 15 days in the month of October and November respectively.

343. The surface winds gradually change from south-westerlies to north-easterlies Frequency of Depressions/Cyclonic Storms is one or two in a few years. The frequency of transient easterly waves is 4 – 6 in this season.

344. **Rainfall**. Being in rain shadow area, this region does not get sufficient rainfall during SW Monsoon season, however, rainfall activity increases during Post-Monsoon season. The region gets around 21% of its annual rainfall during SW Monsoon season and 47% during Post-Monsoon season. Mean rainfall for October and November is 160 mm and 130 mm respectively and October is the wettest month. Driest month is January with just 7 mm of average monthly rainfall.

Coastal Tamil Nadu

345. Coastal Tamil Nadu exhibits tropical climate, specifically tropical wet and dry climate throughout the year. Being located in the coast, the region does not experience extreme variation in seasonal temperature. The weather is hot and humid for most of the year except for January and February. However, the annual cycle of weather over the coastal region has a uniquely different weather pattern from the rest of the regions in the country. The basic difference is that while the places located outside Tamil Nadu get their maximum rainfall during South West Monsoon season, Tamil Nadu gets its maximum rainfall during North East Monsoon season.

346. **Winter Season (January – February)**. January and February are the coolest period over Coastal Tamil Nadu. Morning fog/mist occurs, but are shallow in nature and hardly persists for more than an hour after sun rise. The state of sky is generally Partly Cloudy during these two months. The passage of troughs in upper tropospheric easterlies along lower latitudes is associated with low level perturbations moving westwards. When such troughs/cyclonic circulations in low level easterlies are located over the southwest Bay off Tamil Nadu coast, light precipitation may occur on 2-4 days. Normally, rain occurs for 3-4 days in January and 1-2 days in February. Surface winds veer from NW (<5Kt) in the early morning hours to NE/ENE after midday (speeds of 10-20 Kt). Speed diminishes thereafter, and wind once again backs to NW after 2200 IST. This regular sequence is due to the land and sea breeze effect superimposed upon the synoptic gradient.

347. **Pre – Monsoon Season (March – May)**. During this season there is a gradual rise in the surface temperatures and this is the hottest period of the region. High temperature is a significant phenomenon in May when on many days, temperature is recorded in excess of 40 °C. The hottest part of the year is latter part of May to early June, known locally as Agni Nakshatram ("fire star") or as Kathiri Veyyil. The sky condition generally remains Partly Cloudy in the morning hours becoming Cloudy to Mainly Overcast with the passage of easterly waves or trough in Easterlies and marked Peninsular discontinuity. The position and vertical extent of the peninsular discontinuity determines the weather south of 20°N latitude. Few thunderstorms occur over the region in the evening/night during March and April when the peninsular discontinuity shifts eastward to 80-81°E under the influence of Western Disturbances (WDs) and/or low-level disturbances in easterlies. The intensity and frequency of thunderstorms increases in May. This is so because due to increase in the strength of westerlies at lower levels before the onset of monsoon which make the peninsular discontinuity to shift eastward. In April the sea breeze sets in early, the thunderstorm also occurs in 1200-1500 h. However, as the pressure gradient increases in May, the sea breeze sets in late.

348. Cyclonic storms originating in the Bay of Bengal form between 10°-15°N but generally move towards the north or north-eastwards on a re-curving track. Some of them move north-westwards and affect North Tamil Nadu coast but more generally, the Andhra Pradesh coast. Under such conditions, heavy rainfall accompanied by squally surface winds occurs over the region.

349. Surface winds in March and April are generally south-southwesterly to southerly after midnight, and strong south-easterlies after midday. In the month of May, surface winds are generally southwesterly between 0200h and

midday, south-easterlies thereafter, and southerly around midnight. Wind speeds are usually in excess of 15 Kt. The maximum temperatures over the region reaches before 1430h with a sharp fall in temperatures immediately following the strengthening of the sea breeze.

350. **Southwest Monsoon Season (June – September)**. Over Chennai, the onset of SW monsoon takes place around 29-30 May and within one week the entire TN comes under the influence of SW monsoon. Since Tamil Nadu lies in the rain shadow region, rainfall is scanty and is more characterized by change in wind direction rather than increase in rainfall. However, coastal region exhibits a high frequency of convective activity during afternoon, evening and night due to sea breeze. The activity is more whenever the monsoon is weak. Peak activity is witnessed late at night at the head of the retreating land/sea breeze front. In September, as the pressure gradient decreases, thunderstorm takes place earlier. The thunderstorm activity increases from June (08 days) to October (15 days).

351. The formation of low latitude off-shore vortices over SW Bay early in the monsoon season and their north-westward movement close to the coast results in a day or two of very heavy precipitation. The surface winds generally remain west - south westerly during the day, becoming south-south westerly/southerly in the evening.

352. **Northeast Monsoon Season (October – December)**. The withdrawal of SW Monsoon from Chennai takes place around 20 October. The retreating Northeast Monsoon sets in immediately thereafter over Tamil Nadu and is active till mid-December. The north-northeasterly wind flow over Tamil Nadu has a trajectory which fetches it from Myanmar and Bangladesh after a long travel over the Bay of Bengal, during which the air flow picks up moisture. This is a season of the maximum rainfall over TN. About 55% of the annual rainfall over Coastal Tamil Nadu occurs during these three months.

353. **Rainfall**. There is progressive increase in the amount of mean rainfall from April to November with a slight drop in September. The rainfall amount decreases in December and subsequently from January to March, coastal Tamil Nadu receives very less amount of rainfall. The increase in mean monthly rainfall is predominantly due to weak monsoon conditions in the month of August. The amount of rainfall is the highest in the month of November (357 mm). The least is in the month of March (10 mm).

354. **Climatological Data of Major Cities of Tamil nadu.**

Chennai

Month	Jan	Feb	Mar	Apr	May	Jun	Jul	Aug	Sep	Oct	Nov	Dec
Daily Mean Temp (^{0}C)	25	27	29	31	33	32	30	30	30	28	27	25
Max Temp (^{0}C)	29	31	33	35	37	37	35	35	34	32	30	29
Highest Max (^{0}C)	35	37	41	43	45	43	41	40	39	39	35	33
Minimum Temp (^{0}C)	21	22	24	27	28	27	26	26	26	25	23	22
Lowest Min (^{0}C)	14	15	17	20	21	21	21	21	21	17	15	14
Rainfall (mm)	25	3	3	14	34	56	104	127	148	316	400	177
No. of Rainy Days	1	1	1	1	2	4	7	8	7	11	11	6
Total Annual rainfall	1,407 mm											

Madurai

Month	Jan	Feb	Mar	Apr	May	Jun	Jul	Aug	Sep	Oct	Nov	Dec
Max Temp (^{0}C)	31	33	36	37	38	37	37	36	35	33	31	30
Highest Max (^{0}C)	39	39	42	42	45	42	41	40	40	40	38	37
Minimum Temp (^{0}C)	20	21	23	26	26	26	26	25	25	24	23	21
Lowest Min (^{0}C)	16	11	17	19	18	18	19	21	19	19	17	17
Rainfall (mm)	9	11	18	60	81	34	57	94	121	186	147	51
No. of Rainy Days	1	1	1	3	4	2	3	5	7	10	7	3
Total Annual rainfall	869 mm											

Tiruchirappalli

Month	Jan	Feb	Mar	Apr	May	Jun	Jul	Aug	Sep	Oct	Nov	Dec
Max Temp (^{0}C)	30	33	36	37	38	37	36	36	35	33	30	30
Highest Max (^{0}C)	36	40	42	43	43	44	41	41	41	39	37	36
Minimum Temp (^{0}C)	21	21	23	26	27	27	26	26	25	24	23	21
Lowest Min (^{0}C)	14	14	16	18	19	18	20	21	21	19	17	14
Rainfall (mm)	13	4	5	30	67	38	61	70	153	154	168	81
No. of Rainy Days	1	1	1	2	4	3	3	4	7	9	8	5
Total Annual rainfall	844 mm											

Coimbatore

Month	Jan	Feb	Mar	Apr	May	Jun	Jul	Aug	Sep	Oct	Nov	Dec
Max Temp (^{0}C)	31	34	36	37	35	32	32	32	33	32	30	30
Highest Max (^{0}C)	35	39	41	43	41	38	36	36	38	37	34	34
Minimum Temp (^{0}C)	19	20	22	24	24	23	22	22	22	22	21	19
Lowest Min (^{0}C)	12	13	16	18	16	18	16	16	17	16	14	12
Rainfall (mm)	7	4	26	44	55	24	25	36	53	157	134	33
No. of Rainy Days	1	1	1	3	3	3	3	3	3	8	7	2
Total Annual rainfall	600 mm											

Telangana

355. Telangana is situated on the Deccan Plateau, in the central stretch of the eastern seaboard of the Indian Peninsula. The region is drained by two major rivers, with about 79% of the Godavari River catchment area and about 69% of the Krishna River catchment area, but most of the land is arid. Telangana is also drained by several minor rivers such as the Bhima, the Maner, the Manjira, the Musi, and the Tungabhadra

356. Telangana is a semi-arid area and has a predominantly hot and dry climate. Summers start in March, and peak in mid-April with average high temperatures in the 37–38 °C range. The monsoon arrives in June and lasts until Late-September. A dry, mild winter starts in late November and lasts until early February with little humidity and average temperatures in the 22–23 °C range. The annual rainfall is between 900 and 1500mm in northern

Telangana and 700 to 900mm in southern Telangana. About 75 - 80% of the annual rainfall is received from the southwest monsoons.

357. **Winter Season (December - February)**. It is also known as cold season. It's a season of mild winter. However, the region is occasionally affected by cold wave condition in which temperature drops to less than 10 ^{0}C.

358. **Pre-Monsoon Season (March - May)**. Temperatures remain consistently high and hence the period is called hot season. Increased insolation and rise in temperature with high occurrence of thunderstorms are the main features of the season. Heat wave condition prevails at isolated places during mid-April to mid-May. The most important features that affect the region is the tropical cyclone. The cyclone generally forms over south-east Bay of Bengal and move North-west ward. Some of these cyclones re-curve north or north-east ward before they affect Andhra coast. However, a few of the cyclones re-curve out-skirting the Andhra coast. On such occasions, Telangana receives good amount of rainfall, though, no serious impact is felt over the region. A few cyclones continue to follow the north-west ward path and make landfall over Andhra coast. In such situation, Telangana, along with Andhra Pradesh faces the severity of the cyclones. However, severity is less than that over Andhra Pradesh.

359. **Southwest Monsoon Season (June to September)**. This season accounts for about 75% of annual rainfall over Telangana. With the onset, the summer heat spell is broken and the temperatures significantly drop over Telangana. Strong surface winds, low clouds and rain/thundershower are the main aviation weather hazards during this season. The main synoptic feature that brings good rainfall activity to the region is the movement of monsoon depression/Low across south Odisha/Chhattisgarh.

360. **Post Monsoon Season (October – November)**. It is the cool season or retreating monsoon season and is associated with pleasant weather condition. This season is also marked by formation of cyclonic storms in Indian seas. However, the frequency and intensity of the cyclones are less than that of pre-monsoon season. The cyclones affect the region when they continue to follow north-west ward path and hit Andhra coast.

361. **Rainfall**. Maximum and Minimum amount of rainfall takes place in August and December respectively. Averages rainfall in August is close to 300 mm, while in December it is less than 2 mm. Average annual rainfall is 1095.5 mm. As mentioned earlier, about 855 mm of rainfall is received in the monsoon months which constitutes about 78% of the yearly rainfall. About 12% of rainfall is received in post monsoon months and pre-monsoon showers add about 07%. Number of rainy days during June, July, August and September are 15, 24, 26 and 18 days respectively.

362. **Rainfall Data: Hyderabad.**

Month	Jan	Feb	Mar	Apr	May	Jun	Jul	Aug	Sep	Oct	Nov	Dec
Monthly Rainfall (mm)	11	8	24	33	27	148	155	296	156	109	24	4
Rainy Days	2	2	3	7	9	15	24	26	18	10	4	1

Andhra Pradesh

363. Andhra Pradesh is a coastal state with the Bay of Bengal to its east. Physio-graphically, the state can be divided into coastal and the comparatively drier Rayalaseema region. The state is bordered by Telangana, Chhattisgarh, and Odisha in the north, the Bay of Bengal in the East, Tamil Nadu to the south and Karnataka to the west. Andhra Pradesh has got a coastline of around 974 km, which gives it the 2nd longest coastline in the nation after Gujarat. Two major rivers, the Godavari and the Krishna runs across the state. A small enclave 30 km^2, the Yanam district of Puducherry, lies in the Godavari Delta in the north east of the state.

364. The Eastern Ghats are a major dividing line separating coastal plains and peneplains in the state's geography. The Eastern Coastal Plains comprise the area of coastal districts up to the Eastern Ghats as their border along the Bay

of Bengal, with variable width. These are, for the most part, delta regions formed by the Krishna, Godavari, and Penna rivers. The Eastern Ghats are discontinuous, and individual sections have local names. The Ghats become more pronounced towards the south and extreme north of the coast. These consist of the Papikonda range, the Simhachal hill range, the Yarada hills, the Nallamala Hills, the Papi hills, the Seshachala hills, and the Horsley hills. Peneplains, part of Rayalaseema, slope towards the east, with the Eastern Ghats as their eastern border.

365.   The climate of Andhra Pradesh state is generally hot and humid. The summer season (Pre-monsoon) in this state generally extends from March to June. The coastal areas have higher temperatures than the other parts of the state. The summer is followed by the monsoon season, which starts during June and continues till September. This is the season for heavy tropical rains in Andhra Pradesh and maximum of the annual rainfall is received in this season. About one third of the total rainfall in Andhra Pradesh is brought by the North-East Monsoons around the month of October in the state. Since the state has quite a long coastline, the winter season in Andhra Pradesh is pleasant. This is the time when the state attracts most of its tourists.

366.   **Temperature**. Practically, there is no cold month in Andhra Pradesh. January is the month of lowest temperature with mean temperature 26 ^{0}C and minimum of the order of 12 ^{0}C. With the northward shifting of the solar equator, temperature starts raising by February. Gradual rise continues up to May with mean maximum temperature of 37 ^{0}C in the month. There is slight relief in the month of June due to the pre-monsoon showers. Maximum temperature often rises up to the order of 45 to 46 ^{0}C during April to June and heat wave condition occurs on many days during these months. With arrival of monsoon, both mean and maximum temperatures fall significantly in July and thereafter, there is gradual fall till January.

367.   **Rainfall**. Monsoon sets in over Andhra Pradesh in the first half of June. The state receives about 61% of annual rainfall during monsoon season. About 29% of annual rainfall is received from pre-monsoon rain. A large portion of the June rainfall (75 mm) is actually received from pre-monsoon shower before the onset of monsoon.

Month	Jan	Feb	Mar	Apr	May	Jun	Jul	Aug	Sep	Oct	Nov	Dec
Mean Temp (^{0}C)	26	28	32	34	37	35	33	31	31	29	27	25
Mean Max (^{0}C)	30	33	37	39	42	37	35	34	33	32	31	29
Highest Max (^{0}C)	35	39	43	46	46	46	41	39	37	38	34	33
Mean Min (^{0}C)	19	20	23	26	30	29	28	27	26	24	21	19
Lowest Min (^{0}C)	12	15	17	22	26	19	20	19	19	19	15	13
Rainfall (mm)	9	5	5	7	19	75	76	103	92	101	66	12
Rainy days	3	1	1	2	3	13	13	17	16	14	8	3
Annual rainfall	570 mm											

368.   **Tropical Cyclones**. Andhra Pradesh is one of the most vulnerable States in India to Cyclones which are associated with violent winds, heavy to extremely heavy rainfall and devastating floods. With its coastline of 974 km, the state is more vulnerable to cyclones compared to other natural disasters. All-together, 184 cyclones of all categories including depressions crossed the coast from 1891 to 2019. AP is at risk of at least one cyclone each year on an average with maximum frequency during October and November. Cyclones with moderate to severe intensity occur every two to three years, which results in huge damage to the state. According to the Government of AP, 2.9 million people are vulnerable to cyclones as 3.3 million people are located within 5 km distance from the coastline. Thus, a storm surge with small wave length from the sea level results in effecting thousands of people in addition to the strong winds of the cyclones.

369.   The very severe cyclonic storm 'HUDHUD' is an example that created significant damage to the four districts of the state namely Visakhapatnam, Vizianagaram, Srikakulam and East Godavari. The cyclone made a landfall on 12th October, 2014 near Visakhapatnam with winds exceeding 185 kmph. Another very severe cyclonic storm

'TITLI' caused extensive damage over north Andhra Pradesh and adjoining Odisha districts on 10 Oct 2018. The cyclone was associated with very strong wind 140-150 Kmph gusting to 165 Kmph), heavy rainfall, flood storm surge (about one metre above astronomical tide) caused extensive damage.

370. During the period from 1971 to 2018, 55 tropical cyclones affected Andhra Pradesh. Out of these, 01 was depression, 18 were Cyclonic Storm, 17 were Severe Cyclonic Storm and 19 were very Severe Cyclonic Storm. From 2019 to 2022, a total of 05 tropical systems affected the state. The most severe of them was Extremely Severe Cyclonic Storm 'Fani' which made land fall near Puri on 03 May 2019. Although no fatalities occurred in Andhra Pradesh, Srikakulam and Vizianagaram districts reported an economic loss of 586.2 million. The other storms were a deep depression on 06 October 2020, Very Severe Cyclonic Storm 'Nivar' on 25 November 2020, Cyclonic Storm 'Gulab' on 26 Sep 2021 and Cyclonic Storm Asani on 08 May 2022 (Made landfall over Andhra Pradesh as Deep Depression).

Lakshadweep

371. Lakshadweep is an archipelago of twelve atolls, three reefs and five submerged banks, with a total of about thirty-six islands and islets. The reefs are in fact also atolls, although mostly submerged, with only small unvegetated sand cays above the high-water mark. The submerged banks are sunken atolls. Almost all the atolls have a northeast–southwest orientation with the islands lying on the eastern rim, and a mostly submerged reef on the western rim, enclosing a lagoon. It has ten inhabited islands, 17 uninhabited islands, attached islets, four newly formed islets and five submerged reefs.

372. Lakshadweep enjoys a pleasant tropical climate with summer temperatures ranging from 22° C to 33° C and winter temperatures from 20° C to 32° C. Southwest monsoons from June to September bring plenty of rainfall to the islands. Months from October to may are pleasant with slight rise in temperature during March and April. Cool sea breeze and abundant shade provided by the canopy of coconut palms make the climate pleasant even during peak summer.

373. Lakshadweep is the region around which most of tropical cyclones of Arabian Sea originates. This part of the Arabian Sea is the warmest and that might be the main region for formation of storms over this region. In general, Sea surface temperature over Arabian Sea is less than that over the Bay of Bengal. Hence, frequency of formation of cyclonic storm over Arabian Sea is less than Bay of Bengal. The intensity of the cyclones of Arabian Sea is also less than that of Bay of Bengal. However, in recent years the frequency and intensity of Arabian Sea cyclones have increased. It has been revealed that the sea surface temperature over Arabian Sea has shown an increasing trend compared to many othe parts of the oceans.

374. Pre-monsoon months (April and May) and post monsoon month October are the months when tropical cyclones form over Arabian Sea. These cyclonic storms are, however, not so hazardous for Lakshadweep because they are close to the islands only during their formation stage. As they intensify, the move north or north-west ward after giving some moderate to heavy rainfall and moderately strong winds over the islands.

375. **Average Temperature and Rainfall**. Daily and monthly variation in temperature over Lakshadweep is very less. Rainfall activity increases with the onset of Monsoon in June. The rainfall during October to December accounts for the north-east monsoon current. February to April are the driest months and there is sudden increase in the month of May due to pre-monsoon showers. June is the wettest month.

376. **Climatological Data.**

Month	Jan	Feb	Mar	Apr	May	Jun	Jul	Aug	Sep	Oct	Nov	Dec
Daily Mean Temp (^{0}C)	27	27	28	29	29	28	27	27	27	27	27	27
Max Temp (^{0}C)	27	27	28	29	29	28	28	28	27	28	28	28
Minimum Temp (^{0}C)	26	26	27	28	28	27	26	26	26	26	27	27
Rainfall (mm)	13	4	13	17	149	311	210	172	132	149	90	59
No. of Rainy Days	2	1	1	3	10	18	17	15	14	13	9	5

Andaman & Nicobar (A&N) Islands

377. The Andaman and Nicobar Islands is a union territory of India consisting of 571 islands, of which 37 are inhabited, at the junction of the Bay of Bengal and the Andaman Sea. The territory is about 150 km (93 mi) north of Aceh in Indonesia and separated from Thailand and Myanmar by the Andaman Sea. It comprises two island groups, the Andaman Islands and the Nicobar Islands, separated by the 150 km wide Ten Degree Channel (on the 10°N parallel), with the Andaman islands to the north of this latitude, and the Nicobar islands to the south. The Andaman Sea lies to the east and the Bay of Bengal to the west. The island chains are thought to be a submerged extension of the Arakan Mountains.

378. A&N Islands experience a typical tropical maritime climate by virtue of close proximity to the equator and being isolated islands in the Bay of Bengal. The usual definition of seasons which are applicable to the Indian mainland does not apply to this region. In terms of rainfall and other considerations, one can say that A&N has only two seasons, namely the rainy season (May to mid-December) and the dry season (Mid December to April). However, keeping in view the seasons as applicable to the country in general, the annual cycle of weather can be classified as Winter, Pre-Monsoon, Monsoon and Post-Monsoon seasons. The region remains under the sway of SW Monsoon for five months, which is the longest period in the country. The northern islands, that is, Port Blair and adjacent islands receives more rainfall as these islands have some hilly terrain. The southern islands being mainly plain areas experience less convective rain and hence, annual rainfall is less than the northern islands.

379. The southern islands are located in the south-east Bay of Bengal, though the sea portion to the east of the A&N islands is named as Andaman Sea. The sea surface temperature (SST) of Bay of Bengal is higher than the adjacent seas. During pre-monsoon and post-monsoon SST of Bay of Bengal remains higher than most of the seas of the world. Hence, the Sea is prone to Tropical Cyclone formation during these seasons, though, some cyclones also form over south Bay during winter season. Most of the tropical cyclones form over south-east Bay or south Andaman Sea. These cyclones generally move north-west ward, often re-curve towards north or north-east and hit Andhra, Odisha, West Bengal, Bangladesh or Myanmar coasts and cause heavy loss to properties and lives. The cyclones, however, do not cause damage to the islands in general as the systems remain close to the islands during the pre-mature stage. One odd cyclone, when heads northward from south Andaman Sea may prove to be damaging over northern islands.

380. **Winter Season (January-February)**. The period Jan – Feb is referred to as the winter season. There is no significant change in the day and night temperatures during this period in comparison to rest of the year. This season is the dry weather season as these islands receive lesser rainfall compared to the other months and hence, considered the best season for tourists.

381. During this season, an extended low-pressure area lies over South Bay and adjoining Indian Ocean, West of Andaman & Nicobar Islands. On an average, 3 – 4 Easterly Waves affect these islands during this season, which

is less when compared to other seasons. Winds at surface and lower levels are generally north-easterly / easterly in January, but backs to northerly in February. Feeble lows moving across Andaman Sea is one of the major synoptic features affecting weather over the region. At times, the weather is also affected by perturbation from ITCZ close to the equator. Whenever the easterly wave is well marked and embedded with a cyclonic circulation at lower levels close to the Nicobar Islands, the region experiences thunderstorm.

382. **Pre-Monsoon Season (March – April).** Characteristics that are generally associated with this season in fact extend up to mid-May. Unlike the rest of the country, this island does not witness any rapid rise in surface temperature during this season. The Bay Islands show only a marginal 1 – 2°C increase in the surface temperature in comparison to the preceding winter season. The highest temperature for the season is recorded generally in May, which is of the order of 36.4°C.

383. The pressure field remains diffused over the country during this season. A high-pressure cell forms over the Andaman Sea / SE Bay. During Apr and beginning of May, perturbations are seen in the wind field over Andaman Sea / SE Bay. These perturbations intensify into Depressions / Cyclonic Storms. They show an initial WNW ward movement and thereafter recurve to move N / NE ward.

384. **SW Monsoon Season (May – September).** Onset of the SW monsoon occurs over the Andaman & Nicobar Islands by the third week of May. Monsoon sets in over Nicobar Islands by 15 May. All the known monsoon systems affect this region either singly or in combination. Andaman & Nicobar region receives the maximum amount of rainfall during this season.

385. The NW – SE oriented axis of monsoon trough (AMT) is seen extending from the seasonal heat low along the Indo-Gangetic plains with its SE tip dipping towards SE Bay region. During the beginning of the season, formation of monsoon low over North Bay results in decrease in rainfall activity over A&N islands. However, during the latter part of the season, these systems form closer to the Andaman Islands and affect the weather over the northern islands.

386. Close proximity of the ITCZ results in active monsoon condition over the region. Location of Low-Level Jets (LLJ) over the region is also associated with good monsoon activity. LLJ that form over Kerala coast at lower levels, at times move eastward and can be marked over this region.

387. **Post-Monsoon Season (October – December).** SW monsoon withdraws from the islands by middle of Oct and almost simultaneously the NE monsoon sets in. The island continues to receive significant amount of precipitation during this period also.

389. During this season, the ITCZ shifts southward. An extended low-pressure area lies over South Bay of Bengal. Cyclonic storms and depressions form over the S Bay region. A few of these systems form over SE Bay close to the A&N Islands. They move in a WNW / NW ward and cross the TN or AP coast. The middle level ITCZ circulations form over the Andaman Sea and adjoining region. These result in increased precipitation over these islands. Easterly waves have significant influence over the weather of this region during the season. Some of the tropical cyclones that form over South Bay or Andaman Sea originate from the perturbation in the easterly waves.

390. **Rainfall (Nicobar Islands).** The mean number of rainy days during each month varies between 6 - 9 days during Jan to March, 12 days in April and there after it shows an abrupt increase to between 20 – 24 days from May to November. In December, the mean number of rainy days decreases to 14. The mean monthly rainfall also shows a similar trend. Mean monthly rainfall is less than 70 mm from Jan to March. It starts increasing in the month of April when the mean rainfall exceeds 100 mm and abruptly increases between May to November in which months the mean rainfall is above 300 mm. Rainfall decreases to below 200 mm in the month of December. Total annual rainfall over the Nicobar Islands is about 2,640 mm.

391. **Rainfall and Temperature Data (Port Blair).** Rainfall activity over the northern islands is more than over Nicobar islands. Climatological data in respect temperature and rainfall of Port Blair are shown below.

Month	Jan	Feb	Mar	Apr	May	Jun	Jul	Aug	Sep	Oct	Nov	Dec
Daily Mean Temp (^{0}C)	27	27	28	29	28	27	27	27	27	27	27	27
Max Temp (^{0}C)	30	31	32	33	31	30	30	29	29	30	31	30
Minimum Temp (^{0}C)	23	23	23	25	25	25	24	24	24	24	25	24

CHAPTER 32

CLIMATE OF NEIGHBOURING COUNTRIES

Bangladesh

1. Bangladesh is a coastal south Asian country which is bounded by West Bengal and Meghalaya to the north, Assam, Tripura and Myanmar to the east, West Bengal to the west and Bay of Bengal to the south. Bangladesh is a low-lying and mainly riverine country located in South Asia with a coastline of 580 km (563 Km) on the northern littoral of the Bay of Bengal. The delta plain of the Ganges (Padma), Brahmaputra (Jamuna), and Meghna Rivers and their tributaries occupy 79 percent of the country. Four uplifted blocks (including the Madhupur and Barind Tracts in the centre and northwest) occupy 9 percent and steep hill ranges up to approximately 1,000 m high occupy 12 percent in the southeast (the Chittagong Hill Tracts) and in the northeast. Straddling the Tropic of Cancer, Bangladesh has a tropical monsoon climate characterized by heavy seasonal rainfall, high temperatures, and high humidity. Natural disasters such as floods and cyclones accompanied by storm surges periodically affect the country.

2. **Climate**. Bangladesh has sub-tropical monsoon type climate with hot and humid summer and cool and dry winter. The average Temperature during winter season is 11-20 ^{0}C and during summer 21 to 38 ^{0}C. Rainfall during summer ranges from 1,100 mm to 3,400 mm.

3. **Seasons**. Bangladesh has four seasons. Winter: December to February; Pre-monsoon: March to May; Monsoon: June to September and Post-Monsoon: October to November. The monsoon season actually extends up to mid of October. The winter season is pleasantly cool. However, from mid-February, it is quite warm. Pre-monsoon is actually the hottest season of the year and during this period there is some rainfall accompanied by Norwester and hailstorm and often with recurrence of violent squally winds. April and May are the hottest months. The highest temperature ranging from 44 to 45 0 C is attained in the northern and northwestern districts. Over rest of the country, it ranges from 41 to 43 0 C. The post monsoon months of October and November are transition months from summer to winter and it is quite hot in October.

4. **Rainfall**. Rainfall over the country during winter is very scanty and does not exceed 5% of the yearly total. The driest month of the season is December when the northern and the western districts get hardly 3-10 mm of rainfall.

The coastal districts of Barisal, Noakhali, Chittagong and Chittagong Hill Tracts get 15-30 mm of rain. With the progress of the season, the rainfall increases over the whole of the country. The rainfall during the pre-monsoon is about 15% though it varies from region to region. The rainfall during the summer monsoon is about 75% of the yearly total. The rainfall during post monsoon period is about 5% though, it may be more if any depression or cyclone affects the country.

5. **Cyclones**. Bangladesh is very prone to tropical storm. The country is affected periodically by tropical storms which causes heavy damage to the country. Bangladesh is a part of humid tropics, with the Himalayas in the north and the funnel shaped coast touching the Bay of Bengal in the south. This peculiar geography of Bangladesh causes not only the life-giving monsoons but also catastrophic ravages of cyclones. The Bay of Bengal is an ideal breeding ground for tropical cyclones. In Bangladesh, tropical cyclones occur during pre and post monsoon season.

6. The number of severe cyclonic storms that affected Bangladesh over a ten-year period from 1780-1998. Since 1960 onwards, there is no ten-year period when tropical cyclone did not affect the country. During the period 1960-1970, there occurred ten cyclones in Bangladesh averaging one cyclone every year. The human casualties were also tremendous. During the 12 November 1970 cyclone alone, five lakh people died. Though the frequency has decreased recently than during 1960-70, it still has a significant number. During 29 April 1991 another cyclone of very severe intensity hit Bangladesh, where human casualty was one lakh thirty-eight thousand. In the last few years, human casualties have declined because of taking protective measures. 1997 Cyclone of Bangladesh was almost of the same intensity as in 1970 or 1991 but the casualty has been much less. In the coastal areas, some 2500-cyclone shelters have been built.

7. **Climatological Data: Temperature (^{0}C) and Rainfall (mm).**

National Average

Month	Mean Temp	Mean Max Temp	Mean Min Temp	Rainfall
January	19	25	13	9
February	21	28	15	26
March	26	32	20	52
April	28	33	23	130
May	29	33	25	277
June	29	32	26	459
July	28	31	26	523
August	28	31	26	420
September	28	31	25	318
October	28	31	24	160
November	24	29	19	42
December	20	26	14	10
Annual	26	30	21	2,426

Dhaka

Month	Jan	Feb	Mar	Apr	May	Jun	Jul	Aug	Sep	Oct	Nov	Dec	Annual
Rainfall	12	20	69	120	258	397	386	326	264	158	26	8	2,044
Max Temp	25	29	35	35	33	31	32	31	32	30	29	26	31
Min Temp	10	14	18	26	24	26	27	26	26	23	17	12	21
Mean Temp	19	21	26	29	29	29	29	29	29	27	23	20	26

Chittagong

Month	Jan	Feb	Mar	Apr	May	Jun	Jul	Aug	Sep	Oct	Nov	Dec	Annual
Rainfall	5	12	59	86	240	589	687	538	285	246	55	8	2,810
Max Temp	25	28	32	32	33	30	31	31	32	31	29	27	30
Min Temp	12	16	20	24	25	25	26	25	26	24	19	15	21
Mean Temp	20	22	26	28	29	28	28	28	28	27	25	21	26

Sylhet

Month	Jan	Feb	Mar	Apr	May	Jun	Jul	Aug	Sep	Oct	Nov	Dec	Annual
Rainfall	15	49	82	278	580	826	719	596	453	254	21	4	3,877
Max Temp	--	--	--	32	32	31	25	32	32	30	28	25	--
Min Temp	--	--	--	22	23	25	26	25	25	23	17	12	--
Mean Temp	17	20	24	27	27	27	28	28	28	26	23	20	25

Raj Shahi

Month	Jan	Feb	Mar	Apr	May	Jun	Jul	Aug	Sep	Oct	Nov	Dec	Annual
Rainfall	12	12	29	49	120	298	350	276	256	116	17	5	1,540
Max Temp	25	29	35	36	34	34	31	32	32	31	29	26	31
Min Temp	11	14	19	23	24	27	26	27	26	25	18	14	21
Mean Temp	19	21	26	30	30	30	29	29	29	27	24	19	26

Barisal

Month	Jan	Feb	Mar	Apr	May	Jun	Jul	Aug	Sep	Oct	Nov	Dec	Annual
Rainfall	10	14	45	82	187	419	429	389	326	203	24	5	2,139
Max Temp	26	25	32	35	34	32	31	31	31	32	29	25	30
Min Temp	15	15	23	25	26	26	26	26	26	25	19	13	22
Mean Temp	19	22	27	29	29	29	28	29	29	28	24	20	26

Khulna

Month	Jan	Feb	Mar	Apr	May	Jun	Jul	Aug	Sep	Oct	Nov	Dec	Annual
Rainfall	8	8	33	66	171	371	385	270	232	156	13	3	1,716
Max Temp	26	27	33	35	34	31	31	31	31	27	29	25	30
Min Temp	15	15	23	26	26	27	26	25	26	16	19	12	22
Mean Temp	20	23	28	30	30	29	29	29	29	28	25	21	27

Myanmar

8. Myanmar is bordered in the northwest by the Chittagong Division of Bangladesh and the Mizoram, Manipur, Nagaland and Arunachal Pradesh states of India. Its north and northeast border is with the Tibet Autonomous Region and Yunnan for a Sino-Myanmar border. It is bounded by Laos and Thailand to the southeast. Myanmar has 1,930 km of contiguous coastline along the Bay of Bengal and Andaman Sea to the southwest and the south.

9. In the north, the Hengduan Mountains form the border with China. Hkakabo Razi, located in Kachin State, is at an elevation of 5,881 metres is the highest point in Myanmar. Many mountain ranges, such as the Rakhine Yoma, the Bago Yoma, the Shan Hills and the Tenasserim Hills exist within Myanmar, all of which run north-to-south from the Himalayas. The mountain chains divide Myanmar's three river systems, which are the Irrawaddy, Salween (Thanlwin), and the Sittaung rivers. The Irrawaddy River, Myanmar's longest river at nearly 2,170 kilometres, flows into the Gulf of Martaban.

10. Myanmar's climate is influenced by many factors. It's climate pattern is highly influenced by the south-west monsoon. About 95% of its annual average rainfall is received from the Southwest Monsoon from May to October, with spatial and temporal variability. The Central Dry Zone has the least benefit from the Southwest Monsoon, due to high mountains surrounding the area (rain shadow effect).

11. The Southwest Monsoon starts sweeping Myanmar from the south by 18 May and gradually progresses north ward and covers the entire country by 06 June.

12. The reversal of the wind patterns, from November to February (Northeast Monsoon), brings dry, cool climate to Myanmar. Although the period is typically dry, weather disturbances (Tropical Storms) from the Bay of Bengal and Andaman Sea, on the Southwest and South of the country, respectively, could bring rainfall to coastal zones and other areas. Disturbances/remnants of typhoons from the South China Sea could bring rainfall especially to

the Eastern parts. Disturbances over the Bay of Bengal and Andaman Sea, and localized thunderstorms, could also bring rainfall to the normally dry and warm months of March and April.

13. Tropical cyclones affect the country mainly in the months from April to May and October to December. Number of storms that affected Myanmar from 1877 to 2010 are shown below: -

Month	No. of Storms	Month	No. of Storms	Month	No. of Storms
January	2	May	23	September	0
February	2	June	1	October	15
March	0	July	0	November	16
April	20	August	0	Dec	8

14. **Climate Zones**. Myanmar has been divided into three climate zones, which are the dry central zone, the coastal zone and the hilly zone. The coastal zone receives the highest average annual rainfall. The hilly and Central Dry zones, respectively, receive much less rainfall compared to the coastal zone.

15. **Seasons and Rainfall**. Myanmar has three seasons: the cool, relatively dry northeast monsoon season (late October to mid-February), the hot, dry inter-monsoonal season (mid-February to mid-May), and the southwest monsoon season (mid-May to late October). The coastal regions and the western and south-eastern ranges receive more than 5,000 mm of rainfall annually, while the delta regions receive about 2,500 mm. The central region is not only away from the sea but also on the drier, lee side of the Rakhine Mountains. Precipitation gradually decreases northward until in the region's dry zone it amounts to only 500 to 1,000 mm per year. The Shan Plateau, because of its elevation, usually receives between 1,900 and 2,000 mm rainfall.

16. **Rainfall: The Central Dry Zone (Meiktila Township, Mandalay Region)**. Rainfall in Meiktila demonstrates bimodal rainfall pattern (dual peak; observable in stations located in the Central Dry Zone). With annual average rainfall of about 800mm, rainfall is normally heaviest in May and September.

17. **Rainfall: The Coastal Region (Kyaukpyu Township, Rakhine Region)**. Receiving most of the rainfall from the Southwest Monsoon and from tropical systems over the Bay of Bengal, Kyaukpyu is the wettest place of the country with average annual rainfall of about 4,655mm. Like many areas with dominance of Southwest Monsoon influence, Kyaukpyu receives significant rainfall from May to October, with the maximum in July.

18. **Rainfall: The Hilly Zone (Kengtung Township, Shan State)**. Kengtung receives lesser rainfall compared to townships in the coastal zone, but more compared to townships in the Central Dry Zone. Located inland, it receives less rainfall from the Southwest Monsoon but benefits from rainfall from remnants of severe weather disturbances from the South China Sea. Maximum rainfall of about 230 mm is received in the month August.

19. **Temperature**. Elevation and distance from the sea affect temperature in the various regions. Although Myanmar generally is a tropical country, temperatures are not uniformly high throughout the year. The daily temperature range is greater than that in nearly all other parts of Southeast Asia, but no locality has a continental type of climate (i.e., one characterized by large seasonal differences in average temperature). Mandalay, in the centre of the dry zone, has some of the greatest daily temperature ranges, which span about 12 °C annually. In broader perspective, however, average daily temperatures show little variation, ranging from 26 °C to 28 °C between Sittwe (Akyab) in the Rakhine region, Yangon near the coast, and Mandalay in the northern part of the central basin. At Lashio, on the Shan Plateau, the average daily temperature is somewhat cooler, 22 °C.

20. **Temperature: The Central Dry Zone (Meiktila Township, Mandalay Region)**. The average annual maximum temperature is around 33°C. April records the warmest daytime temperature, averaging at 38.2°C. December has the coolest month with average temperature around 28.8°C. On the other hand, minimum or night time temperature is warmest in May at an average of around 25°C and in April at around 25°C. Coolest nights are in January and

December, at 14.6°C and 16°C, respectively. Highest maximum temperature 46°C was recorded at Meiktila on 13 April 2010 and the lowest minimum temperature recorded was 19°C on 03 February 2007 and 7 December 1992.

21. **Temperature: The Coastal Region (Kengtung Township, Shan State).** The warmest average daytime temperature is in April (33.5°C). The diurnal variation in April is significant, with its average night time temperature at 17.8°C. The lowest average maximum temperature is in December, at 25.5°C, and January at 27.1°C. Both are also the coolest months in terms of average night time temperature (January at 10°C and December at 11°C). February registers as among the months with coolest night time temperature, averaging at 11.1°C. Highest night time temperature is in the wet months of June (21.7°C), July (21.5°C), August (21.3°C). Daytime temperature, for these months, averages at 30.7°C (June), 29.5°C (July) The most extreme maximum temperature observed was 39.7°C on 19 September 1988 and the coolest minimum temperature was in 24 December 1983 (10°C).

22. **Temperature: The Hilly Zone (Maylamyine Township, Mon State).** The warmest days in the township are normally in March and April, averaging at 35.5°C and 35.6°C, respectively. On the average, the days in January are the coolest, at 18.4°C. Extremely warm daytime temperature, of 37.7°C was recorded twice in Mawlamyine, both in April 1992. Coolest daytime temperature is 21°C (18 October 1992). On the other hand, night time temperatures are warmest in April (24.7°C) and in May (24.3°C). Warmest night time temperature recorded in 30 years was 29.20°C (19 April 1991); coolest was in 26 December 1999, at 8°C

Bhutan

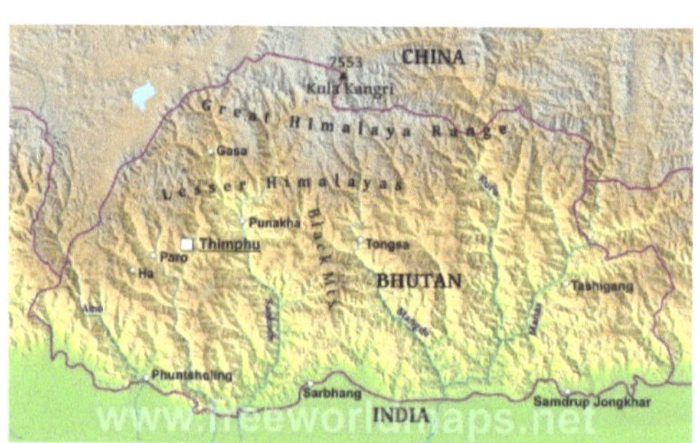

23. Bhutan is on the southern slopes of the eastern Himalayas, landlocked between the Tibet Autonomous Region of China to the north and the Sikkim, West Bengal, Assam to the west and south, and Arunachal Pradesh to the east. The land consists mostly of steep and high mountains crisscrossed by a network of swift rivers that form deep valleys before draining into the Indian plains. In fact, 98.8% of Bhutan is covered by mountains, which makes it the most mountainous country in the world. Elevation rises from 200 m in the southern foothills to more than 7,000 m. Bhutan's northern region reaches up to glaciated mountain peaks with an extremely cold climate at the highest elevations. Most peaks in the north are over 7,000 m above sea level. The highest point is 7,570 m tall Gangkhar Puensum, which has the distinction of being the highest unclimbed mountain in the world. The lowest point, at 98 m, is in the valley of Drangme Chhu, where the river crosses the border with India.

24. **Climate.** Bhutan's climate varies with elevation, from subtropical in the south to temperate in the highlands and polar-type climate with year-round snow in the north. Bhutan experiences five distinct seasons: summer, monsoon, autumn, winter and spring. Western Bhutan has the heavier monsoon rains; southern Bhutan has hot humid summers and cool winters; central and eastern Bhutan are temperate and drier than the west with warm summers and cool winters. South-west monsoon affects the country from June to September and contributes the maximum of the annual rainfall.

25. **Synoptic Features effecting Climate.** The main synoptic features that effect the climate of Bhutan are similar to that of other parts of the eastern Himalaya region like the states of Sikkim and Arunachal Pradesh of India.

The winter (December to Mid-February) season is cold and comparatively dry. The spring season (Mid-February to may) is the hot season with thunderstorm activities during afternoon/evening hours on many days (mainly towards the southern areas). The south-west monsoon season (June to September) brings abundant of rainfall to the country. Most of the annual rainfall is received during this season. The autumn season (October to November) exhibits pleasant temperature profile with some rainfall during October and scanty rainfall during November.

a) **Winter Season.** Mainly dry continental northerly to north-easterly winds prevail over the country during this season. Weather remains dry except for the periods of passage of WDs. The WDs, which are mainly induced systems and in occluded states affect the country during mid-December to March. Under their influence, Bhutan experiences rainfall in the low altitude areas and snowfall over the high-altitude regions. The troughs extending from a parent westerly system moving across Tibet or China also cause rainfall and snowfall over the country during winter season. Cold wave conditions prevail over the country mainly over the mountainous areas and higher plateaus when the sky clears up rapidly after the passage of the westerly system and cessation of precipitation.

b) **Spring Season.** This is the season of gradual rise in temperature. By April/May, Bhutan experiences dry westerly to north-westerly winds. April/May are the hottest months of the year. Like the western foot-gill region of Assam, Bhutan experiences severe thunderstorm in April and May mainly over the southern plain and adjoining hilly regions. Passage of WDs along the middle latitudes, occasionally enhances the thunderstorm activity.

c) **Monsoon Season.** Bhutan receives abundance of rainfall during this season. The southerly monsoon current is pre-dominant. The monsoon winds, after encountering the southern hills gets lifted up and cause thunderstorm, though of lesser intensity and frequency than the previous season. Occasionally, movement of WDs across higher latitudes enhances rainfall activities. The position of the AMT plays an important role in monsoon activity. East ward extension of the AMT up to NE states is favourable for good monsoon activity over Bhutan. Also, during the Break Monsoon condition, when the AMT passes along the foot-hill, Bhutan experiences heavy rainfall. During the period when a monsoon system forms over north Bay of Bengal, rainfall activity decreases over the country.

26. **Climate over Plains.** In the thin **flat southern region,** in cities like Gelephu and Phuentsholing, the climate is **subtropical.** Winter is very mild, but sometimes it can get very cold at night. Spring, from March to May, is hot, especially in April and May, when the temperature can reach as high as 40 °C. However, the frequent afternoon/evening thunderstorms subside the heat effect on many days. From June to mid-October, it's the monsoon season, with **abundant rainfall, at times torrential,** which are enhanced by the proximity to the mountains, and sultry heat. Total annual rainfall can exceed 3,000 mm. The autumn months (October to November) have pleasant weather conditions. The winter season (December to February) is cold and mostly dry.

27. Average temperatures profile of Gelephu are given in the following table: -

Temp (^{0}C)	Jan	Feb	Mar	Apr	May	Jun	Jul	Aug	Sep	Oct	Nov	Dec
Mean	17	19	22	26	27	27	28	28	27	25	21	18
Maximum	23	25	29	32	32	31	31	31	31	30	27	25
Minimum	10	12	15	20	22	24	25	25	24	21	15	11

28. **Climate over Hills and Mountains.** In **hilly and mountainous areas** up to 1,500 meters (5,000 feet), the climate becomes **progressively colder** with relatively dry winters and warm, rainy summers. The summer thunderstorm activity and monsoon rainfall activities are less in compared to the southern plain region. In **Punakha**, located at 1,300 m ASL, the average temperature ranges from 11 °C in January to 24 °C in July. The summer is quite hot and humid despite the altitude. The average temperatures of Punukha are given below: -

Temp (^{0}C)	Jan	Feb	Mar	Apr	May	Jun	Jul	Aug	Sep	Oct	Nov	Dec
Mean	11	13	17	19	22	25	24	24	23	21	17	13
Maximum	17	19	22	25	27	29	28	28	27	26	23	20
Minimum	5	8	11	14	17	20	20	20	20	16	10	6

29.   In the capital **Thimphu which is** located in a valley at an altitude varying between 2,200 and 2,600 m, the average temperature ranges from 6 °C in January to 21/22 °C in the period from June to August. Winter is dry and sunny, but snowfall occurs sometimes. The average temperatures of Thimphu as given below: -

Thimphu

Temp (^{0}C)	Jan	Feb	Mar	Apr	May	Jun	Jul	Aug	Sep	Oct	Nov	Dec
Mean	6	9	12	15	18	21	22	22	20	16	11	8
Maximum	15	17	19	22	25	27	27	27	26	24	20	17
Minimum	-2	0	4	8	12	15	17	16	15	9	3	-1

30. **Rainfall**. The characteristics of rainfall is similar to that of the Indian states of Sikkim and Arunachal. Main source of rainfall for the country is the south-west monsoon followed by the spring season (pre-monsoon) during which mainly convective rainfall takes place and then followed by the winter season when the country, mainly the northern high-altitude region receives rainfall under the influence of the westerly systems.

31. **Rainfall Data: Thimphu**.

Rainfall	Jan	Feb	Mar	Apr	May	Jun	Jul	Aug	Sep	Oct	Nov	Dec
Mean (mm)	5	10	20	30	50	90	145	120	70	40	0	5
Rainy days	1	1	4	6	9	14	19	17	12	5	0	1
Yearly Rainfall	590 mm											

Nepal

32. Nepal is landlocked by China's Tibet Autonomous Region to the north and India on other three sides. West Bengal's narrow *Siliguri Corridor* separate Nepal and Bangladesh. Nepal has a very high degree of geographic diversity and can be divided into three main regions: Terai, Hilly, and Himal. The Terai region, covering 17% of Nepal's area, is a lowland region with some hill ranges and is culturally more similar to parts of India. The Hilly region, encompassing 68% of the country's area, consists of mountainous terrain without snow and is inhabited by various indigenous ethnic groups. The Himal region, covering 15% of Nepal's area, is home to several high mountains including Mount Everest.

33. The country can be divided into five physiographic regions. From south to north, and with increasing altitude, these are the following: (a) Terai, (b) Siwaliks, (c) Middle Mountains, (d) High Mountains, and (e) High Himalaya

(Kansakar et al. 2004). Terai is a low-lying plain, and northwards, middle mountains and hills are elongated east–west with a series of complex valleys breaks, such as Kathmandu, Pokhara, Dang, and Surkhet.

34. Nepal has three categories of rivers: the largest systems (Koshi, Gandaki/Narayani, Karnali/Goghra, and Mahakali), second category rivers (rising in the Middle Hills and Lower Himalayan Range), and third category rivers (rising in the outermost Siwalik foothills and mostly seasonal). These rivers can cause serious floods and pose challenges to transportation and communication networks.

35. Based on average temperature and precipitation, the climatic condition over Nepal is classified in eight zones. They are, Tropical Savanah, Arid Steep Cold, Temperate with Dry Winter & Hot Summer, Temperate with Dry Winter & Warm Summer, Cold Climate with Dry Winter & Warm Summer, Cold Climate with Dry Winter & Cold Summer, Polar Tundra and Polar Frost climate.

36. The climate of Nepal is affected by two major weather systems, summer monsoon circulation (June to September) and westerly circulation (November to May). The influence of these two circulation systems is different, with summer precipitation greater in the southeast and westerly-derived winter precipitation greatest in the northwest. Nepal as a whole receives approximately 80 % of its annual precipitation during the summer monsoon.

37. Nepal can broadly grouped into three climatic regions – The plain belt to the south, the lower mountain region north of the plains and the high mountainous region to the north. and the Himalaya. Weather conditions over the plains to the south are similar to that of Indo-Gangetic plains. The eastern sector of this region receives maximum rainfall. Over the lower mountains, temperature is lower and rainfall is less than the plain areas. The higher mountainous region is characterized by very cold temperature and the high peaks remain submerged in snow throughout the year.

38. **Main Synoptic Features Affecting Weather of Nepal**. There is similarity in the weather conditions over Nepal with the Himalayan regions of India. Conditions over the southern plain region resembles with that of north Indo Gangetic plains. Various synoptic features affecting the weather of Nepal in various months are discussed below.

    a)   **Mid-March to Mid-June**.

        i.   These are the hot and dry months when the seasonal heat establishes over north India. The heat low, which makes it's appearance towards the end of February over Indian Peninsular region shifts gradually north ward with the progress of the summer season in India. The heat low is actually marked over the hottest area where the air pressure is the lowest. During April/May it can be located over eastern Indian region indicating maximum insolation over the region. The effect of the hotness is also experienced over the eastern plain region of Nepal.

        ii.   As the summer season in India progresses, the heat low gradually shifts west ward to Rajasthan and adjoining region by June. In this period the temperature over the western part also rises leading to heat wave like condition over the southern plains. During this month, the southern plains and the lower mountainous regions of Nepal experience thunderstorm activities during the occasions of slight moisture incursion from Bay of Bengal or Arabian Sea due to movement of WDs across higher latitude or location of a low level anticyclone over north Bay or West Bengal. Thunderstorm activities are maximum towards the eastern sector.

        iii.   The normal track of the WDs shifts much to the north of Nepal. However, some of the WDs still affect the country, especially the northern mountainous region.

    b)   **Mid-June to September**.

        i.   The south-west monsoon starts affecting Nepal from west from Mid of June and covers the entire country by end of June or early July.

ii.   Mainly the south-easterly to easterly Bay of Bengal branch of monsoon affects the country when the AMT passes through Indo Gangetic plain or to it's south. Nepal has the merits of getting better monsoon rainfall during monsoon season as the monsoon current undergoes orographic lifting by the hills of Nepal. During the Break Monsoon Conditions when the AMT passes along the foothills, Nepal experiences heavy rainfall, sometime resulting in over flowing of the rivers flowing southward, which eventually cause flood in southern plain areas and north Gangetic plains.

iii.   The monsoon systems (Low/Depression) that forms over north/head Bay of Bengal moves north-west or west-north-west ward during the monsoon seasons. Many of the systems tends to re-curve north or north-east ward under the influence of the moving WDs across northern latitudes. In such occasions, the monsoon systems reach up to the foot-hills and causes active monsoon condition over the affected areas. The monsoon systems generally weaken over the hills.

iv.   Withdrawal of monsoon from the country begins from the western end during end of September and completely withdraws before mid-October.

c)   **Mid-October to November**.

i.   After withdrawal of monsoon, the sky condition improves rapidly and temperature stars falling gradually. Toward mid of November the lower and higher mountainous regions start experiencing cold climate conditions, while the conditions become pleasant over the southern plains.

ii.   Northerly to north-easterly continental winds start prevailing over the country after withdrawal of monsoon.

iii.   The track of WDs starts shifting south ward and by November some places over north mountainous region receives rain/snow under the influence of the WDs, though, still following a track much to the north of Nepal. By mid of November the track is close to Nepal and rain/snowfall activity increases over the northern region.

d)   **December to Mid-March**.

i.   This is the cold season and the main synoptic feature affecting the country is the westerly systems which follow a track very close to the country. The WDs cause good snowfall activity over the mountainous regions during the season. East-north-east ward movement of Induced Westerly Systems from Rajasthan also cause rainfall over the southern plains & low mountain regions and snowfall over the northern mountains and plateaus.

ii.   Mainly north-westerly dry & cold continental winds prevail over the country during the season. Prolonged prevalence of dry & cold winds results in cold wave condition. The dry spells are broken on passage of WDs and cold wave condition ceases. However, after the disturbances move away east ward, the sky clears rapidly and again results in cold wave condition for two to four days.

39. **Climatological Data (Five Major Cities)**.

(Temperature in ^{0}C (Maximum, Mean & Minimum) and Rainfall in mm).

Dipayal

Element	Jan	Feb	Mar	Apr	May	Jun	Jul	Aug	Sep	Oct	Nov	Dec
Mean Temp	13	16	20	25	28	30	29	29	27	24	19	15
Max Temp	22	25	30	34	36	36	34	34	33	31	27	23

Min Temp	5	7	11	15	20	23	24	24	22	16	10	6
Rainfall	39	52	36	39	87	161	239	223	167	46	6	20
Annual	1,118 mm											

Pokhara

Element	Jan	Feb	Mar	Apr	May	Jun	Jul	Aug	Sep	Oct	Nov	Dec
Mean Temp	13	16	20	23	24	26	26	26	25	22	18	14
Max Temp	20	22	27	30	30	31	30	30	29	27	24	21
Min Temp	7	9	13	16	18	21	22	22	21	17	12	8
Rainfall	23	35	60	128	359	669	940	866	641	140	18	22
Annual	3,900 mm											

Kathmandu

Element	Jan	Feb	Mar	Apr	May	Jun	Jul	Aug	Sep	Oct	Nov	Dec
Mean Temp	11	13	17	20	22	24	24	24	23	20	16	12
Max Temp	19	21	25	28	29	29	28	29	28	27	24	20
Min Temp	2	5	8	12	16	19	20	20	19	13	8	4
Rainfall	14	19	34	61	124	236	363	331	200	51	8	13
Annual	1,455 mm											

Dhankuta

Element	Jan	Feb	Mar	Apr	May	Jun	Jul	Aug	Sep	Oct	Nov	Dec
Mean Temp	12	14	18	21	22	23	23	23	23	20	17	14
Max Temp	18	19	23	26	27	27	27	27	27	25	23	19
Min Temp	7	9	13	16	18	20	20	20	19	16	12	9
Rainfall	12	18	25	46	108	147	256	180	133	44	12	9
Annual	991 mm											

Biratnagar

Element	Jan	Feb	Mar	Apr	May	Jun	Jul	Aug	Sep	Oct	Nov	Dec
Mean Temp	16	19	23	27	28	29	29	29	28	26	22	18
Max Temp	23	26	31	34	33	33	32	33	32	32	29	25
Min Temp	9	11	16	20	23	25	26	26	25	21	15	11
Rainfall	12	13	13	53	186	302	531	378	299	92	6	7
Annual	1,891 mm											

Pakistan

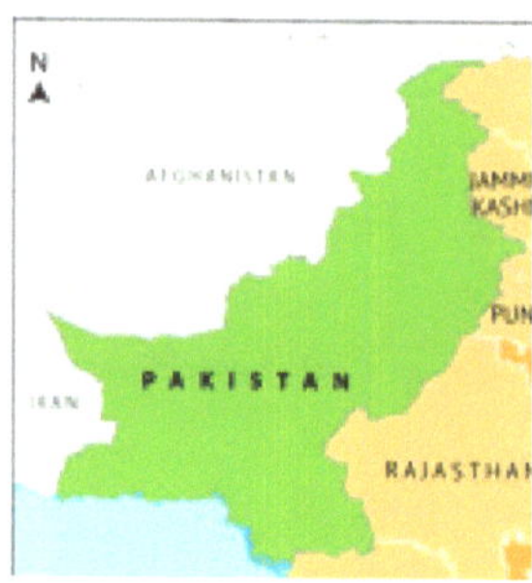

40. Pakistan is bordered by India to the east, Afghanistan to the west, Iran to the southwest, and China to the northeast. It is separated from Tajikistan by Afghanistan's narrow Wakhan Corridor in the north, and also shares a maritime border with Oman. There are five distinct geographic regions:

    a) The Thar Desert and Lower Indus Valley, located in the southernmost province of Sindh, consists largely of arid valleys and rocky hills that extend into neighbouring India. Farming is successful only in the irrigated areas nearest to the Indus River.

    b) The Balochistan Plateau is a broad, arid tableland that lies between 1,000 and 3,000 feet above sea level in the western province of Balochistan. The plateau is encircled by rugged mountains and covers nearly one-half of the country's territory.

    c) The Indus Basin features the largest contiguous irrigation system in the world. "Punjab," the name of the province in which much of the basin is located, means "five waters" in Persian, referring to the five major rivers (Indus, Jhelum, Chenab, Ravi, and Sutlej) in the basin. The province of Punjab comprises the northeastern quarter of Pakistan.

    d) The Northwest Frontier is a region of barren mountains sheltering rich irrigated valleys. The provincial capital of Peshawar is situated on an ancient trade route that leads through the Khyber Pass and into Afghanistan.

    e) The Far North offers Pakistan's most spectacular scenery with towering snow-capped mountains, deep narrow valleys, and glaciers. The world's second highest mountain, K–2, is located in the Far North, as are a dozen other peaks of more than 25,000 feet elevation, including Nanga Parbat, Gasherbrun, and Rakaposhi.

41. Pakistan lies in the temperate zone. The climate is generally arid, characterized by hot summers and cool or cold winters. There are wide variations between extremes of temperature at given locations. Also, the country has a diverged climatological condition from north to south and from west to east. For example, the coastal area along the Arabian Sea is usually warm, whereas the frozen snow-covered ridges of the Karakoram Range and of other mountains of the far north are very cold year-round and are not accessible by general public.

42. Pakistan has four seasons: a cool, dry winter from December to February; a hot, dry spring from March to May; the summer rainy season, or southwest monsoon period, from June to September; and the retreating monsoon period from October to November. The onset and duration of these seasons vary somewhat according to location.

43. The climate in the capital city of Islamabad varies from an average daily low of 2° C in January to an average daily high of 40° C in June. Half of the annual rainfall occurs in July and August, averaging about 255 mm in each of those two months. The other months of the year have significantly less rain, amounting to about fifty mm per month. The spring season is associated with thunderstorm and hailstorm.

44. Pakistan's largest city, Karachi, is more humid than Islamabad but gets less rain. Only July and August average more than twenty-five mm of rain in the Karachi area and the remaining months are exceedingly dry. The temperature

is also more uniform in Karachi than in Islamabad, ranging from an average daily low of 13° C during winter to an average daily high of 34° C on summer days. Although the summer temperatures do not get as high as those in Punjab, the high humidity causes great discomfort.

45. Most areas in Punjab experience fairly cool winters, often accompanied by rain. By mid-February the temperature begins to rise and the summer heat sets in in April. The south-west monsoon sets in by early July and withdraws by beginning of September. Month of June is oppressively hot. Maximum temperature may shoot up to 46 ^{0}C or more during June. Cool season commences in October.

46. **Main Synoptic Features Affecting Weather of Pakistan.**

   a) **March to June.**

   i. There is general rise in the temperature over the country, rise becoming rapid from mid-April to June over south and central region. Heat wave conditions commonly prevails over Balochistan and Punjab sectors. With the arrival of the monsoon current temperature starts decreasing by July.

   ii. Due to prevailing high insolation the atmosphere remains highly instable. Incursion of a little moisture from the Arabian Sea or from a westerly system moving across north Iran, there is thunderstorm activity over Balochistan and Punjab regions. During April and May, movement of an induced WD across central Pakistan may result in fairly widespread thunderstorm or dust storm with hail at some places.

   iii. During the season tropical cyclones form over south-east Arabian Sea and mostly move north-west ward. In one od case a storm may recurve north ward and affect Gujarat and south Pakistan. In such case parts of south Pakistan and adjoining central Pakistan may get affected with heavy rain and squally winds.

   iv. A few WDs moving across north of Pakistan affects the weather over north Pakistan.

   b) **July to August (Monsoon Season).**

   i. The seasonal heat low of south-west monsoon season is located over north Balochistan/south Punjab sector and the adjoining areas of Rajasthan. An anti-cyclone gets established above the CYCIR associated with the heat low from about 7,000' ASL. Hence, monsoon activity is minimum over the area of the heat low. Hence, Pakistan is not so much affected by the monsoon current.

   ii. Occasionally, a few monsoon depressions/lows moving west-north-west ward may cause rainfall over central Pakistan when the systems are located over west Rajasthan.

   iii. The east-west ridge of Tibetan high which is marked over Tibet at middle and upper tropospheric levels, sometimes become well pronounced and extends up to central Pakistan and further to the west. It is observed in such situation that monsoon rainfall extends up to Pakistan (Central Pakistan).

   iv. Occasionally one odd monsoon system located over north-west India may interact with a westerly system moving across higher latitudes. In such situation, there is increase in clouding over north Pakistan and some places experience rainfall.

   c) **October to November.**

   i. There is fall in temperature after August over the entire country. By end of October, weather condition over the country becomes pleasure with clear sky, warm day and cool night. By November the extreme northern region starts exhibiting winter climate with cold and dry weather conditions.

   ii. The track of WDs starts shifting south ward and by November, a few WDs affect the north Pakistan with rainfall over plains/plateaus and snowfall over the high mountainous areas.

ii.   Tropical Cyclones form over Arabian Sea in these two months also. Like Pre the summer months, one odd storm may re-curve north or north-north-east ward from central Arabian Sea and effect south Pakistan.

d)   December to February/Mid-March.

i.   This is winter season with cold and dry weather conditions. Towards extreme north, temperatures are very low. The cold and dry north-westerly continental winds often result in cold wave condition over parts of central and north India. South Pakistan, however, remains pleasantly warm in day and cool at night.

ii.   There is increase in the frequency of WDs and with their tracks shifting further to the south. North Pakistan often remains affected by these disturbances.

iii.   This is also the season when Induced WDs move across central Pakistan and affect central and north Pakistan.

iv.   The approach of the WDs, beside causing precipitation, also arrest cold wave conditions. However, after passing away of the systems the sky clears up rapidly, cold and dry continental winds again sets in and cold wave condition prevails for a few days.

47. **Climatological data: Major Cities of Pakistan.**

Quetta

Element	Jan	Feb	Mar	Apr	May	Jun	Jul	Aug	Sep	Oct	Nov	Dec
Mean Temp	5	8	13	18	23	27	29	28	23	17	11	6
Max Temp	11	14	19	25	31	35	36	35	31	25	19	13
Min Temp	-2	1	6	10	15	20	23	21	15	8	3	-1
Rainfall	50	45	50	30	9	3	17	16	1	4	6	30
Rainy Days	4	5	6	5	4	2	5	4	2	2	1	3
Annual	260 mm											

Peshawar

Element	Jan	Feb	Mar	Apr	May	Jun	Jul	Aug	Sep	Oct	Nov	Dec
Mean Temp	11	14	19	24	29	32	32	31	29	24	18	18
Max Temp	18	20	25	30	36	39	37	35	34	31	24	20
Min Temp	5	8	13	17	22	26	27	26	23	17	11	6
Rainfall	30	45	75	50	25	8	45	70	22	11	13	25
Rainy Days	5	7	11	10	7	4	8	9	5	4	3	4
Annual	410 mm											

Lahore

Element	Jan	Feb	Mar	Apr	May	Jun	Jul	Aug	Sep	Oct	Nov	Dec
Mean Temp	12	16	21	27	32	33	31	30	29	26	19	14
Max Temp	18	22	28	34	39	39	35	35	34	32	27	21
Min Temp	6	10	15	19	25	27	27	26	24	19	12	7
Rainfall	25	30	40	20	20	35	200	165	60	12	4	14
Rainy Days	4	5	6	5	5	5	11	11	5	2	2	3
Annual	630 mm											

Islamabad

Element	Jan	Feb	Mar	Apr	May	Jun	Jul	Aug	Sep	Oct	Nov	Dec
Mean Temp	11	13	18	24	29	31	30	29	27	23	17	13
Max Temp	17	20	25	31	36	38	35	34	33	30	25	20
Min Temp	4	7	12	17	21	24	25	24	22	16	9	5
Rainfall	65	95	110	80	55	80	275	290	100	35	25	45
Rainy Days	5	7	10	10	8	8	16	16	9	4	3	4
Annual	1,250 mm											

Jacobabad

Element	Jan	Feb	Mar	Apr	May	Jun	Jul	Aug	Sep	Oct	Nov	Dec
Mean Temp	15	19	24	30	36	37	35	33	32	28	22	17
Max Temp	21	25	31	37	44	44	40	37	36	34	29	23
Min Temp	9	12	17	23	29	30	30	29	27	22	16	10
Rainfall	3	7	10	2	2	5	35	25	11	2	1	4
Rainy Days	0	1	1	1	1	0	2	1	1	0	0	0
Annual	110 mm											

Hyderabad

Element	Jan	Feb	Mar	Apr	May	Jun	Jul	Aug	Sep	Oct	Nov	Dec
Mean Temp	18	21	26	31	34	34	32	31	31	30	25	19
Max Temp	24	28	33	39	41	40	37	35	36	36	31	25
Min Temp	12	14	19	23	26	28	27	27	25	23	18	13
Rainfall	1	4	5	6	4	14	55	60	20	2	2	2
Rainy Days	0	1	0	0	0	1	2	3	1	0	0	0
Annual	180 mm											

Karachi

Element	Jan	Feb	Mar	Apr	May	Jun	Jul	Aug	Sep	Oct	Nov	Dec
Mean Temp	19	22	26	29	31	32	31	30	30	29	25	21
Max Temp	26	29	32	35	36	35	34	32	33	35	32	28
Min Temp	12	15	19	24	27	29	28	27	26	23	18	13
Rainfall	5	8	10	5	0	7	80	55	20	1	2	6
Rainy Days	0	1	0	0	0	1	3	2	1	0	0	1
Annual	205 mm											

48.   **Climatological Data: Extremes.**

Quetta

Month	Maximum Temperature		Minimum Temperature		Monthly Rainfall	
	Highest (^{0}C)	Date	Lowest (^{0}C)	Date	Highest (mm)	Year
January	24	28/1987	-15	09/1967	178	1982
February	27	26/1953	-17	01/1970	189	1982
March	31	17/2022	-8	12/1973	232	1982
April	35	26/2015	-4	02/1965	159	1992
May	39	11/2000	0	03/1989	40	1963
June	41	04/2005	5	18/1951	61	2007
July	42	10/1998	9	07/1955	164	1956
August	41	09/1970	3	23/1949	208	2022
September	38	01/1970	-1	30/1962	62	1994
October	34	01/1998	-1	29/1949	69	1982
November	29	06/2008	-13	30/1962	72	2006
December	25	05/2015	-17	21/1950	162	1982

Peshawar

Month	Maximum Temperature		Minimum Temperature		Monthly Rainfall	
	Highest (^{0}C)	Date	Lowest (^{0}C)	Date	Highest (mm)	Year
January	27	30/2007	-4	07/1970	150	1999
February	31	24/2016	-2	03/1905	236	2007
March	37	26/1892	-1	01/1905	223	1978
April	42	29/1941	5	08/1918	267	2008
May	48	31/1941	11	02/1881	131	1901
June	50	18/1995	13	08/1949	98	1881
July	50	05/1920	18	23/1968	409	2010
August	48	09/1915	19	27/1954	451	1892
September	43	02/1920	12	30/1991	127	2012
October	39	05/2006	6	30/1916	203	1996
November	35	03/1979	0	30/1912	111	1959
December	29	03/1979	-2	13/1937	145	1967

Lahore

Month	Maximum Temperature		Minimum Temperature		Monthly Rainfall	
	Highest (^{0}C)	Date	Lowest (^{0}C)	Date	Highest (mm)	Year
January	28	23/1952	-2	17/1935	132	2022
February	33	27/1953	-1	02/1905	117	1990
March	41	25/1892	3	05/1945	167	1978
April	46	29/1941	8	02/1903	141	1983
May	48	30/1944	14	14/1977	111	1885
June	48	08/1929	18	18/1977	291	2023
July	48	06/1901	20	04/1974	478	1981
August	44	08/1911	19	10/1980	640	1996
September	43	04/1905	17	23/1972	525	1954
October	41	07/1931	8	31/1949	155	1985
November	35	05/1943	2	30/1962	78	1997
December	31	01/1899	-2	23/1910	112	1967

Islamabad

Month	Maximum Temperature		Minimum Temperature		Monthly Rainfall	
	Highest (^{0}C)	Date	Lowest (^{0}C)	Date	Highest (mm)	Year
January	28	30/2007	-3	28/1984	219	1992
February	31	24/2016	-2	24/1984	298	2013
March	36	18/2022	2	04/2003	286	2015
April	41	30/1999	4	07/1994	231	1983
May	46	12/2001	10	09/1997	186	1987
June	47	21/1994	15	13/1999	230	2008
July	44	03/2012	18	09/2023	1038	2001
August	41	12/1987	15	01/2017	664	2013
September	37	02/2023	11	25/1994	445	2014
October	37	01/2009	5	31/1984	197	2021
November	31	01/1996	0	30/1995	93	2020
December	29	07/1998	-4	25/1984	162	1990

Karachi

Month	Maximum Temperature		Minimum Temperature		Monthly Rainfall	
	Highest (^{0}C)	Date	Lowest (^{0}C)	Date	Highest (mm)	Year
January	33	16/1965	0	21/1934	89	1995
February	37	28/2016	3	11/1950	96	1979
March	43	31/2022	7	09/1979	130	1967
April	44	16/1947	12	29/1967	53	1935
May	48	09/1938	18	04/1989	33	1939

June	47	18/1979	22	03/1940	110	2007
July	43	18/2020	22	22/1938	429	1967
August	42	09/1964	20	07/1984	367	2020
September	43	30/1951	18	23/1994	316	1959
October	43	01/1951	10	30/1949	98	1956
November	39	02/1994	6	29/1938	83	1959
December	35	03/2011	1	14/1986	64	1980

Sri Lanka

49. Sri Lanka is an island nation in the Indian Ocean, southeast of the Indian subcontinent with a coast line of 1,340 Km long. The main island of Sri Lanka has an area of 65,268 sq km and is the twenty-fifth largest island in the world. Dozens of offshore islands account for the remaining 342 sq km area. The largest offshore island, Mannar Island, is linked to Adam's Bridge, a land connection to the Indian mainland, which is now mostly submerged with only a chain of limestone shoals remaining above sea level.

50. **Climate**. Due to the location of Sri Lanka, within the tropics between 5° 55' to 9° 51' North latitude and between 79° 42' to 81° 53' East longitude, the climate of the island could be characterized as tropical.

51. **Topography.** The central part of the southern half of the island is mountainous with heights more than 2.5 Km. The core regions of the central highlands contain many complex topographical features such as ridges, peaks, plateaus, basins, and valleys. The remainder of the island is practically flat except for several small hills that rise abruptly in the lowlands. These topographical features strongly affect the spatial patterns of winds, seasonal rainfall, temperature, relative humidity and other climatic elements, particularly during the monsoon season.

52. **Rainfall**. Rainfall in Sri Lanka has multiple origins. Monsoonal, Convectional and depressional rain accounts for a major share of the annual rainfall. The mean annual rainfall varies from under 900mm in the driest parts (south-eastern and north-western) to over 5000 mm in the wettest parts (western slopes of the central highlands).

53. The rainfall pattern is influenced by the monsoon winds of the Indian Ocean and Bay of Bengal and is marked by four seasons. The first is from mid-May to October (Monsoon), when winds originate in the southwest, bringing moisture from the Indian Ocean. When these winds encounter the slopes of the Central Highlands, they unload heavy rains on the mountain slopes and the southwestern sector of the island. Some of the windward slopes receive up to 2500 mm of rain per month, but the leeward slopes in the east and northeast receive little rain. The second season occurs in October and November, the inter-monsoonal months. During this season, periodic squalls occur and sometimes tropical cyclones bring rains to the southwest, northeast, and eastern parts of the island. During the third season, December to mid-March NE Monsoon), monsoon winds come from the northeast, bringing moisture from the Bay of Bengal. The north-eastern slopes of the mountains may be inundated with up to 1250 mm of rain during these months. Another inter-monsoonal period occurs from mid-March until mid-May, with light, variable winds and evening thundershowers.

54. **Temperature**. Regional differences observed in air temperature over Sri Lanka are mainly due to altitude, rather than to latitude. The mean monthly temperatures differ slightly depending on the seasonal movement of the Sun, with some modified influence caused by rainfall. The mean annual temperature in Sri Lanka manifests largely homogeneous temperatures in the low lands and rapidly decreasing temperatures in the highlands. In the lowlands, up to an altitude of 100 m to 150 m, the mean annual temperature varies between 26.5 °C to 28.5 °C, with an annual temperature of 27.5 °C. In the highlands, the temperature falls quickly as the altitude increases. The mean annual temperature of Nuwaraeliya, at 1800 m sea level is 15.9 °C. The coldest month with respect to mean monthly temperature is generally January, and the warmest months are April and August. The mean annual temperature varies from 27°C in the coastal lowlands to 16°C at NuwaraEliya in the central highlands (1900m above mean sea level).

55. **Seasons**. The Climate of Sri Lanka is dominated by the topographical features of the country and the Southwest and Northeast monsoons wind regimes. There are four seasons in Sri Lanka, which are, First Inter-monsoon Season (March - April), Southwest -monsoon Season (May - September), Second Inter-monsoon Season (October-November) and Northeast -monsoon Season (December - February).

56. **First Inter-monsoon Season (March - April)**. Warm and uncomfortable conditions, with thunderstorm-type rain, particularly during the afternoon or evening, are the typical weather conditions during this season. The distribution of rainfall during this period shows that the entire South-western sector at the hilly region receives 250 mm of rainfall, with localize area on the South-western slops experiencing rainfall in excess of 700 mm (Keragala 771 mm). Over most parts of the island, the amount of rainfall varies between 100 and 250 mm, with exception being the Northern Jaffna Peninsula (Jaffna- 78 mm, Elephant pass- 83 mm).

57. **Southwest -monsoon Season (May - September)**. Windy weather during this monsoon eases off the warmth that prevailed during the First Inter monsoon season. Southwest monsoon rains are experienced at any times of the day and night, sometimes intermittently mainly in the Southwestern part of the country. Amount of rainfall during this season varies from about 100 mm to over 3000 mm. The highest rainfall received in the mid-elevations of the western slopes (Ginigathhena- 3267 mm, Watawala- 3252 mm, Norton- 3121 mm). Rainfall decreases rapidly from these maximum regions towards the higher elevation, and in Nuwara-eliya, it drops to 853 mm. The variation towards the Southwestern coastal area is less rapid, with the Southwestern coastal belt experiencing between 1000 mm to 1600 mm of rain during this 5-month long period. Lowest figures are recorded from Northern and South-eastern regions.

58. The thunderstorm-type of rain, particularly during the afternoon or evening, is the typical climate during this season. But unlike in the First Inter-monsoon season, the influence of weather systems like depression and cyclones in the Bay of Bengal is common during the this Inter-monsoon season. Under such conditions, the whole country experiences strong winds with wide spread rain, sometimes leading to floods and landslides. The second Inter-monsoon period of October – November is the period with the most evenly balanced distribution of rainfall over Sri Lanka. Almost the entire island receives in excess of 400 mm of rain during this season, with the Southwestern slopes receiving higher rainfall of 750mm to 1200 mm range.

59. The dry and cold wind blowing from the Indian land-mass will establish a comparatively cool, but dry weather over many parts making the surrounding pleasant and comfortable weather except for some rather cold morning hours. Cloud-free skies provide days full of sunshine and pleasant and cool night. During this period, the highest rainfall figures are recorded in the north-eastern slopes of the hill region and the Eastern slopes of the Knuckles/Rangala range. The maximum rainfall is experience at Kobonella (1281 mm), and the minimum is in the Western coastal area around Puttalam (Chilaw- 177 mm) during this period.

ACKNOWLEDGEMENT

1. I worked as a professional in Aviation Meteorology for the Indian Air Force (IAF) over a span of three decades, including about two decades of experience in analysing and forecasting. I am highly obliged to the Met Services department of IAF for entrusting me with responsibilities and providing ample opportunities to enhance my professional knowledge through various training programs. I am grateful to the Met Officers who patiently guided me and honed my skills in the fields of analysis and forecasting while working under their leadership. Some of the IAF Officers who were instrumental in my development are Wg Cdr AP Serrao (Retd), Gp Capt RK Mehajan (Retd), Gp Capt SK Thakur, Wg Cdr GAM Vamsi, and Gp Capt Sanjay Trivedi (Retd).

2. The IMD (Indian Meteorological Department) homepage has proven to be the greatest single source of information for me. The information gleaned from this platform has greatly contributed to the compilation of this book. I extend my sincere gratitude to this esteemed organization for providing such comprehensive and vital information.

3. I also express my sincere gratitude to Dr. Sanjay O'Neill Shaw, Scientist-F, RMC Guwahati; Mr Mukul Kumar Saikia, IAS, Principal Secretary, Karbi Anglong Autonomous Council, Assam and Mr Dipak Malla Buzar Baruah, Scientist-E & Director (IT), NIC Assam who spent their valuable time to read the book patiently and adorn it with their generous words of praise, adding a touch of encouragement that truly brightened my authorial journey. It is the support and encouragement from such personalities that make the journey of creating and publishing a book truly rewarding.

REFERENCES

Met Office, Govt of UK: www.metoffice.gov.uk/weather/atmosphere/maiden-julian-oscillation

Weather fronts   www.metoffice.gov.uk/weather/learn-about/weather/atmosphere/global-circulation-patterns

El-Nino and Its impact on Indian Monsoon and Economy: International Journal of Current Research Vol. 8, Issue, 09

El Nino and its Impact on Summer Monsoon: A Presentation by Neha Singh

Tripsavvy: www.tripsavvy.com/major-mountain-ranges-in-india-4687498

statista: www.statista.com/statistics/1298391/number-named-storms-north-indian-ocean

Climate Research Lab CCCR, IITM: www.climate.rocksea.org/tropical-cyclones-indian-ocean

Science Direct: Clear Air Turbulence: www.sciencedirect.com/tropics/earth-and-planetary-science/clear-air-turbulence

SKY brary: https://skybrary.aero/articles/clear-air-turbulence-cat

Britanica: Tropical cyclone: www.britanica.com/science/tropical-cyclone

Tornado: www.britanica.com/science/tornado

ITCZ: www.britanica.com/science/intertropical-convergence-zone

NASA: earth observatory: https://earthobservatory.nasa.gov/images/703/the-intertropical-convergence-zone

India Meteorological Department: Image Archive (imd.gov.in)

Climate of Jharkhand: Climatological Summaries of States-No. 17

Climate of Bihar: Climatological Summaries of States-No. 18

Climate of Goa: Climatological Summaries of States-No. 25

Climate of HP: Climatological Summaries of States-No. 15

	Climate of Maharashtra
	Climate of Orissa
	Climate of Sikkim: Climatological Summaries of States-No. 24
	Climate of Tripura, Meteorological Centre, Agartala, IMD
	Climate of West Bengal
	Northeast Monsoon of South Asia: No: MoES/IMD/Synoptic Met/02
	(2022)/27
Clear Air Turbulence (CAT) in India and Neighbourhood:	GR Gupta, MM Nayasthani and RK Verma, Met Office, New Delhi
Thunderstorm:	E Phillip Krider

NOAA: National Weather Service

National Ocean Service: Waterspout: https://oceanservice.noaa/facts/waterspout.html

Centers for Disaster Control and Prevention: Natural Disaster and Severe Weather: Lightning Safety Tips

ThoughtCo.:	www.thoughtco.com/how-thunderstorm
National Geography:	https://education.nationalgeographic.org/resource/hail-ee
	https://education.nationalgeographic.org/environment/article/tornado
Own your weather:	ownyourweather.com/what-is-a-dust-storm
WorldAtlas:	www.worldatlas.com/what-is-a-tornado.html
Microsoft Bing: Tornado:	www.bing.com/images/search?view=detailV2&ccid=LfByqL1
Timeanddate:	www.timeanddate.com/astronomy/optical-phenomenon.html
ENCYCLOPEDIA.COM:	www.encyclopedia.com/science/encyclopedias-almanacs-transcripts-and-maps/atmospheric-optical-phenomena#A
Geography:	https://geography.name/global-wind-and-pressure-patterns
RMetS:	International Journal of Climatology: Variability of monsoon low-level jet and associated rainfall over India: Yesubabu Vishwanadhapalli, Hari Prasad Desai, Sanjeev Dwivedi, Venkatratnam Madineni, Sabique Langodan, Ibrahim Hoteit
AccuWeather:	Leh Ladakh, India Monthly Weather

Department of Meteorology, Srilanka:	https://meteo.gov.lk/index.php?option=com_content&view+article&id=94&Itemid=310=en
Climate Profile:	Climate Variabilities, Extremes and Trends in Central, Coastal and Hilly Zones, Department of Meteorology and Hydrology, Myanmar: Prepared by RIMES
Climate of Bangladesh:	Dr AM Choudhury

Meghalaya State Climate Change Action Plan: Govt of Meghalaya publication.

ICAO Publication: Annexure 14 to the Convention of International Civil Aviation:

Vol 1, Aerodrome Design and Operations: 7th edition

Cloud Atlas:	WMO: https://cloudatlas.wmo.int/en/home.html
South West Monsoon:	YP Rao
Cyclone Preparedness and Response Plan: Govt of AP:	https://apsdma.ap.gov.in/latestupdate_pdf/Cyclone_Plan.09062020.pdf
Climate of Bhutan: Bhattadev University:	https://bhattadevuniversity.ac.in>studyMaterial

ResearchGate: Climatic Impact of El Nino/La Nina on Indian Monsoon: A New Perspective

New Climatic Classification of Nepal: https://www.researchgate.net/publication/28003490

Climate Classification of Pakistan: Saifullah Khan1, Mahmood-Ul-Hasan: www.econ-environ-geol.org

Pakistan Meteorological Department: Home Page - https://www.pmd.gov.pk/en/

Pakistan's Monthly Climate Summary July 2023

Open Data Nepal: https://opendatanepal.com/dataset/monthly-climate-normal-for-20-different-stations-in-nepal-1980-2010/resource/bbae0b33-6de1-430d-9f00-c18bfcc52500

Action Plan for Climate Change of various states

Climate Data: https://en.climate-data.org/